RADIATIVE PROCESSES IN ATOMIC PHYSICS

RADIATIVE PROCESSES IN ATOMIC PHYSICS

VLADIMIR P. KRAINOV
HOWARD R. REISS
BORIS M. SMIRNOV

A Wiley-Interscience Publication
JOHN WILEY & SONS, INC.
New York • Chichester • Weinheim • Brisbane • Singapore • Toronto

This text is printed on acid-free paper.

Library of Congress Cataloging in Publication Data:

Kraĭnov, V. P. (Vladimir Pavlovich), 1938–
Radiative processes in atomic physics / by Vladimir P. Krainov, Howard R. Reiss, Boris M. Smirnov.
p. cm.
Includes index.
ISBN 0-471-12533-4 (cloth : alk. paper)
1. Atomic spectroscopy. 2. Radiative transitions I. Reiss, H.R. (Howard Robert), 1929– . II. Smirnov, B. M. (Boris Mikhaĭlovich), 1938– . III. Title.
QC454.A8K73 1997
539.2—dc21 96-47468
CIP

Printed in the United States of America

10 9 8 7 6 5 4 3 2 1

CONTENTS

PREFACE

This book has evolved from lectures by the authors on radiative processes in atomic physics. The intent of the book is to help students and investigators in this field to extend and to make contemporary their knowledge of this important branch of atomic physics. It is envisioned that both advanced students and active researchers in this field will find it useful. Much material is contained herein that is not to be found elsewhere. Radiative processes, while constituting a fundamental area in atomic physics, have not been the focus of many books.

A novel method has been adopted for the presentation of the subject matter. The material is presented as a succession of problems. These problems are stated succinctly, solved using basic principles of atomic physics, and then the results are discussed in qualitative terms. It has been our experience that a possible initial discomfort with this unfamiliar structure gives way to an appreciation of its important advantages: First, different aspects of a single topic are treated in separate problems, which makes possible a progressive deepening of the understanding of the subject. Second, by considering limited cases of a general topic, it is possible to simplify the underlying mathematics so as to highlight the fundamental concepts. Third, although some of these problems build progressively on the results of those that precede it, it is also possible to enter into the subject matter at any point. Fourth, there is the very important feature that the problem/solution format reinforces in the reader the ability to analyze the content of a physical problem and to apply the suitable mathematical and physical tools to solve it. Finally, the qualitative discussion of the outcome of the solution aids in the development of physical intuition.

It is presumed that the reader is already acquainted with the mathematical apparatus of quantum mechanics. These matters are not explicitly developed in the body of the book, but a set of appendices is provided to give a concise overview of those methods of quantum mechanics most often employed in the treatment of radiative problems in atomic and molecular physics. The appendices review the basic tools of the quantum treatment of angular momentum, including rotation functions, Clebsch–Gordan coefficients, and the Wigner $3j$ and $6j$ symbols. There are summaries of aids to calculation in atomic physics, such as the density matrix method and fractional parentage coefficients. Two sections are devoted to quantities that arise in strong-field, nonperturbative problems. Finally, there are several sections that present some of the basic data of atomic physics (such as ionization potentials) as well as a compilation of the parameters of the hydrogen atom, including angular and radial wave functions, expectation values of various positive and negative powers of r for the lower-lying states, and oscillator strengths and lifetimes for radiative transitions.

V. P. Krainov
H. R. Reiss
B. M. Smirnov
Moscow and Washington
1996

RADIATIVE PROCESSES IN ATOMIC PHYSICS

1

RADIATIVE TRANSITIONS OF ATOMIC ELECTRONS

We devote our attention here to the study of interaction processes between the radiation field and atomic electrons. As a result of these processes, the atomic electrons will experience transitions between states available to them. If the radiation field is regarded as a collection of the elementary radiation quanta—or photons—then the number of photons changes as a consequence of such transitions.

In this book we consider transitions in which the valence electrons of an atom or molecule participate or in which vibrational or rotational molecular states change (see Appendix G). In the case of valence electron transitions, a characteristic transition energy can be constituted from fundamental atomic constants: electron mass m, Planck constant $\hbar$, and electron charge e. The value of this basic energy unit can be estimated by combining these atomic constants in such fashion as to produce the dimensions of energy. This process yields $me^4/\hbar^2$. Vibrational and rotational molecular transition energies are smaller in an essential way than this estimate, since the molecular mass is large as compared to the electron mass.

The atomic electron velocity v corresponding to atomic energies is found from $mv^2 \sim me^4/\hbar^2$, so v is of the order of $e^2/\hbar$. Comparing this velocity with the velocity of light, c, we find that $v/c \sim e^2/\hbar c \approx \frac{1}{137} \ll 1$, so that one can use the nonrelativistic theory of radiation. The applicability of nonrelativistic theory leads to essential simplifications in our general analysis.

The interaction between the electromagnetic field and the atomic system is described by the interaction Hamiltonian (Ref. 1)

$$\widehat{V} = -\frac{e}{2mc}(\widehat{\mathbf{p}}\cdot\mathbf{A} + \mathbf{A}\cdot\widehat{\mathbf{p}}) + \frac{e^2}{2mc^2}\mathbf{A}^2,$$

where $\mathbf{A}(\mathbf{r}, t)$ is the vector potential of the electromagnetic radiation field, and $\mathbf{p}$ is the momentum operator of an atomic electron that participates in the radiation transition. We presume that the radiation wavelength is large compared to a typical atomic dimension (i.e., the Bohr radius). This statement can be written as $c/\omega \gg \hbar^2/me^2$, where ω is the circular frequency of the radiation. This inequality can be rearranged to read $\hbar\omega/mc^2 \ll e^2/\hbar c$, so it is again equivalent to a nonrelativistic condition since $e^2/\hbar c \ll 1$ as cited above. Since the radiation frequency ω is of the order of the frequency of an atomic transition $me^4/\hbar^3$, then we see that the *dipole approximation* is applicable (see, e.g., Ref. 2). That is, since the vector potential (and electric field) is nearly the same over the entire radius of the atom, we may set $\mathbf{A}(\mathbf{r}, t) \approx \mathbf{A}(t)$. In consequence of this approximation, the Hamiltonian written above for the interaction between the radiation field and the atomic electron can be written in the simpler form:

$$\widehat{V} = -\mathbf{E} \cdot \mathbf{D}. \tag{1.1}$$

Here $\mathbf{E}$ is the electric field strength of the radiation field being considered, and $\mathbf{D}$ is the dipole moment operator of the atomic electron. We see that the dipole approximation is a consequence of the inequality $v/c \ll 1$.

Through most of this book, we shall also assume that the energy of interaction between the radiation field and the atomic system is small compared to the transition energy. This justifies the description by perturbation theory of radiation and absorption processes involving photons of the radiation field. Perturbation theory is applicable if the electric field strength E is small compared to a characteristic internal electric field of the atom. We can take this internal field to be of the order of the Rydberg energy divided by the product of the Bohr radius and the electron charge, or of the order of $m^2e^5/\hbar^4$. Hence, we presume that $E \ll m^2e^5/\hbar^4$. When this inequality holds true, then the operator (1.1) used for investigation of radiative transitions between states of an atom can be considered to be a small perturbation.

Thus the most important emission or absorption processes involve a single photon, if that is permitted by quantum selection rules and by energy conservation. Unless very high-powered lasers are involved, the rates of processes in which more than one photon participate are small compared to single-photon rates, and this smallness is determined by the size of the small parameter of perturbation theory. These circumstances allow one to make essential simplifications in the solution of problems.

The interaction Hamiltonian (1.1) will be used through most of this book. Succeeding terms in the expansion in powers of v/c (quadrupole approximation, *et seqq.*) are important only if the selection rules forbid dipole transitions.

The electric field can be represented as the sum over monochromatic waves:

$$\mathbf{E} = \sum_{\omega} \mathrm{Re}[\mathbf{E}_\omega \exp(i\omega t)],$$

where $\mathbf{E}_\omega$ is the complex strength of the electric field component of frequency ω. We denote by $\delta\omega$ the difference between neighboring wave frequencies and presume that all differences are the same, a situation usually realized in laser radiation. The width of the spectrum is designated by $\Delta\omega$. We presume that the value of $\mathbf{E}_\omega$ changes only slightly within this width, and we also impose the inequality $\Delta\omega \gg \delta\omega$. That is, we require that there be many frequency components in the spectrum. If the phases of the amplitudes $\mathbf{E}_\omega$ are statistically independent, then we can neglect the interactions between different waves when we treat the most prominent radiative transitions in lowest-order perturbation theory. This interaction contributes only to the next higher order of perturbation theory. Another situation in which the interaction between component waves can be neglected arises when the frequency difference $\delta\omega$ between neighboring modes is sufficiently large, that is, when $\delta\omega \gg w$, where w is the transition rate. (See Problem 1.1.)

Radiative transitions will be considered in all but the last two chapters within the framework of time-dependent perturbation theory. We first consider radiative transitions in which only one photon is emitted or absorbed, so that we employ first-order perturbation theory.

The Appendices at the end of this book contain a summary of many basic results, such as the properties of angular momentum in quantum theory and the ways in which angular momenta can couple. These Appendices will be referred to frequently.

1.1 PRINCIPAL QUANTITIES OF SINGLE-PHOTON TRANSITIONS

Problem 1.1. Find the transition rate for an atomic system for the absorption of one photon and also give the expression for the intensity of the absorption.

SOLUTION. Our goal is the calculation of the radiative transition rate in the context of first-order perturbation theory. This rate is small due to the small magnitude of the interaction between the radiation field and the atom, as discussed in the introduction to this chapter. Then the inverse quantity, the transition lifetime, or elapsed time between successive transitions will be large as compared to characteristic atomic times. The atomic time scale can be estimated as the period of the motion of an electron around the atomic core, which is $\hbar^3/(me^4)$.

In the introduction to this chapter, it was pointed out that the interaction between an atomic electron and each monochromatic component of the electromagnetic wave can be considered separately. Therefore we shall examine the operator for the interaction of an atomic electron with a single component of the field, which is

$$\widehat{V} = -\mathbf{D} \cdot \mathrm{Re}[\mathbf{E}_\omega \exp(i\omega t)]. \tag{1.2}$$

The Hamiltonian of an atomic system in the absence of the electromagnetic field will be denoted as $\widehat{H}_0$. This Hamiltonian can describe, for example, an electron in the effective mean potential of the atomic core. It is this electron that changes its state in the radiative transition. If the radiative transition is one involving a molecule, then the quantity $\widehat{H}_0$ will refer to the entire molecule, describing its vibrations or rotations, if they change in the process of the radiative transition of the molecule. We do not detail here the form of $\widehat{H}_0$; that will be done in subsequent problems.

We denote by ψ_k the system of eigenfunctions of the Hamiltonian $\widehat{H}_0$, where ε_k are the corresponding energy eigenvalues. These functions obey the time-independent Schrödinger equation $\widehat{H}_0\psi_k = \varepsilon_k\psi_k$. The expansion of the wave function of the system in terms of the eigenfunctions ψ_k is

$$\Psi = \sum c_k\psi_k(\mathbf{r})\exp(-i\varepsilon_k t/\hbar). \tag{1.3}$$

Here $\mathbf{r}$ is the set of coordinates of the atomic electron, or those coordinates that determine that vibrational or rotational state of the molecule which changes in the radiation transition. We insert Eq. (1.3) for the wave function of the atomic electron into the Schrödinger equation,

$$i\hbar\frac{\partial\Psi}{\partial t} = \widehat{H}\Psi,$$

where $\widehat{H} = \widehat{H}_0 + \widehat{V}$ is the total Hamiltonian of the system. Then we multiply the resulting equation by ψ_m^*, integrate over the coordinates of the atomic electron, and use the orthogonality conditions for the eigenfunctions ψ_k of the unperturbed atomic electron. We thus obtain for the coefficients c_k the equation

$$i\hbar\frac{dc_m}{dt} = \sum_k V_{mk}c_k\exp(i\omega_{mk}t), \tag{1.4}$$

where $\omega_{mk} = (\varepsilon_m - \varepsilon_k)/\hbar$, and V_{mk} is the matrix element of the time-dependent interaction operator, Eq. (1.2), taken between states of the atomic electron. Using first-order perturbation theory, we suppose that the perturbation is not present until the time $t = 0$, before which the electron was in the state with index 0. Then in zero approximation we have $c_k^{(0)} = \delta_{k0}$.

We take the radiation field to be linearly polarized, which means that $\mathbf{E}_\omega$ is a real quantity. Furthermore, we replace $\mathbf{E}_\omega$ by $\mathbf{E}$. This replacement does not change the essential character of the results. When we substitute the explicit form of the expression (1.2) for the interaction operator,

$$\widehat{V} = -\mathbf{D}\cdot\mathbf{E}\cos\omega t,$$

we obtain from Eq. (1.4) the first approximation

$$c_k^{(1)} = -\frac{\mathbf{D}_{k0}\cdot\mathbf{E}}{2\hbar}\left\{\frac{1-\exp[i(\omega_{k0}-\omega)t]}{\omega_{k0}-\omega} + \frac{1-\exp[i(\omega_{k0}+\omega)t]}{\omega_{k0}+\omega}\right\}. \tag{1.5}$$

The strongest transitions correspond to those states k for which the resonance condition $\omega_{k0} \approx \omega$ is fulfilled. In this case, we need to retain only the first term in Eq. (1.5). The probability for transition to the state k is

$$\left|c_k^{(1)}\right|^2 = \frac{E^2(\mathbf{D}_{k0}\cdot\mathbf{s})^2}{\hbar^2}\frac{\sin^2[(\omega_{k0}-\omega)t/2]}{(\omega_{k0}-\omega)^2}, \tag{1.6}$$

where **s** is the unit vector in the direction of the electric field vector, the so-called field polarization vector. For small a and large t, the function in Eq. (1.6) of the form $\sin^2(at)/a^2$ can be replaced by $\pi t\delta(a)$, where $\delta(a)$ is the delta function. The delta function has the properties that $\delta(a) = 0$ when $a \neq 0$, $\delta(a) = \infty$ at $a = 0$, and

$$\int_{-\infty}^{\infty} \delta(a)\, da = 1.$$

The function $\sin^2(at)/(\pi a^2 t)$ has essentially all of the properties just listed for $\delta(a)$, and is, in fact, often employed as a representation of the delta function when $t \longrightarrow \infty$. Taking account of these considerations, we obtain the transition rate

$$w = \frac{1}{t}\left|c_k^{(1)}\right|^2 = \frac{E^2}{2\hbar^2}(\mathbf{D}_{k0} \cdot \mathbf{s})^2 \delta(\omega_{k0} - \omega). \tag{1.7}$$

This formula is valid even when t is not extremely large, as long as $wt \ll 1$ and perturbation theory is applicable.

We now take into account the multimode properties of the radiation. To sum Eq. (1.7) over all modes of the field, we make the identification

$$\sum_{\omega} \rightarrow \frac{1}{\delta\omega}\int d\omega,$$

where $\delta\omega$ is the frequency difference of neighboring modes (see above), and the quantity $1/\delta\omega$ is the energy density of the states. In addition, we express the rate w in terms of the number of field photons n_ω in the state with frequency ω, instead of the electric field strength **E**. We recall that the quantity

$$\frac{(E^2/8\pi) + (H^2/8\pi)}{2} = \frac{E^2}{8\pi}$$

represents the mean value of the energy density of the electromagnetic field in a single mode of the field. In arriving at this result, we replace the time-averaged value of $\cos^2 \omega t$ by $\frac{1}{2}$, and we take into account that the electric field strength $|\mathbf{E}|$ is equal to the magnetic field strength $|\mathbf{H}|$ in a monochromatic plane wave. We can thus write the energy density of modes in the frequency range $[\omega, \omega + d\omega]$ as $(E^2/8\pi)(d\omega/\delta\omega)$. From another point of view, this quantity is also given by

$$\frac{\hbar\omega n_\omega 2d\mathbf{k}}{(2\pi)^3} = \frac{\hbar n_\omega \omega^3 d\omega}{\pi^2 c^3},$$

where **k** is the photon wave vector, $d\mathbf{k}/(2\pi)^3$ is the number of states contained in a unit volume and the range of the frequencies ω considered (Ref. 1), n_ω is the number of photons in a single state, and the factor 2 takes into account the two polarization states of the photon (Ref. 3). The wave number k is connected with the photon frequency ω by the relation $\omega = ck$, where c is the speed of light. These considerations establish the connection between the electric field strength E and the corresponding number of photons n_ω.

From the last two expressions we find

$$E^2 = \left(\frac{8\hbar\omega^3}{\pi c^3}\right)\delta\omega n_\omega.$$

When we substitute E^2 into Eq. (1.7), and integrate over the frequency range in the neighborhood of the transition frequency, we obtain the rate for absorption of photons with polarization **s** as

$$w = \frac{4\omega^3 n_\omega}{\hbar c^3}(\mathbf{D}_{k0}\cdot\mathbf{s})^2 \qquad (\omega = \omega_{k0}). \tag{1.8}$$

Such an integration, employing the delta function property $\int_{-\infty}^{\infty}\delta(a)\,da = 1$, is correct when the frequency range $\Delta\omega$ (the width characteristic of the radiation spectrum) is large compared to typical values of the difference $\omega_{k0} - \omega$, which, in turn, is of the order of t^{-1}. Since $wt \ll 1$, we obtain the inequality $\Delta\omega \gg w$.

Thus, the result (1.8) is valid for a weak field with a broad frequency spectrum and at a time of observation that is not extremely large. We also require the incoherence of waves with different frequencies.

Equation (1.8) is derived for the case $n_\omega \gg 1$. We have implied that the electric field strength does not change as a result of the radiation transition. Such a change does take place in that the number of photons is diminished by one upon the absorption of a photon. Nevertheless, Eq. (1.8) is correct at any value of n_ω. Indeed, from the nature of photon absorption, it follows that the photon absorption probability is proportional to the number of photons in the volume under consideration. Thus, the relation (1.8) is also correct when one introduces the quantum properties of the electromagnetic field.

We find the intensity of the absorption by multiplying Eq. (1.8) by $\hbar\omega$ to obtain

$$I = \frac{4\omega^4 n_\omega}{c^3}(\mathbf{D}_{ko}\cdot\mathbf{s})^2.$$

Although this expression does not contain the Planck constant, it becomes classical only when $n_\omega \gg 1$. To go to the classical limit, we replace the matrix element of the dipole moment, $\mathbf{D}_{k0}$, by the Fourier component of the time-dependent dipole moment corresponding to the transition frequency $\omega_{k0} = (\varepsilon_k - \varepsilon_0)/\hbar$ (see, e.g., Ref. 4). This is the form taken by the Bohr correspondence principle in this case.

We can average Eq. (1.8) over the polarization direction **s** by taking the solid angle average of $\overline{\cos^2\theta}$, where θ is the angle between the vectors **D** and **s**. Since this average value is $\overline{\cos^2\theta} = \frac{1}{3}$, we obtain the transition rate in the form

$$w_{0k} = w(0, n_\omega \to k, n_\omega - 1) = \frac{4\omega^3}{3\hbar c^3}|\mathbf{D}_{k0}|^2 g_k n_\omega. \tag{1.9}$$

Here we introduce the factor g_k, which is the statistical weight of the final state. That is, g_k is the number of substates for this final level. Unlike Eq. (1.8), the expression (1.9) does not depend on the field polarization.

The quantity $1/w$ represents the time between two consecutive photon absorption events. The time for the absorption of one photon is much less; it is determined by the uncertainty principle, and is of the order of $1/\omega_{k0}$.

It can be seen from Eq. (1.6) that at very small times the transition rate is proportional to t^2. However, this rate is so small that it is experimentally unobservable. On the other hand, the solution we have obtained will be incorrect at times so large that $w_{k0}t \geq 1$. In this case, the rate will depart from linear dependence. (See Chapter 2.)

Problem 1.2. Determine the photon emission transition rate between states of an atom.

Remark: Unlike the case of photon absorption, radiation from an excited atomic state consists of two parts. One is the so-called spontaneous transition rate, corresponding to radiation in the absence of the external field. Spontaneous emission occurs due to the interaction of the charged atomic electron with the electromagnetic vacuum field (see Refs. [1, 4–6]).The second part, corresponding to production of photons stimulated by the external field, is called induced radiation. Thus the total rate for the emission of radiation from the excited state k is equal to

$$w_{k0} = w(k, n_\omega \to 0, n_\omega + 1) = A_{k0} + B_{k0} n_\omega, \tag{1.10}$$

where the A term refers to spontaneous emission and B to induced emission. The induced emission rate is proportional to the number of photons in the cavity n_ω since these photons produce the transition being considered. The quantities A_{k0} and B_{k0} are called Einstein coefficients. In this problem, the Einstein coefficients will be derived. We presume that a gas of atomic electrons is in thermodynamic equilibrium with the radiation field.

SOLUTION. We shall determine the Einstein coefficients starting from the equilibrium condition for atomic electrons in a gas in an enclosure we shall refer to as a cavity. At equilibrium, the mean rate at which excitation transitions take place is balanced by the mean rate at which transitions occur in the opposite direction, that is,

$$N_0 \overline{w}_{0k} = N_k \overline{w}_{k0} = N_0 A_{0k} \overline{n}_\omega. \tag{1.11}$$

The quantities N_0 and N_k are the number densities of atomic electrons in the states 0 and k, respectively. The bar over a quantity denotes an average over photon states, since different numbers of photons n_ω are possible at a fixed temperature in the cavity. The quantity $\overline{n}_\omega$ is the mean number of photons with the frequency ω. The densities can be related to each other by the expression

$$\frac{N_k}{N_0} = \frac{\overline{n}_\omega}{\overline{n}_\omega + 1} \frac{g_k}{g_0}, \tag{1.12}$$

where g_0 and g_k are statistical weights of the states 0 and k, respectively. This relation follows from the fact that the number of photons increases by one in the deexcitation from state k to state 0. Also, the number of electrons in the given state is proportional to the statistical weight of the corresponding state. When we insert Eq. (1.12) into Eq. (1.11), and take (1.10) into consideration, we find, for arbitrary $\overline{n}_\omega$, the relation

$$A_{k0} = B_{k0} = A_{0k}\frac{g_0}{g_k}.$$

Thus the transition rate for a process with the emission of a photon is of the form

$$\overline{w}_{k0} = \overline{w}(k, n_\omega \rightarrow 0, n_\omega + 1) = \frac{4\omega^3}{3\hbar c^3}|\mathbf{D}_{k0}|^2 g_0(\overline{n}_\omega + 1). \tag{1.13}$$

Equation (1.9) has been employed here.

In the derivation of Eq. (1.13), we did not specify any particular distribution function for $\overline{n}_\omega$. In the case of thermal equilibrium, it is the well-known Planck distribution. We have used only the fact that photons are bosons, so that any arbitrary number of them can be in a given state. The fact that thermal equilibrium has been specified has been used only to establish the connection between the rates of the direct and inverse processes. Thus, the formula (1.13) is valid even without averaging. We can rewrite it for the case of radiation of a photon with definite polarization **s** [as in Eq. (1.8)] as

$$w_{k0} = \frac{4\omega^3}{\hbar c^3}|\mathbf{D}_{k0}\cdot\mathbf{s}|^2 g_0(n_\omega + 1). \tag{1.14}$$

The radiation transition rate given by Eq. (1.8) or (1.14) contains a square of a matrix element of the dipole moment operator. This matrix element is nonzero only for transitions with a certain change in the quantum numbers of the atom. The rules giving these possible transitions are called *selection rules*. Some of these rules will be considered below.

The factor $(n_\omega + 1)$ in Eq. (1.14) gives two terms in the transition rate. The term proportional to n_ω gives the rate for induced radiation. In this case, the radiated photon has the same wave vector and the same polarization **s** as the photons causing the transition. The second term in Eq. (1.14), the one independent of n_ω, corresponds to spontaneous radiation, since the transition can occur even in the absence of the external field. That is, it occurs even if $n_\omega = 0$. The spontaneous radiation rate with arbitrary polarization is given by the second term on the right side of Eq. (1.13).

We have found that all three Einstein coefficients A_{0k}, A_{k0}, and B_{k0} are interconnected, and the rates of all radiation transitions can be expressed in terms of one of these coefficients. We have expressed these Einstein coefficients as functions of the parameters of the electromagnetic field and of the atomic state.

Problem 1.3. Derive a classical formula for the intensity of spontaneous dipole radiation from the general quantum mechanical expression (1.14).

SOLUTION. We can state the intensity of spontaneous radiation as the energy radiated per unit time. From Eq. (1.14), this radiation intensity for a photon of polarization **s**, stemming from the radiative transition from an excited state k to a lower state m, is

$$\hbar\omega_{km} w_{km} = \frac{4\omega_{km}^4}{c^3} |\mathbf{D}_{km} \cdot \mathbf{s}|^2 g_m,$$

where $\hbar\omega_{km}$ is the energy difference between the states k and m. When we average over the polarization of the emitted photon, and assume for simplicity that $g_m = 1$, we obtain from Eq. (1.13) the total intensity of spontaneous radiation

$$I_{km} = \frac{4\omega_{km}^4}{3c^3} |\mathbf{D}_{km}|^2.$$

In the classical case we should sum I_{km} over all states m below the state k in order to determine the lifetime of the state k. That is, we require the intensity

$$I_k = \sum_{m,\omega_{km}>0} I_{km}. \tag{1.15}$$

This expression can be simplified. Since the matrix element $\mathbf{D}_{km}$ is proportional to $\exp(i\omega_{km}t)$ according to general principles of quantum mechanics, we can then write

$$\left(\ddot{\mathbf{D}}\right)_{km} = -\omega_{km}^2 \mathbf{D}_{km}.$$

When we employ this in Eq. (1.15), we obtain

$$I_k = \frac{4}{3c^3} \sum_{m,\omega_{km}>0} \left|\left(\ddot{\mathbf{D}}\right)_{km}\right|^2. \tag{1.16}$$

We can estimate the sum in Eq. (1.16) in the classical approximation. From matrix multiplication rules we have

$$\sum_m \left|\left(\ddot{\mathbf{D}}\right)_{km}\right|^2 = \left(\ddot{\mathbf{D}}^2\right)_{kk}.$$

Furthermore, we know that the quasi-classical sum in Eq. (1.16) is dominated by those states m that are close in energy to state k. In fact, the values of the matrix elements $(\ddot{\mathbf{D}})_{km}$ calculated with quasi-classical wave functions decrease quickly with increasing energy difference between states k and m, due to oscillations in the integrand which grow rapidly with energy difference. Now we take into account that states m both above and below state k are very similar to those that lie very close to k. We thus obtain

$$\sum_{m,\omega_{km}>0} \left|\left(\ddot{\mathbf{D}}\right)_{km}\right|^2 = \sum_{m,\omega_{km}<0} \left|\left(\ddot{\mathbf{D}}\right)_{km}\right|^2 = \frac{1}{2}\left(\ddot{\mathbf{D}}^2\right)_{kk}.$$

The diagonal matrix element of an operator representing a physical quantity gives the mean value of that operator. That mean value, in turn, yields the classical value for the physical quantity when the classical limit is taken. When we implement this observation, we obtain from Eq. (1.16) the total radiation intensity of an atom in the classical limit,

$$I_k = \frac{2}{3c^3}\ddot{\mathbf{D}}^2. \tag{1.17}$$

The classical dipole moment **D** in this equation should be regarded as a given function of time.

We conclude that, in the quasi-classical limit, the lifetime of a given state is determined by spontaneous transitions to nearby states.

1.2 OSCILLATOR STRENGTH

Problem 1.4. The oscillator strength for radiative transitions in an atom is introduced by the relation

$$f_{0k} = \frac{2m}{\hbar e^2}\omega_{k0}|(D_z)_{k0}|^2, \tag{1.18}$$

where 0 and k are the initial and final states, ω_{k0} is the radiation frequency corresponding to the transition between these states, and $(D_z)_{k0}$ is the matrix element taken between these states of the projection of the dipole operator. Obtain the formula

$$\sum_k f_{0k} = N,$$

where the sum is taken over all possible states, and N is the total number of electrons in the atom.

SOLUTION. The concept of the oscillator strength arises from a model of the electric properties of matter in which we suppose that the atomic electrons are in equilibrium positions and react elastically to small perturbations. That is, the electrons in this model are caused by a low-intensity electromagnetic field to execute simple harmonic motion around their equilibrium positions. Actually, electrons do not have fixed equilibrium positions in atoms. Instead, it is the statistical distribution of such positions that is realized, as determined by the squared modulus of the wave function. However, in classical language, a portion of the integral of this squared modulus of the wave function corresponds to each spectral frequency, and it is this part that determines the oscillator strength for that frequency.

The oscillator strength f_{0k} has been introduced in Eq. (1.18). The radiation transition rate given in Eq. (1.14) can be written in terms of the oscillator strength as

$$w_{k0} = \frac{2e^2\omega_{k0}^2}{mc^3} f_{0k}g_0(n_\omega + 1). \tag{1.19}$$

The coordinate system is such that the z axis lies along the direction of the polarization vector **s**. Equation (1.19) shows that the oscillator strength is the dimensionless parameter that measures the intensity of the given radiative transition.

The Hamiltonian of a system of atomic electrons consists of the sum of the electron kinetic energies and the potential energies of their interactions with each other and with the atomic nucleus. The commutator of the projection of the momentum operator $\widehat{p}_{jz}$ of the electron bearing the index j, and the coordinate projection z_j in this direction, is given by the standard result

$$\left[\widehat{p}_{jz}, z_j\right] = \hbar/i. \tag{1.20}$$

It follows from this that the commutator of the Hamiltonian $\widehat{H}$ and z_j arises from the kinetic energy part only, with the outcome

$$\left[\widehat{H}, z_j\right] = \frac{1}{2m}\left[\widehat{p}_z^2, z_j\right] = \frac{\hbar}{mi}\widehat{p}_{jz}.$$

For the commutator of the Hamiltonian $\widehat{H}$ and the projection of the total dipole moment, we obtain the expression

$$\left[\widehat{H}, D_z\right] = \frac{e\hbar}{mi}\sum_j \widehat{p}_{jz}.$$

The matrix element of this relation taken between the states 0 and k gives

$$\omega_{k0}(D_z)_{0k} = \frac{ei}{m}\sum_j (\widehat{p}_{jz})_{0k}.$$

When we insert this expression into (1.18) for the oscillator strength, we arrive at

$$f_{0k} = \frac{2i}{e\hbar}\sum_j (\widehat{p}_{jz})_{0k}(D_z)_{k0},$$

which can be rewritten as

$$f_{0k} = -\frac{2i}{e\hbar}\sum_j (\widehat{p}_{jz})_{k0}(D_z)_{0k},$$

upon using the properties of the operators $\widehat{p}_{jz}$ and D_z. We can now express the oscillator strength as half the sum of the two preceding expressions to give the symmetrized form

$$f_{0k} = \frac{i}{\hbar}\sum_{j,j'}\{(\widehat{p}_{jz})_{0k}(z_{j'})_{k0} - (z_{j'})_{0k}(\widehat{p}_{jz})_{k0}\}.$$

When this expression is summed over all final states k, we obtain

$$\sum_k f_{0k} = \frac{i}{\hbar}\sum_{j,j'}\left[\widehat{p}_{jz}, z_{j'}\right]_{00}.$$

Since the momentum operator of each electron commutes with the position operator of any other electron, then the double sum above reduces to the case $j = j'$ only. Furthermore, using the commutation relation (1.20), we find

$$\sum_k f_{0k} = N, \tag{1.21}$$

where N is the total number of electrons in the atomic system. If the sum is restricted only to those transitions in which valence electrons take part, then N is the number of valence electrons in the atom.

The sum in Eq. (1.21) is, in fact, equal to the sum over all the states of the discrete spectrum and the integral over the continuum states. It is seen from (1.21) that this sum is independent of the quantum numbers of the initial atomic state 0. Equation (1.21) is a generalization of the Thomas–Reiche–Kuhn sum rule.

If the state 0 is a highly excited state, then we face the problem of how to calculate the sum semiclassically. We treated an analogous problem in Problem 1.3 for the intensity of the radiative transition. However, that method does not apply in the present case, where we would obtain zero instead of N on the right-hand side of Eq. (1.21) because all commutators are zero in the classical limit.

The semiclassical approach thus does not give the correct result. A partial explanation for this comes from the fact that those terms in the series in Eq. (1.21) corresponding to highly excited states with energies more than ε_0 and those for energies less than ε_0 contribute values of opposite sign to the sum. Each term is much greater than the right-hand side of Eq. (1.21). The correct result arises as a small difference of large terms. For example, we can verify for the harmonic oscillator potential or for the Coulomb potential that the most important terms on the left side of Eq. (1.21) are of the order of n, where n is the principal quantum number of the state 0. As a result of compensation of terms with opposite signs, the value n disappears from the right side of Eq. (1.21).

Problem 1.5. Express the static polarizability of an atom in terms of the oscillator strengths for radiative transitions.

The static polarizability α_0 of the state 0 in the constant electric field E is given by the relation

$$\frac{d\delta\varepsilon_0}{dE} = -\alpha_0 E, \tag{1.22}$$

where $\delta\varepsilon_0$ is the static Stark shift of the energy of the state 0 due to the electric field E.

SOLUTION. The energy shift of the state 0 resulting from the perturbation (1.1) is calculated by second-order perturbation theory to be

$$\delta\varepsilon_0 = \frac{E^2}{\hbar}\sum_k \frac{|(D_z)_{0k}|^2}{\omega_{0k}}.$$

Hence the quantum mechanical expression for the static polarizability of an atom is of the form

$$\alpha_0 = -\frac{2}{\hbar}\sum_k \frac{|(D_z)_{0k}|^2}{\omega_{0k}} = \frac{e^2}{m}\sum_k \frac{f_{0k}}{\omega_{0k}^2}, \tag{1.23}$$

when we make use of Eq. (1.18) for the oscillator strength.

From a classical viewpoint, the static polarizability of a harmonic oscillator with frequency ω is determined by the same Eq. (1.22), where $\delta\varepsilon_0$ is the energy of the oscillator in the electric field. This quantity is clearly of the form

$$\alpha = \frac{e^2}{m\omega^2}.$$

From a comparison of these last two expressions, we see that the polarizability of an atom is given by the sum of the polarizabilities of atomic oscillators, each of which appears with a relative importance measured by the oscillator strength f_{0k}. This explains the origin of the term *oscillator strength*.

We note that when state 0 is a highly excited state, we can repeat all of the conclusions just reached.

Problem 1.6. Calculate sums of the form

$$S_0^{(n)} = \frac{2m}{3e^2\hbar^2}\sum_k (\varepsilon_k - \varepsilon_0)^n |(\mathbf{D})_{k0}|^2,$$

where $n = 0, 1, 2, 3$, and 4. These are examples of what are called *sum rules*. All sums include the complete basis set of eigenstates, that is, the quantity $S_0^{(n)}$ includes a sum over all discrete bound states and an integral over all continuum states.

SOLUTION. We consider first the simplest case, $n = 0$. We have

$$S_0^{(0)} = \frac{2m}{3\hbar^2}\sum_k |(\mathbf{r})_{k0}|^2 = \frac{2m}{3\hbar^2}(r^2)_{00}. \tag{1.24}$$

In the classical limit, the quantity $(r^2)_{00}$ is the mean value $\overline{r^2(t)}$.

For the case $n = 1$, the simplest avenue to the result is to use the solution of Problem 1.4. We set $N = 1$ in Eq. (1.21), and obtain

$$S_0^{(1)} = \frac{1}{3}3\sum_k f_{0k} = 1. \tag{1.25}$$

Now we shall evaluate the sum rule for $n = 2$. We use the relations

$$\sum_k |(\widehat{\mathbf{p}})_{k0}|^2 = (p^2)_{00} = 2m(E - V)_{00}$$

and

$$(\widehat{\mathbf{p}})_{k0} = im\omega_{k0}(\mathbf{r})_{k0}$$

to find

$$S_0^{(2)} = \frac{2m}{3}\sum_k \omega_{k0}^2|(\mathbf{r})_{k0}|^2 = \frac{4}{3}(E - V)_{00}. \tag{1.26}$$

In the classical limit, this expression is equal to $(\frac{4}{3})\overline{T(t)}$, where $\overline{T(t)}$ is the mean value of the electron kinetic energy.

We next consider the sum with $n = 3$. According to the Ehrenfest theorem of quantum mechanics, we have

$$\frac{d}{dt}(\widehat{\mathbf{p}})_{k0} = i\omega_{k0}(\widehat{\mathbf{p}})_{k0} = -(\nabla V)_{k0}.$$

Using these relations, we find

$$\sum_k \omega_{k0}|(\widehat{\mathbf{p}})_{k0}|^2 = i\sum_k (\widehat{\mathbf{p}})_{0k}\cdot(\nabla V)_{k0} = -i\sum_k (\nabla V)_{0k}\cdot(\widehat{\mathbf{p}})_{k0}$$
$$= \frac{i}{2}(\widehat{\mathbf{p}}\cdot\nabla V - \nabla V\cdot\widehat{\mathbf{p}})_{00} = \frac{\hbar}{2}(\Delta V)_{00}.$$

The sum with $n = 3$ is therefore

$$S_0^{(3)} = \frac{2m\hbar}{3}\sum_k \omega_{k0}|\omega_{k0}(\mathbf{r})_{k0}|^2 = \frac{\hbar^2}{3m}(\Delta V)_{00}.$$

Since $\Delta V = 4\pi e\rho$, where ρ is the density of the positive charge producing the potential energy V, we obtain

$$S_0^{(3)} = \frac{4\pi e\hbar^2}{3m}(\rho)_{00}. \tag{1.27}$$

For example, $\rho = Ze\delta(\mathbf{r})$ for a hydrogenlike atom, and so Eq. (1.27) yields the result

$$S_0^{(3)} = \frac{4\pi Ze^2\hbar^2}{3m}|\psi_0(0)|^2.$$

Since $\lim_{r\to 0}\psi_0(r) \sim r^l$, where l is the orbital angular momentum quantum number of the state 0, then we can conclude that $S_0^{(3)} = 0$ for $l \neq 0$.

Finally, we consider the case $n = 4$. Again using the Ehrenfest theorem, we have

$$\sum_k |(\nabla V)_{k0}|^2 = \left((\nabla V)^2\right)_{00} = \sum_k \omega_{k0}^2|(\widehat{\mathbf{p}})_{k0}|^2.$$

We thus obtain

$$S_0^{(4)} = \frac{2m\hbar^2}{3}\sum_k \omega_{k0}^2|\omega_{k0}(\mathbf{r})_{k0}|^2 = \frac{2\hbar^2}{3m}\left((\nabla V)^2\right)_{00}. \tag{1.28}$$

For example, a hydrogenlike atom has $(\nabla V)^2 \sim 1/r^4$, so Eq. (1.28) gives

$$S_0^{(4)} \sim \int |\psi_0(\mathbf{r})|^2 \frac{dr}{r^2}.$$

This integral diverges in the case $l = 0$, while it is finite for $l \neq 0$.

Sum rules are useful for estimating matrix elements in asymptotic limits. For example, it follows from Eq. (1.27) that, when $\varepsilon_k \to \infty$, part of the sum is of the form

$$\int \omega_{k0}^3|(\mathbf{r})_{k0}|^2 d\omega_{k0}.$$

Since this integral converges, we conclude that the matrix element $(\mathbf{r})_{k0}$ has the simple behavior

$$(\mathbf{r})_{k0} \underset{\varepsilon_k\to\infty}{\sim} \omega_{k0}^{-p},$$

where $p > 2$.

Problem 1.7. Using the shell model of an atom, determine the effect on sum rules of the Pauli principle (i.e., electrons cannot occupy states that are already filled).

SOLUTION. On first inspection, we see that the Pauli principle should influence the sum rule (1.21), which states that the sum of the oscillator strengths is equal to the number of electrons in an atom. In fact, when we found this sum rule (see Problem 1.4), we summed (1.21) over all possible states of an electron in the self-consistent field of the atomic core. According to the Pauli principle, we must exclude the filled electron states from this sum. However, we shall show that we can nevertheless neglect the Pauli principle in the calculation of the sum of the oscillators.

The sum of the oscillator strengths for a given electron (in the single-electron approximation) in state j can be written in a form taking into account the Pauli principle by writing the sum as

$$\sum_i f_{ji},$$

where the i index refers to the empty electron states. We showed in Problem 1.4 that

$$\sum_k f_{jk} = 1,$$

where k refers to all possible electron states. It follows from this that

$$\sum_{j,k} f_{jk} = N,$$

and

$$\sum_{j,i} f_{ji} = \sum_{j,k} f_{jk} - \sum_{j,j'} f_{jj'},$$

where j and j' are filled states, and N is the number of electrons. It follows from Eq. (1.18) that the oscillator strengths have the property $f_{jj'} = -f_{j'j}$. Hence we find that

$$\sum_{j,j'} f_{jj'} = 0,$$

so that

$$\sum_{j,i} f_{ji} = N.$$

Thus the Pauli principle does not affect the sum rule for the oscillator strengths. An analogous conclusion can also be reached for odd values of n in the sums

$$\sum_{j} S_j^{(n)}$$

treated in Problem 1.6.

Problem 1.8. In a transition of an atom to a highly excited state, find the dependence of the oscillator strength on the principal quantum number of this state.

SOLUTION. As shown in the solution to Problem 1.4 [see Eq. (1.18)], the oscillator strength is determined by the radial dipole matrix element,

$$f_{nl,n'l'} \sim \left| \int_0^\infty R_{nl}(r) R_{n'l'}(r) r^3 \, dr \right|^2 .$$

Here R_{nl} and $R_{n'l'}$ are radial solutions of the Schrödinger equation for initial (nl) and final highly excited $(n'l')$ states. The other factors in Eq. (1.18) for $f_{nl,n'l'}$, including $\varepsilon_n - \varepsilon_{n'}$ do not depend on n' due to the condition $|\varepsilon_n| \gg |\varepsilon_{n'}|$. Thus, in order to find the dependence of $f_{nl,n'l'}$ on the principal quantum number $n' \gg 1$ of the highly excited state, we must estimate the functions R_{nl} and $R_{n'l'}$, and also we must establish the most important values of r in the integral for the dipole matrix element.

In the solution of this problem, we employ the atomic unit system in which $m = e = \hbar = 1$. These units make possible a substantial decrease in the apparent complexity of equations, with a concomitant improvement in clarity of the meaning. When $n' \gg 1$, the radial wave function $R_{n'l'}$ can be written using the semiclassical approximation in the form

$$R_{n'l'}(r) = \frac{a}{r(k_{n'l'})^{1/2}} \cos \left[\int_{r_1}^{r} k_{n'l'}(r') \, dr' - \frac{\pi}{4} \right],$$

where

$$k_{n'l'} = 2(E_{n'} - V_{l'})^{1/2},$$

and $E_{n'} = -1/(2n'^2)$ is the energy of the term with the principal quantum number n'. This term is hydrogenlike due to the condition $n' \gg 1$. The quantity

$$V_{l'}(r) = -\frac{1}{r} + \frac{l'(l'+1)}{2r^2}$$

is the effective potential for the radial motion of an electron. The index l' is the orbital quantum number of the final state, and r_1 is the left turning point for the corresponding classical motion.

We predicate $n, l \sim 1$ for the initial atomic state. Then we find that the essential region of r for the wave function $R_{nl}(r)$, and hence for the integrand in the radial matrix element of the dipole moment, is of the order $r \sim 1$. Furthermore, it follows from the dipole selection rule (see Section 1.3) that at $l \sim 1$ we also have $l' \sim 1$. In the region $r \sim 1$ we have $k_{n'l'} \sim 1$, and, consequently, $R_{n'l'}(r) \sim a$.

We now wish to calculate the normalization factor a of a radial wave function for a highly excited final state. It is found from the normalization condition

$$\int_0^\infty R^2_{n'l'}(r) r^2 \, dr = 1.$$

In the semiclassical approximation this condition is of the form

$$\frac{a^2}{2} \int_{r_1}^{r_2} \frac{dr}{k_{n'l'}} = 1.$$

From this relation we find

$$\frac{1}{a^2} = \int_{r_1}^{r_2} \frac{dr}{2\{-(1/n'^2) + (2/r) - [l'(l'+1)/r^2]\}^{1/2}},$$

where r_2 is the right turning point for the corresponding classical motion of an electron. Since the important region in the integrand is where $r \sim n'^2 \gg 1$, and $l' \sim 1$, then we can neglect the contribution of the centrifugal potential in the normalization integral. We thus obtain

$$\frac{1}{a^2} = \int_{r_1}^{r_2} \frac{dr}{2[-(1/n'^2) + (2/r)]^{1/2}} = n'^3 \int_{x_1}^{x_2} \frac{dx}{2[(2/x) - 1]^{1/2}} \sim n'^3.$$

Our final result is that $a \sim 1/n'^{3/2}$.

In the essential region of the integrand in the expression for $f_{nl,n'l'}$, which is of the order of $r \sim 1$, we have $R_{nl} \sim 1$ and $R_{n'l'} \sim a$. This leads to the result

$$f_{nl,n'l'} \sim |R_{n'l'}|^2 \sim (n')^{-3}.$$

The coefficient in this dependence is of order unity (in atomic units). For example, the explicit calculation for the ground state of hydrogen ($n = 1,\ l = 0,\ l' = 1$) yields

$$f_{10,n'1} = 1.6(n')^{-3}.$$

1.3 SELECTION RULES AND RATES OF ELECTRON TRANSITIONS IN ATOMS

In this section we consider electron transitions in atoms. We limit ourselves to light atoms, and examine selection rules for quantum numbers of the states involved in transitions. We also consider the connection between radiative transition rates and symmetry properties of the wave function.

Problem 1.9. Consider a resonantly excited atom in a P state interacting with another atom of the same type in an S state. The distance R between the atoms remains fixed in the process of radiating a photon. Assume that the interaction between the atoms may be regarded as a dipole–dipole interaction. Find the selection rules and calculate the radiative transition rate for $P \to S$.

SOLUTION. A quasi molecule consisting of the two atoms in the problem has a symmetry plane perpendicular to the axis connecting the atomic nuclei and positioned at the midpoint of this axis. The wave functions of the system can be even or odd depending on the preservation of sign or change of sign, respectively, upon reflection in the plane of symmetry. We shall compose the wave function of the system with due regard to parity. Then we shall determine the rates of spontaneous transitions that follow from Eq. (1.13).

The wave functions of even (*gerade*, or g) states and odd (*ungerade*, or u) states of the quasi molecule are of the form

$$\Psi^m_{g,u} = 2^{-1/2}(\psi_1\varphi_2^m \pm \psi_2\varphi_1^m),$$

where ψ is the wave function of an atom in its ground (S) state, and φ^m is the wave function of the excited (P) state. The atoms are labeled by the indices 1 and 2, and m is the projection of the angular momentum of the excited P state on the axis connecting the atomic nuclei ($m = -1, 0, +1$). The wave function of the ground-state molecular term is $\psi_1\psi_2$.

Substituting these wave functions into the relation (1.13), we find the spontaneous transition rate for the quasi molecule, averaged over the polarizations of the emitted photon, to be

$$w = \frac{4\omega^3}{3\hbar c^3}\frac{1}{2}|\langle\psi_1\varphi_2^m \pm \psi_2\varphi_1^m|\mathbf{D}_1 + \mathbf{D}_2|\psi_1\psi_2\rangle|^2. \tag{1.29}$$

Here $\mathbf{D}$ is the dipole moment operator of the atom, and ω is the frequency of the atomic transition $P \to S$. It is seen that $w_{um} = 0$ for an odd state. That is, the radiative transition between an odd excited state of a quasi molecule and its ground state is forbidden. This follows from the fact that the wave function of the ground state of the quasi molecule and the total dipole moment operator are even functions with respect to permutation of the atoms, while wave functions of odd excited states of the quasi molecule are odd functions with respect to such a permutation. Therefore the matrix element of such a transition vanishes.

For even excited states of the quasi molecule, we first consider the case in which an atom in a P state has zero projection of the angular momentum on the axis z connecting the atomic nuclei. It is seen that

$$\langle \varphi_1^0 | D_{1x,y} | \psi_1 \rangle = 0.$$

We now introduce the notation

$$D_z = \langle \varphi_1^0 | D_{1z} | \psi_1 \rangle,$$

in terms of which Eq. (1.29) yields

$$w_{g0} = \frac{8\omega^3}{3\hbar c^3} |D_z|^2. \tag{1.30}$$

When we perform the angular integrations in the expression for D_z (see Appendix A), we obtain

$$D_z = D \int_0^\pi \cos^2\theta \sin\theta \, d\theta \frac{3^{1/2}}{2} = \frac{D}{3^{1/2}},$$

where the quantity

$$D = e \int_0^\infty R_p(r) R_s(r) r^3 \, dr$$

is the radial part of the integral in the matrix element of the dipole moment, and $R_p(r)$ and $R_s(r)$ are the radial wave functions. Equation (1.30) can thus be rewritten in the form

$$w_{g0} = \frac{8\omega^3}{9\hbar c^3} D^2. \tag{1.31}$$

The lifetimes of the excited states with $m = +1$ and $m = -1$ are plainly the same. In this case, $D_z = 0$ and

$$D_y = D_x = D \int_0^\pi \sin^2\theta \sin\theta \, d\theta \int_0^{2\pi} \cos\varphi \exp(i\varphi) \left(\frac{3}{8\pi}\right)^{1/2} \frac{d\varphi}{(4\pi)^{1/2}}$$
$$= \frac{D}{6^{1/2}},$$

so that

$$w_{g1} + w_{g,-1} = \frac{8\omega^3}{3\hbar c^3}\left(|D_x|^2 + |D_y|^2\right) = w_{g0}. \tag{1.32}$$

The expressions obtained for the radiative transition rates are valid when $R \ll \lambda$, where R is the interatomic distance and λ is the wavelength of the light ($\lambda = 2\pi c/\omega$). In the opposite case, where $R \gg \lambda$, the decay of the excited atom in the P state is independent of the atom in the S state, and the radiative transition rates are of the form

$$w_0 = \frac{4\omega^3}{9\hbar c^3} D^2 = w_1 = w_{-1}.$$

This result also holds when $R \gg a$, where a is a dimension typical of the atom; but when R is not much larger than a, then the wave functions are strongly distorted because of interactions between the atoms.

Problem 1.10. When two atoms are separated by a distance large as compared to the radius of either atom, show that the interaction potential between the atoms can be written as

$$\widehat{V} = \frac{1}{R^3}\left[\mathbf{D}_1 \cdot \mathbf{D}_2 - 3(\mathbf{D}_1 \cdot \mathbf{n})(\mathbf{D}_2 \cdot \mathbf{n})\right],$$

where the subscripts identify the two atoms, $\mathbf{D}$ is the atomic dipole moment, R is the distance between the atomic nuclei, and $\mathbf{n}$ is the unit vector along the axis connecting the atomic nuclei ($\mathbf{R} = \mathbf{n}R$).

SOLUTION. This form of the interaction operator corresponds to an expansion valid in the limit of large distances between the atoms.

We can represent the Hamiltonian of the electrons of the interacting atoms in the form

$$\begin{aligned} \widehat{H} &= \widehat{H}_0 - \sum_i \frac{Z_1}{|\mathbf{R} - \mathbf{r}_i|} - \sum_k \frac{Z_2}{|\mathbf{R} + \mathbf{r}_k|} + \sum_{i,k} \frac{1}{|\mathbf{R} - \mathbf{r}_i + \mathbf{r}_k|} \\ &= \widehat{H}_0 + \widehat{V}, \end{aligned} \tag{1.33}$$

where $\widehat{H}_0$ is the Hamiltonian of the noninteracting atoms. The index k refers to electrons of the first atom, which has a nucleus of charge Z_1; and the index i refers to the electrons of the second atom, which has a nucleus of charge Z_2; and $\mathbf{r}_i$, $\mathbf{r}_k$ are the radius vectors of each of the electrons with respect to their own nuclei. A standard result of classical electrodynamics (see, e.g., Ref. 7) is that the interaction energy between two dipoles $\mathbf{p}_1$ and $\mathbf{p}_2$ separated by a distance R is

$$W_{1,2} = \frac{1}{R^3}\left[\mathbf{p}_1 \cdot \mathbf{p}_2 - 3(\mathbf{p}_1 \cdot \mathbf{n})(\mathbf{p}_2 \cdot \mathbf{n})\right], \tag{1.34}$$

obtained from a multipole expansion of the total electrostatic energy between the two charge distributions whose dipole moments are $\mathbf{p}_1$ and $\mathbf{p}_2$. This multipole expansion is identical to an expansion of V in Eq. (1.33) in terms of the small quantities r_k/R and r_i/R. When the atomic dipole moment operators $\mathbf{D}_1 = \sum_k \mathbf{r}_k$, $\mathbf{D}_2 = \sum_i \mathbf{r}_i$, are substituted for $\mathbf{p}_1$ and $\mathbf{p}_2$ in Eq. (1.34), we obtain the desired interaction energy operator between two atoms,

$$\widehat{V} = \frac{1}{R^3}\left[\mathbf{D}_1 \cdot \mathbf{D}_2 - 3(\mathbf{D}_1 \cdot \mathbf{n})(\mathbf{D}_2 \cdot \mathbf{n})\right]. \tag{1.35}$$

Problem 1.11. Under the conditions of the preceding problem, calculate the spectral shift of the radiation line compared to its unperturbed value $\hbar\omega = \varepsilon_P - \varepsilon_S$.

SOLUTION. The dipole–dipole interaction between atoms produces a spectral shift in the radiation line. The operator of this interaction is of the form

$$\widehat{V} = \frac{1}{R^3}\left[\mathbf{D}_1 \cdot \mathbf{D}_2 - 3(\mathbf{D}_1 \cdot \mathbf{n})(\mathbf{D}_2 \cdot \mathbf{n})\right],$$

where $\mathbf{n}$ is the unit vector in the direction of the interatomic axis. We shall treat this interaction in the context of first-order perturbation theory, which will be correct for sufficiently large distances between the atoms ($R \gg a$). We treat the distance R as a given quantity. The interaction potential between the atoms in the quasi molecule in an even excited state is of the form

$$U_m^g(R) = \langle \Psi_g^m | \widehat{V} | \Psi_g^m \rangle. \tag{1.36}$$

Obviously, the interaction potential for the quasi molecule in the ground S state is zero, so that the shift of the spectral line comes entirely from (1.36). We note that this shift vanishes when $R \gg \lambda$, since the diagonal matrix element from the dipole operator goes to zero for the excited state φ^m.

The quantity m in the expression (1.36) is the projection of the angular momentum of an atom in a P state onto the axis connecting the atoms. Simplifying Eq. (1.36), we find

$$U_m^g(R) = \frac{1}{R^3}\left(|\langle\psi|\mathbf{D}|\varphi^m\rangle|^2 - 3|\langle\psi|\mathbf{D}\cdot\mathbf{n}|\varphi^m\rangle|^2\right).$$

We introduce the radial matrix element D as in the preceding problem and calculate the angular integrals. When $m = 0$, we obtain

$$|\langle\psi|\mathbf{D}|\varphi^0\rangle|^2 = |\langle\psi|D_z|\varphi^0\rangle|^2 = |\langle\psi|\mathbf{D}\cdot\mathbf{n}|\varphi^0\rangle|^2 = D^2/3,$$

so that

$$U_0^g(R) = -2D^2/3R^3.$$

Similarly, when $m = 1$ and $m = -1$, we find

$$|\langle\psi|\mathbf{D}|\varphi^1\rangle|^2 = |\langle\psi|D_x|\varphi^1\rangle|^2 + |\langle\psi|D_y|\varphi^1\rangle|^2 = D^2/3.$$

We also have the result that

$$|\langle\psi|\mathbf{D}\cdot\mathbf{n}|\varphi^1\rangle|^2 = |\langle\psi|D_z|\varphi^1\rangle|^2 = 0,$$

so that

$$U^g_{\pm 1}(R) = D^2/3R^3.$$

The excited term of odd parity is also shifted. The interaction potential has the same amplitude as the corresponding even-parity interaction potential $U^g_m(R)$, but has the opposite sign. We cannot, however, seek such a shift in the spontaneous transition, since it is forbidden by the dipole selection rule.

The shift of the even excited term is observed as a splitting of the spectral line for the $P \to S$ radiative transition. This splitting is asymmetrical, with the energy of the $m = \pm 1$ terms increasing by half the amount by which the line for $m = 0$ decreases. Also, we see from the solution of the preceding problem that the intensities of the transitions are different. According to Eq. (1.32), the intensity of the $m = \pm 1$ line is twice that of the intensity of the $m = 0$ line.

All of the above considerations are valid if the interaction potential $U_m(R)$ between the atoms is small compared to the transition energy $\hbar\omega$, that is, when the shift of the transition line is small. This is true when $R \gg a$ and allows us to represent the wave function of a quasi molecule as a combination of the wave functions of the noninteracting atoms in the zero order of perturbation theory.

It should be noted that the shift of the spectral line is inversely proportional to the cube of the distance R between the atoms. That is, it is large compared to the usual van der Waals interaction, which is a phenomenon of second-order perturbation theory, and is proportional to the inverse sixth power of R.

Problem 1.12. Find the selection rules for the single-electron radiative dipole transition between states of a light atom.

The atomic level scheme of a light atom is based on the picture that the orbital angular momenta of the electrons couple to form the total angular momentum L of the atom, and the spins of the electrons couple to form the total spin momentum S (the so-called LS coupling). Relativistic effects, including spin-orbit interactions, are smaller and are not included here.

SOLUTION. We designate by M_L the projection of the total angular momentum on a fixed direction, and M_S is the projection of the atomic spin on the same direction. We are interested in the final states of a system when the matrix element

$$\langle LM_LSM_S\alpha|D_q|L'M_L'S'M_S'\alpha'\rangle$$

is nonzero, where α, α' are those quantum numbers of the initial and final states, respectively, that are necessary to define the states in addition to the angular momentum quantum numbers. The quantity D_q is a component of the dipole moment vector operator in spherical vector coordinates. (See Appendix C.) These components are

$$D_0 = D_z, \qquad D_{\pm 1} = (D_x \pm iD_y)/\sqrt{2}.$$

In the general form D_q for $q = -1, 0, 1$, we have

$$D_q = erY_{1q}(\Omega),$$

where r, Ω are spherical electron coordinates. We assume that only one electron takes part in the radiative transition.

Since the dipole operator does not depend on spin, it is nonzero only when the selection rules

$$S' = S, \qquad M_S' = M_S$$

are fulfilled. From Eq. (C.9), we have

$$\langle LM_L\alpha|D_q|L'M_L'\alpha'\rangle \sim \langle L1, M_Lq|L'M_L'\rangle\langle L1, 00|L'0\rangle.$$

The Clebsch–Gordan coefficient $\langle L1, M_Lq|L'M_L'\rangle$ is nonzero only when $M_L = M_L'$, $M_L' \pm 1$. Thus, selection rules for the radiative transition being examined are of the form (for single-electron transitions)

$$L - L' = \pm 1, \qquad S - S' = 0, \qquad M_L - M_L' = 0, \pm 1, \qquad M_S - M_S' = 0. \quad (1.37)$$

It can be shown in the case of many-electron atoms that the rule

$$L - L' = 0$$

should be added to Eq. (1.37), as well as the requirement that the parity of the state should change in the transition. (This was fulfilled automatically above.) See Problem 1.17 for details.

Problem 1.13. A light atom undergoes a radiative transition from a state with orbital angular momentum L and spin S to a state with orbital angular momentum L' and spin S'. Fine structure of the atom in the initial state is determined by the total angular momentum J and its projection M on a fixed direction. Find the relative probability of the atom to be in the final state J' after the transition.

SOLUTION. As in the preceding problem, we presume LS coupling to hold true. That is, states are determined not only by the total angular momentum J but also by the orbital angular momentum L and the spin angular momentum S. Since the dipole moment D_q is an orbital quantity, its matrix elements are diagonal with respect to S. From Eq. (1.13) the relative transition probability to the given final state is proportional to the quantity

$$\sum_{q,M'} |\langle JMLS\alpha|D_q|J'M'L'S\alpha'\rangle|^2,$$

where $q = -1, 0, 1$, and J, M, L, S are, respectively, the total angular momentum, its projection, and the orbital and spin angular momentum quantum numbers of the atom in its initial state. Further, α and α' are all other quantum numbers necessary to describe the initial and final states. We assume that the transition is a single-electron process.

We have $D_q = erD^1_{q0}$, where $D^1_{q0} = Y_{1q}(\Omega)$ is the rotation function. Using the formula (D.12) for the matrix elements of D_q, we obtain

$$\begin{aligned}\sum_{q,M'} &|\langle JMLS\alpha|D_q|J'M'L'S\alpha'\rangle|^2 \\ &= (2L'+1)(2J'+1) \\ &\quad\times \sum_{q,M'} \langle L'1,00|L0\rangle^2 \langle 1J',qM'|JM\rangle^2 \begin{Bmatrix} L' & 1 & L \\ J & S & J' \end{Bmatrix}^2 |\langle L'\alpha'\|er\|L\alpha\rangle|^2 \\ &= (2L'+1)(2J'+1)\langle L'1,00|L0\rangle^2 \begin{Bmatrix} L' & 1 & L \\ J & S & J' \end{Bmatrix}^2 |\langle L'\alpha'\|er\|L\alpha\rangle|^2.\end{aligned}$$

Selection rules on $L \to L'$ were established in Problem 1.12. The same rules follow also from the Clebsch–Gordan coefficient $\langle L'1, 00|L0\rangle$ in the expression above. From the same expression, the selection rules on J follow from the properties of the Wigner $6j$ symbol

$$\begin{Bmatrix} L' & 1 & L \\ J & S & J' \end{Bmatrix}.$$

This symbol is nonzero if $J - J' = 0, \pm 1$ (except for the transition $J = 0 \to J' = 0$).

The relative probability for transition to the state with a given angular momentum J' is of the form

$$\frac{(2J'+1)\begin{Bmatrix} L' & 1 & L \\ J & S & J' \end{Bmatrix}^2}{\sum_{J'}(2J'+1)\begin{Bmatrix} L' & 1 & L \\ J & S & J' \end{Bmatrix}^2} = (2L+1)(2J'+1)\begin{Bmatrix} L' & 1 & L \\ J & S & J' \end{Bmatrix}^2,$$

obtained with the use of relation (D.4). We thus find

$$w(J, L \to J', L') = (2L+1)(2J'+1)\begin{Bmatrix} L' & 1 & L \\ J & S & J' \end{Bmatrix}^2 w(J, L \to L'), \qquad (1.38)$$

where w is the rate for the radiative transition between the states given in the parentheses attached to the w.

We now wish to average this expression over the total angular momentum J of the initial state. The probability of finding an atom in the initial state with a given value of J is proportional to the statistical weight of this state. That is, it is equal to

$$\frac{2J+1}{(2S+1)(2L+1)},$$

which is the ratio of the statistical weight of this state to the statistical weight of the term $(SL\alpha)$. We then find that

$$w(J, L \to L') = \frac{2J+1}{(2S+1)(2L+1)} w(L \to L'). \tag{1.39}$$

From Eqs. (1.38) and (1.39) we obtain

$$w(J, L \to J', L') = \frac{(2J+1)(2J'+1)}{2S+1} \begin{Bmatrix} L' & 1 & L \\ J & S & J' \end{Bmatrix}^2 w(L \to L'). \tag{1.40}$$

We now sum this quantity over the initial total momentum J to obtain

$$\begin{aligned} w(L \to J', L') &= \sum_J w(J, L \to J', L') \\ &= \frac{2J'+1}{(2L'+1)(2S+1)} w(L \to L'). \end{aligned} \tag{1.41}$$

This formula is obvious in that it says that the sum of the intensities of all the lines of a spectral multiplet with the same final level is proportional to the statistical weight of this final level, which is

$$\frac{2J'+1}{(2L'+1)(2S+1)}.$$

For example, consider the transition with $L' = L - 1$. When we make the appropriate quantum number entries in the Wigner $6j$ coefficient in the expression for $w(J, L \to J', L')$ and then make the simplifying assumption that $J \gg 1$, we obtain the ratios

$$\begin{aligned} &w(JL \to J-1, L-1) : w(JL \to J, L-1) : w(JL \to J+1, L-1) \\ &\quad = \left[(J+L)^2 - S^2\right]^2 : 2\left|\left[(J+L)^2 - S^2\right]\left[S^2 - (J-L)^2\right]\right| : \left[(J-L)^2 - S^2\right]^2. \end{aligned}$$

Thus, among the lines of a multiplet, the most intense is the line with $\Delta J = \Delta L$, the so-called principal line. The line with $\Delta J = \Delta L + 1$ is approximately J^2 times weaker, since at $J \gg 1$ we have $J \approx L$. Finally, the line with $\Delta J = \Delta L + 2$ is approximately J^4 times weaker than the principal line. These last two lines are called satellites. The greater the value of the total angular momentum J, the stronger the principal line is as compared to the satellite lines.

If the condition $J \gg 1$ cannot be fulfilled, the analysis is generally greatly complicated, except that relatively simple expressions can nevertheless be obtained for alkali atoms. We consider the atom to consist of a single electron in the field of an atomic core. We then have $S = \frac{1}{2}$. The term $L\alpha$ splits into two subterms with $J = L \pm \frac{1}{2}$, and the term $(L-1)\alpha'$ also splits into two subterms, but with $J' = L - \frac{1}{2}$ and $J' = L - \frac{3}{2}$. The selection rules on J allow three transitions from the term $L\alpha$ to the term $(L-1)\alpha'$. Their relative probabilities are found from the formula obtained above to be in the ratios

$$w\left(J = L + \tfrac{1}{2} \to J' = L - \tfrac{1}{2}\right) : w\left(J = L - \tfrac{1}{2} \to J' = L - \tfrac{1}{2}\right)$$
$$: w\left(J = L - \tfrac{1}{2} \to J' = L - \tfrac{3}{2}\right)$$
$$= (L+1)(2L-1) : 1 : (L-1)(2L+1).$$

The fourth transition with $J = L + \frac{1}{2} \to J = L - \frac{3}{2}$ is forbidden since it would require $\Delta J = 2$, in contradiction with the selection rules on J found above. If we suppose that $L \gg 1$, then we see that $\Delta J = \Delta L = 1$ for the first and third transitions. Thus, both these lines are principal lines, as they have comparable intensities. We can see this also from the result obtained for the relative intensities of these lines. For the second line, $\Delta J = \Delta L - 1 = 0$, so this is a satellite line with an intensity approximately L^2 times weaker than the principal lines.

We can generalize this example as well as the previous one, where transitions from a fixed initial state of a multiplet with a definite value of J were considered. The conclusion can be reached that when one takes account of all possible transitions between states of two multiplets with $J \gg 1$, the most intense are lines with $\Delta J = \Delta L$, which are principal lines. The greater the difference $\Delta J - \Delta L$, the weaker is the intensity of the corresponding line.

In conclusion, we note that, physically, averaging over initial or final states of a multiplet is realized, for example, when we consider radiation in a gas with a temperature much greater than the fine structure spacing, but much less than the energy distance between the multiplets considered.

Problem 1.14. Solve Problem 1.13 when $J, J' \sim L, L' \gg 1$.

SOLUTION. We can simplify the expression for the $6j$ symbol in Eq. (1.40) using the general expression (D.7). We shall estimate the quantity

$$\begin{Bmatrix} L' & 1 & L \\ J & S & J' \end{Bmatrix}.$$

Suppose first that $J,\ J',\ L,\ L' \gg S \gg 1$. We now examine some particular cases.

a. The case $L' = L - 1$ has the three possibilities

$$\begin{Bmatrix} L-1 & 1 & L \\ J & S & J-1 \end{Bmatrix}^2 = \frac{(L+J+S)^2(L+J-S)^2}{(8LJ)^2},$$

$$\begin{Bmatrix} L-1 & 1 & L \\ J & S & J \end{Bmatrix}^2 = \frac{2(L+J+S)(L+J-S)(L+S-J)(J+S-L)}{(8LJ)^2},$$

$$\begin{Bmatrix} L-1 & 1 & L \\ J & S & J+1 \end{Bmatrix}^2 = \frac{(L+S-J)^2(J+S-L)^2}{(8LJ)^2}.$$

The ratio of the rates for these transitions are given in Problem 1.13.

b. The case $L' = L$, which is realized in many-electron atoms, gives

$$\begin{Bmatrix} L & 1 & L \\ J & S & J-1 \end{Bmatrix}^2 = \frac{2(L+J+S)(L+J-S)(L+S-J)(J+S-L)}{(8LJ)^2},$$

$$\begin{Bmatrix} L & 1 & L \\ J & S & J \end{Bmatrix}^2 = \frac{4(J^2+L^2-S^2)^2}{(8LJ)^2},$$

$$\begin{Bmatrix} L & 1 & L \\ J & S & J+1 \end{Bmatrix}^2 = \frac{2(L+J+S)(L+J-S)(L+S-J)(J+S-L)}{(8LJ)^2}.$$

Consequently, we obtain the ratios of transition rates

$$\begin{aligned} &w(J, L \to J-1, L) : w(J, L \to J, L) : w(J, L \to J+1, L) \\ &\quad = \left[(J+L)^2 - S^2\right]\left[S^2 - (J-L)^2\right] : 2(J^2 + L^2 - S^2)^2 \\ &\qquad : \left[(J+L)^2 - S^2\right]\left[S^2 - (J-L)^2\right]. \end{aligned}$$

c. The case $L' = L + 1$ contains the possibilities

$$\begin{Bmatrix} L+1 & 1 & L \\ J & S & J-1 \end{Bmatrix}^2 = \frac{(S+L-J)^2(S+J-L)^2}{(8LJ)^2},$$

$$\begin{Bmatrix} L+1 & 1 & L \\ J & S & J \end{Bmatrix}^2 = \frac{2(L+S+J)(L+J-S)(L+S-J)(J+S-L)}{(8LJ)^2},$$

$$\begin{Bmatrix} L+1 & 1 & L \\ J & S & J+1 \end{Bmatrix}^2 = \frac{(L+J+S)^2(L+J-S)^2}{(8LJ)^2}.$$

This gives the transition rate ratios

$$\begin{aligned} &w(J, L \to J-1, L+1) : w(J, L \to J, L+1) : w(J, L \to J+1, L+1) \\ &\quad = \left[S^2 - (L-J)^2\right]^2 : 2\left|\left[(J+L)^2 - S^2\right]\left[S^2 - (J-L)^2\right]\right| : \left[(L+J)^2 - S^2\right]^2. \end{aligned}$$

In this case, the line with $\Delta J = \Delta L = 1$ is the principal line, a satellite with $\Delta J = \Delta L - 1 = 0$ is approximately J^2 times weaker, and a satellite with $\Delta J = \Delta L - 2 = -1$ is J^4 times weaker than the principal line.

We have completed the treatment of transitions with large spin momentum $S \gg 1$. Now we consider small values of S, such that $S \sim 1$, with the conditions that $J, J', L, L' \gg 1$. We need to retain ± 1 terms when they are in combination with S, but we can neglect S in combinations with $J + L$. We again examine several particular cases.

a. When $L' = L - 1$, we obtain

$$w(J, L \to J-1, L-1) : w(J, L \to J, L-1) : w(J, L \to J+1, L-1)$$
$$= (J+L)^4 : [(J+L)^2(L+S-J)(J+S-L+1)]$$
$$: [(S+L-J)(S+L-J-1)(S+J-L+1)(S+J-L+2)].$$

b. The case $L' = L$ gives

$$w(J, L \to J-1, L) : w(J, L \to J, L) : w(J, L \to J+1, L)$$
$$= [(J+L)^2(J+S-L)(L+S-J+1)] : (J+L)^4$$
$$: [(J+L)^2(J+1+S-L)(L+S-J)].$$

c. Finally, when $L' = L - 1$, the results are

$$w(J, L \to J-1, L+1) : w(J, L \to J, L+1) : w(J, L \to J+1, L+1)$$
$$= [(J+S-L-1)(J+S-L)(L+S-J+1)(L+S-J+2)]$$
$$: |[(J+L)^2(L+1+S-J)(J+S-L)]| : (J+S)^4.$$

Problem 1.15. Find the relative transition probabilities for multiplets in the transitions ${}^2S \to {}^2P$, ${}^2P \to {}^2S$, ${}^2P \to {}^2D$, and ${}^2D \to {}^2P$.

SOLUTION. All the transitions under consideration have spin of $S = \frac{1}{2}$. First we treat the ${}^2S \to {}^2P$ transition. Here, the relevant quantum numbers are $L = 0$, $L' = 1$, and $J = \frac{1}{2}$, $J' = \frac{1}{2}$ or $\frac{3}{2}$. According to Eq. (1.40), we obtain

$$\frac{w(\frac{1}{2}, 0 \to \frac{1}{2}, 1)}{w(\frac{1}{2}, 0 \to \frac{3}{2}, 1)} = \frac{\begin{Bmatrix} 1 & 1 & 0 \\ \frac{1}{2} & \frac{1}{2} & \frac{1}{2} \end{Bmatrix}^2 2}{\begin{Bmatrix} 1 & 1 & 0 \\ \frac{1}{2} & \frac{1}{2} & \frac{1}{2} \end{Bmatrix}^2 4} = \frac{1}{2}.$$

Now we consider the transition ${}^2P \to {}^2S$. The quantum numbers here are $L = 1$, $L' = 0$, and $J = \frac{3}{2}$ or $\frac{1}{2}$, $J' = 0$. There are thus no multiplets in this case.

The next transition we consider is ${}^2P \to {}^2D$. The orbital angular momentum quantum numbers are $L = 1$, $L' = 2$. If $J = \frac{1}{2}$, then we must have $J' = \frac{3}{2}$, and there is no multiplet. However, if $J = \frac{3}{2}$, then $J' = \frac{5}{2}$ or $\frac{3}{2}$. Then Eq. (1.40) leads to

$$\frac{w(\frac{3}{2}, 1 \to \frac{3}{2}, 2)}{w(\frac{3}{2}, 1 \to \frac{5}{2}, 2)} = \frac{4}{6} \frac{\begin{Bmatrix} 2 & 1 & 1 \\ \frac{3}{2} & \frac{1}{2} & \frac{3}{2} \end{Bmatrix}^2}{\begin{Bmatrix} 2 & 1 & 1 \\ \frac{3}{2} & \frac{1}{2} & \frac{5}{2} \end{Bmatrix}^2} = \frac{1}{9}.$$

Thus, the principal line corresponds to the transition with $\Delta J = \Delta L = 1$. It is nine times stronger than the satellite line.

Now we turn to the last transition, ${}^2D \rightarrow {}^2P$. The orbital angular momentum quantum numbers are $L = 2$, $L' = 1$. First we consider the case where $J = \frac{3}{2}$ in the initial state. Then the final state can have either $J' = \frac{1}{2}$ or $J' = \frac{3}{2}$. For the ratios of the transition rates, we obtain

$$\frac{w(\frac{3}{2}, 2 \rightarrow \frac{1}{2}, 1)}{w(\frac{3}{2}, 2 \rightarrow \frac{3}{2}, 1)} = \frac{2}{4} \frac{\begin{Bmatrix} 1 & 1 & 2 \\ \frac{3}{2} & \frac{1}{2} & \frac{1}{2} \end{Bmatrix}^2}{\begin{Bmatrix} 1 & 1 & 2 \\ \frac{3}{2} & \frac{1}{2} & \frac{3}{2} \end{Bmatrix}^2} = \frac{15}{2}.$$

The principal line with $\Delta J = \Delta L = -1$ is 7.5 times stronger than the satellite line. If $J = \frac{5}{2}$ in the initial state, then only $J' = \frac{3}{2}$ is possible, and there is no multiplet. We have thus considered all the available options for the transitions being examined.

In conclusion, note that we observed in Problems 1.13 and 1.14 that the principal lines are substantially stronger than the satellite lines when $J \gg 1$. We have now found in the present problem that even at moderate values of J there is a large factor in the ratios of the intensities of principal lines to satellite lines.

Problem 1.16. The total angular momentum F of an atom arises from a combination of the total atomic electron angular momentum J and the nuclear spin I. Calculate relative probabilities for radiative transitions with the change $J \rightarrow J'$ in the quantum numbers of the atom between states with hyperfine structure $F \rightarrow F'$.

SOLUTION. Hyperfine splitting of atomic levels occurs as a result of interactions of the atomic electrons with the nuclear spin when that spin is nonzero. This problem is analogous to Problem 1.13. Specifically, in Problem 1.13, the total angular momentum J was a result of the combination of the orbital and spin angular momenta L and S, but the transition operator did not depend on the spin. In the problem we consider now, the total angular momentum F is a combination of the electron total angular momentum J and the nuclear spin I, but the transition operator does not depend on the nuclear spin. Therefore, in analogy to Problem 1.13, we find that the selection rules on F are $F' = F,\ F \pm 1$.

As a consequence of the similarity of the two problems, we can rewrite the results of Problem 1.13 with the changes in notation

$$J \rightarrow F, \quad L \rightarrow J, \quad S \rightarrow I.$$

From Eq. (1.40), the radiative transition rate for the stated change in hyperfine structure is given by the expression

$$w(J, F \rightarrow J', F') = \frac{(2F+1)(2F'+1)}{(2I+1)} \begin{Bmatrix} J' & 1 & J \\ F & I & F' \end{Bmatrix}^2 w(J \rightarrow J').$$

The total transition rate from the given hyperfine state is given by Eq. (1.39) as

$$w(J, F \to J') = \sum_{F'} w(J, F \to J', F')$$
$$= \frac{2F + 1}{(2J + 1)(2I + 1)} w(J \to J').$$

The radiative transition rate from all components of the hyperfine multiplet to the given hyperfine state is

$$w(J \to J', F') = \sum_{F} w(J, F \to J', F')$$
$$= \frac{2F' + 1}{(2J' + 1)(2I + 1)} w(J \to J').$$

The statistical factors in these expressions have the simple explanation as the relative probabilities for filling the given states in the hyperfine structure.

We can treat as negligibly small the interactions between electron and nuclear spins. Therefore, in the transitions considered here, all selection rules from previous problems with respect to electron total momentum J and electron parity remain valid. In particular, electric dipole transitions between components of the hyperfine structure of the same term are forbidden since they have the same parity.

In closing, we note that transitions of the type $J = 0 \to J' = 0$, forbidden in the dipole approximation, have the forbidden analog

$$F = 0 \to F' = 0$$

in the present problem.

Problem 1.17. Clarify the differences in selection rules for dipole radiative transitions in many-electron atoms as compared to single-electron light atoms.

SOLUTION. The dipole moment is an orbital vector. Hence, matrix elements of all components of this vector can be nonzero only for transitions in which orbital angular momenta of the L electron shell change by ± 1 or 0, that is, $L \to L,\ L \pm 1$. Also, there is the additional selection rule that forbids transitions between states that both have $L = 0$. This rule is a consequence of the spherical symmetry of states with $L = 0$, so that there does not exist a vector quantity representing the matrix element of the dipole moment vector.

The dipole moment operator

$$\mathbf{D} = e \sum_i \mathbf{r}_i$$

is an odd operator. Hence matrix elements of this operator between states of the same parity vanish. Thus we obtain the so-called Laporte rule: Transitions between states with the same parity vanish.

The Laporte rule forbids transitions with $\Delta L = 0$ in single-electron atoms, since in the case of a single electron the orbital quantum number L determines the parity of the state through the parity of $(-1)^L$. This means that $\Delta L = 0$ would require unchanged parity in a dipole transition. However, for many-electron atoms the total electron orbital angular momentum L does not have a direct connection with parity, so the condition $\Delta L = 0$ can be realized together with the requirement of a change of parity in a dipole transition.

1.4 SELECTION RULES AND RATES FOR VIBRATIONAL AND ROTATIONAL TRANSITIONS IN MOLECULES

The selection rules and rates considered in this section are for radiative dipole transitions in molecules. Appendix G describes many of the properties of vibrational and rotational molecular states we shall require here.

Problem 1.18. Consider a molecule consisting of two identical atoms. Show that radiative dipole transitions between vibrational states in such a molecule are forbidden.

SOLUTION. We shall prove that the matrix element of the dipole transition operator vanishes. We first determine the dipole moment of the molecule, considering the nuclei to be in fixed positions, and viewing these nuclei as sources of a potential field. The molecule is symmetrical with respect to a plane that is a perpendicular bisector of the straight line connecting the nuclei. The electron density in the molecules is divided symmetrically by this plane. The electron density also has axial symmetry with respect to the line joining the nuclei. Therefore the electron density ρ is invariant with respect to inversion of all electrons. The dipole moment is thus

$$\mathbf{D} = \int \sum_i e\mathbf{r}_i \rho \, d\mathbf{r}_1 \ldots d\mathbf{r}_n = -\int \sum_i e\mathbf{r}_i \rho \, d\mathbf{r}_1 \ldots d\mathbf{r}_n = 0$$

for this case of fixed internuclear distance. Therefore the matrix element of this operator between vibrational states is zero, and there are no vibrational transitions. This conclusion holds true whether the two nuclei are of the same isotope or not. It is important only that the electric charges of the nuclei be the same. However, this statement is invalidated if molecular rotation influences the electron state. Then the inversion symmetry of the electron wave function is lost if the two nuclei have different masses, that is, if they are of different isotopes of the given element.

On a more subtle level, the conclusion about vanishing of the dipole moment can be violated even for identical nuclei because of the very weak interaction of the nuclear spins with the electrons. This interaction disrupts the symmetry of the electron wave function due to the fact that the two nuclear spins may be differently oriented, and they will thus have different effects on the electron densities associated with each of the nuclei. The interaction of the electrons with the total nuclear spin will lead to a weak mixing of electron states with opposite parity, and makes dipole transitions possible.

Problem 1.19. Calculate the rates for radiative transitions between rotational states of a linear dipole molecule, and determine the selection rules.

SOLUTION. For the sake of simplicity, we consider first only those terms in which the total molecular spin is zero. We denote by **J** the total angular momentum of the molecule in the initial state. It is composed of the orbital electron momenta and the rotational angular momentum of the nuclei. The projection of the total angular momentum on a fixed axis is denoted by M. The projection of the orbital angular momentum on the molecular axis is conserved due to the axial symmetry of the molecule. We label it Λ. Since the rotational angular momentum is perpendicular to the molecular axis, then the quantity Λ also represents the projection of the total angular momentum of the molecule onto its axis. All other quantum numbers necessary to define the initial state are designated by α. Analogous quantities for the final state of the molecule are distinguished by a prime.

We wish to consider the transition $JM \to J'M'$ between rotational states of the molecule for a given electron state, that is, for fixed quantum numbers Λ and α. The problem reduces to the calculation of the matrix element of the dipole moment operator. From Eq. (C.11), the matrix element of the component D_q (where q is a spherical component) of the dipole moment vector in the rest system is of the form

$$\langle J'M'\Lambda\,\alpha|D_q|JM\Lambda\,\alpha\rangle.$$

It can be related to the analogous matrix element in the coordinate system in which the z axis is along the direction of the molecular axis. The connection is given by

$$\langle J'M'\Lambda\,\alpha|D_q|JM\Lambda\,\alpha\rangle = \sqrt{\frac{2J'+1}{2J+1}}\langle J'1, M'q|JM\rangle\langle J'1, \Lambda 0|j\Lambda\rangle\langle\Lambda\,\alpha|D_z|\Lambda\,\alpha\rangle.$$

The index q takes on the values $0, \pm 1$.

It is clear that the matrix element of the dipole moment operator does not depend on the rotational quantum numbers in the coordinate system associated with the molecular axis. It is determined only by the electronic state of the molecule. Thus this matrix element is diagonal for rotational transitions and is equal to the mean dipole moment of the molecule,

$$\overline{D} = \langle\Lambda\,\alpha|D_z|\Lambda\,\alpha\rangle.$$

Selection rules for dipole rotational transitions follow from the properties of the Clebsch–Gordan coefficients contained in the expression above as

$$J - J' = 0, \pm 1; \qquad M - M' = q = 0, \pm 1. \tag{1.42}$$

Since the energies ε_J of the rotational states are determined by the well-known formula

$$\varepsilon_J = BJ(J+1),$$

then the spontaneous transition from the state with angular momentum J to the lower state is possible only with $J' = J - 1$.

Using Eq. (1.13) for the rate of radiative processes, we obtain

$$w(J \to J' = J-1) = \frac{4\omega_{JJ'}^3 \overline{D}^2}{3\hbar c^3} \frac{2J'+1}{2J+1} \langle J'1, \Lambda 0 | J\Lambda \rangle^2 \sum_{M',q} \langle J'1, M'q | JM \rangle^2.$$

This expression is averaged over polarizations of the emitted photon and integrated over the angle of emission. When we carry out the sum with the help of the rule (B.4) and use explicit values of the Clebsch–Gordan coefficients, we find the spontaneous transition rate to be

$$\begin{aligned} w(J \to J-1) &= \frac{4\omega_{J,J-1}^3 \overline{D}^2}{3\hbar c^3} \frac{2J-1}{2J+1} \langle J-1\,1, \Lambda 0 | J\Lambda \rangle^2 \\ &= \frac{4\omega_{J,J-1}^3 \overline{D}^2}{3\hbar c^3} \frac{J^2 - \Lambda^2}{J(2J+1)} = \frac{32 B^3 \overline{D}^2}{3\hbar c^3} \frac{J^2(J^2 - \Lambda^2)}{2J+1}. \end{aligned} \tag{1.43}$$

Clearly, this result is valid for the transition considered only if the vibrational and electronic states of the molecule do not change. (See Problem 1.20.) Also, it is obvious that in a diatomic molecule composed of identical atoms, symmetry considerations require $\overline{\mathbf{D}} = 0$. The above-cited transition rates then vanish. (See also Problem 1.18.)

If the spin of the molecule is nonzero, the results must be modified somewhat. We consider two limiting cases. If the spin interaction with the molecular axis (due to spin interactions of the electrons of both atoms) is large compared to the spacing of rotational levels, then rotation does not destroy this coupling. Thus the projection of the total spin onto the molecular axis is conserved. We denote this quantity as Σ. Then the projection of the total angular momentum is $\Omega = \Lambda + \Sigma$. In this case, Eq. (1.43) need be modified only by the replacement of Λ by Ω.

In the opposite limiting case, when the interaction of the spin with the molecular axis is small, rotation destroys the coupling of the spin with the axis. We can then introduce the conserved sum $\mathbf{K}$ of the orbital and rotational angular momenta. The total angular momentum $\mathbf{J}$ is also conserved and is given by $\mathbf{J} = \mathbf{K} + \mathbf{S}$, where $\mathbf{S}$ is the spin vector of the molecule. Each rotational level splits into a multiplet with $2S + 1$ components, which have angular momenta ranging from $J = K - S$ to $J = K + S$. If we do not specify the component of the multiplet, then the total rotational transition rate is obtained from Eq. (1.43) by replacing

J with K. However, if we are interested in the relative probabilities for the individual lines of the multiplet, then we obtain expressions analogous to those of Problem 1.13. Specifically, in the present case, the angular momenta $\mathbf{K}$ and $\mathbf{S}$ are added in analogy to the addition of momenta $\mathbf{L}$ and $\mathbf{S}$ in the fine structure of light atoms. There is no coupling of the angular momenta $\mathbf{K}$ and $\mathbf{S}$ with the molecular axis.

Problem 1.20. Find selection rules for radiative transitions between vibrational states of a molecule. The vibrational motion is to be considered in the harmonic oscillator approximation.

SOLUTION. Radiative transitions are possible between those vibrational states for which the dipole moment matrix element $\langle v|\overline{\mathbf{D}}|v'\rangle$ is nonzero. Here, v and v' are vibrational quantum numbers, and $\overline{\mathbf{D}}$ is the dipole moment averaged over that part of the electron configuration that does not change during the vibrational transition. The electron state is thus presumed to be given. The distance between nuclei in the parameter $\overline{\mathbf{D}}$ is regarded as a given parameter. Such an approach is based on the adiabatic approximation, justified by the fact that the motion of the nuclei is much slower than the electron motion.

We assume that the amplitude of the vibrations of the nuclei is small as compared to the distance between them. We can then employ the expansion

$$\overline{\mathbf{D}} = \overline{\mathbf{D}}_0 + \sum_i \left(\frac{\partial \overline{\mathbf{D}}}{\partial Q_i}\right)_0 Q_i + \frac{1}{2}\sum_{i,k}\left(\frac{\partial^2 \overline{\mathbf{D}}}{\partial Q_i \partial Q_k}\right)_0 Q_i Q_k + \ldots ,$$

where the Q_i are normal coordinates, and the index i enumerates the type of vibration. The quantity $\overline{\mathbf{D}}_0$ represents the dipole moment of the molecule at the equilibrium configuration of the nuclei. Derivatives of the normal coordinates are also evaluated at the equilibrium configuration.

We consider the molecular vibrations within the framework of the harmonic oscillator model. Then the matrix element of the normal coordinate Q_i in the second term on the right-hand side of the last expression is nonzero only for transitions with a change of the vibrational quantum number by one. This matrix element is of the form

$$\langle v|Q_i|v-1\rangle = \left(\frac{\hbar v}{2\mu_i \omega_i}\right)^{1/2},$$

where μ_i is the reduced mass of the molecule for the given type of vibrations, and ω_i is the frequency of this vibration in mode i. It is seen from the above formula for the mean dipole moment that transitions with a change of vibrational quantum number by two are possible. They correspond to the third term on the right-hand side of the expression for the mean value of the dipole moment.

We now compare the expressions for the radiative transition rates with a change of vibrational quantum number of two to those with a change of one. The ratio of these radiative transition rates is given by

$$\begin{aligned}
\frac{w(v \to v-2)}{w(v \to v-1)}
&= \left| \frac{\langle v|\overline{\mathbf{D}}v-2\rangle}{\langle v|\overline{\mathbf{D}}|v-1\rangle} \right|^2 \\
&= \left[\frac{1}{2} \sum_{i,k} \left(\frac{\partial^2 \overline{\mathbf{D}}}{\partial Q_i\, \partial Q_k} \right)_0 \langle v|Q_i Q_k|v-2\rangle \right]^2 \left[\sum_i \left(\frac{\partial \overline{\mathbf{D}}}{\partial Q_i} \right)_0 \langle v|Q_i|v-1\rangle \right]^{-2} \\
&= \frac{\hbar(v-1)}{8} \left[\sum_{i,k} \left(\frac{\partial^2 \overline{\mathbf{D}}}{\partial Q_i\, \partial Q_k} \right)_0 \frac{1}{(\mu_i \mu_k \omega_i \omega_k)^{1/2}} \right]^2 \left[\sum_i \left(\frac{\partial \overline{\mathbf{D}}}{\partial Q_i} \right)_0 \frac{1}{(\mu_i \omega_i)^{1/2}} \right]^{-2},
\end{aligned}$$

where we have used the rule of matrix multiplication that

$$\langle v|Q_i Q_k|v-2\rangle = \langle v|Q_i|v-1\rangle\langle v-1|Q_k|v-2\rangle.$$

Let us estimate the value of ω_i. The potential energy $\mu_i \omega_i^2 r^2/2$ of harmonic oscillations is of the order of the electron energy $\varepsilon_{\mathrm{el}}$ when the oscillation amplitude r is of the order of the distances a_{el} between atoms in a molecule. It follows from this that

$$\hbar\omega_i \sim (m/\mu_i)^{1/2} \varepsilon_{\mathrm{el}},$$

where m is the electron mass. The derivatives of the dipole moment are of the order of atomic quantities. Hence, we obtain the estimate for the ratio of probabilities

$$\frac{w(v \to v-2)}{w(v \to v-1)} \sim \left(\frac{m}{\mu} \right)^{1/2} v.$$

Numerically, this ratio is of the order of 10^{-2}–10^{-3}.

Thus the most probable radiative transitions between vibrational states take place with a change of vibrational quantum number of one. Transitions with a change of two in the vibrational quantum number are much less probable if the vibrational quantum number v is not very large. The ratio of the rates becomes of order unity when the quantum number v is of the order of $(\mu/m)^{1/2} \gg 1$. When this is true, the vibrational energy is of the order of $\varepsilon_{\mathrm{el}}$, which would imply that the harmonic oscillator approximation is not valid. Consequently, such quantum numbers are not present in practice.

Transitions with $v \to v-2$ also take place in the first-order term in the expansion of the dipole moment $\overline{\mathbf{D}}$ if we take into account the anharmonicity of the oscillations of the nuclei. To estimate the effect, we introduce the anharmonic term αQ^3 into the Hamiltonian describing vibrations. The value of α is of the order of an atomic unit. For simplicity, we shall consider only one type of vibration, so we may omit the indexes i and k. In first-order perturbation theory, the correction to the harmonic wave function $\psi_{v-2}^{(0)}$ of the state with the quantum number $v - 2$ is of the form

$$\psi_{v-2}^{(1)} = \alpha \sum_{v'} \frac{\langle v' | Q^3 | v - 2 \rangle}{\varepsilon_{v-2}^{(0)} - \varepsilon_{v'}^{(0)}} \psi_{v'}^{(0)}.$$

We now single out the term with $v' = v - 1$. Then we find that the matrix element $\langle v | Q | v - 2 \rangle$ is of the order of

$$\alpha \frac{\langle v - 1 | Q^3 | v - 2 \rangle}{\hbar\omega} \langle v | Q | v - 1 \rangle \sim \frac{\alpha}{\hbar\omega} \frac{\hbar^2 v^2}{\mu^2 \omega^2}.$$

Since $\hbar\omega \sim (m/\mu)^{1/2} \varepsilon_{\text{el}}$, we estimate the ratio of the radiative transition rates to be

$$\frac{w^{\text{anharm}}(v \to v - 2)}{w(v \to v - 1)} = \left| \frac{\langle v | Q | v - 2 \rangle}{\langle v | Q | v - 1 \rangle} \right|^2 \sim \left(\frac{\alpha \hbar^2 v^2}{\hbar\omega\mu^2\omega^2} \right)^2 \frac{\mu\omega}{\hbar v}$$

$$\sim \frac{\alpha^2 \hbar v^3}{\mu^3 \omega^5} \sim v^3 \left(\frac{m}{\mu} \right)^{1/2}.$$

This ratio is of the order of unity at the quantum numbers $v \sim (\mu/m)^{1/6}$, when the energy of the vibrational states, $v\hbar\omega \sim (m/\mu)^{1/3} \varepsilon_{\text{el}}$ is still small compared to the electron energies ε_{el}. We find that the corrections to the transitions $v \to v - 2$ due to anharmonicity of the nuclear vibrations are much more significant than the corrections due to the dependence of the mean dipole moment on the distance between the nuclei at large vibrational quantum numbers. However, both corrections are of the same order of magnitude for small quantum numbers v.

We have found that the rates for transitions with $v \to v - 2$ and $v \to v - 1$ are of the same order of magnitude if the correction to the wave function

$$\psi_{v-2}^{(1)} \sim v^{3/2} \left(\frac{m}{\mu} \right)^{1/4} \psi_{v-1}^{(0)}$$

due to anharmonicity is comparable to the unperturbed harmonic wave function $\psi_{v-2}^{(0)}$. The harmonic approximation is inapplicable under such circumstances, and such quantum numbers v are indeed absent even though the corresponding energies are still small compared to the typical electron energies ε_{el}.

The conclusion can thus be reached that, in the range where the vibrational quantum number v is a "good" quantum number, the most effective radiative transitions take place with a change in v of one.

Problem 1.21. Obtain the selection rules for rotational quantum numbers in vibrational-rotational transitions of a linear molecule. Assume that the spacing of rotational terms is small as compared to the spacing of the vibrational terms, and find the relative transition probability to the state with a given rotational angular momentum. (Take $\Lambda = 0$).

SOLUTION. This problem is a generalization of Problem 1.19, where the rotational transition rate without a change in the vibrational quantum number was determined. In that case, the matrix element of the projection of the dipole moment on the molecular axis was equal to the mean dipole moment of the molecule at the distance $r = r_0$ between the nuclei, where r_0 is the equilibrium distance. Such a matrix element is zero for transitions associated with a change in the vibrational state as a consequence of the orthogonality of vibrational wave functions. For this reason, we should write the next term of the expansion of the mean dipole moment (in the sense of integration over electron coordinates) for small differences $Q = r - r_0$. The term proportional to Q leads to matrix elements arising from the linear harmonic oscillator coordinate. It is well known that such matrix elements are nonzero for transitions between neighboring vibrational states only. Thus the selection rule $v' - v = \pm 1$ is valid for the vibrational quantum number v.

Having established the change of vibrational quantum number, we now consider the change of the rotational state. The dipole moment operator of the molecule is directed along its axis. We define **n** to be the unit vector along this molecular axis. In analogy to Problem 1.19, the matrix element of the dipole moment is proportional to the quantity

$$\langle JM|n_q|J'M'\rangle.$$

That is, n_q bears the same relationship to **n** that D_q does to **D**. The spontaneous transition rate of the molecule can be written in the form

$$w(v, J, M \rightarrow v', J', M') \sim |\langle JM|n_q|J'M'\rangle|^2,$$

where v, J, M are, respectively, the vibrational quantum number, the total angular momentum of the linear molecule, and the projection of the angular momentum on a fixed axis in space. The primed quantities v', J', M' are the analogous quantum numbers for the final state of the molecule.

The total transition rate for transitions to all rotational states is written as an inverse lifetime, or

$$\sum_{J'M'q} w(v, J, M \rightarrow v', J', M') = \frac{1}{\tau}.$$

Since a normalization condition is applicable,

$$\sum_{J'M'q} |\langle JM|n_q|J'M'\rangle|^2 = \sum_q \langle JM|n_q^2|JM\rangle = 1,$$

then the last two expressions lead to

$$w(v, J, M \rightarrow v', J', M') = \frac{1}{\tau}|\langle JM|n_q|J'M'\rangle|^2,$$

$$q = M' - M.$$

Now we calculate the matrix element of the operator

$$n_q = Y_{1q}(\theta, \varphi).$$

Using the relation (C.9), we obtain

$$\Phi_{JM}(J', M') = |\langle JM|n_q|J'M'\rangle|^2 = \frac{2J'+1}{2J+1}\langle J'1, M'q|JM\rangle^2 \langle J'1, 00|J0\rangle^2,$$

where q takes the values 0, -1, $+1$. The second factor in this expression is nonzero only when $J' = J \pm 1$. Therefore the only transitions possible are those with a change in the rotational quantum number by one. Values of the function $\Phi_{JM}(J', M')$ for various values of J' and M' are given in Table 1.1.

The function Φ_{JM} satisfies the sum rules

$$\sum_{M'} \Phi_{JM}(J-1, M') = \frac{J}{2J+1},$$

$$\sum_{M'} \Phi_{JM}(J+1, M') = \frac{J+1}{2J+1}.$$

They are obtained using the relation (B.3). It is seen also that

$$\sum_{J'M'} \Phi_{JM}(J', M') = 1.$$

Since energy increases with increasing vibrational quantum number v, then it is obvious that a spontaneous radiative transition will lead to the quantum number $v' = v - 1$. If the rate for this transition is averaged over projections M of the molecular angular momentum, then we obtain the spontaneous rate for emission of a photon with an increase or decrease by one of the rotational quantum number J as

$$w(v, J \to v-1, J+1) = \frac{J+1}{2J+1}\frac{1}{\tau},$$

$$w(v, J \to v-1, J-1) = \frac{J}{2J+1}\frac{1}{\tau}.$$

TABLE 1.1. Relative Probabilities $\Phi_{JM}(J', M')$ for Molecular Vibrational-Rotational Transitions

	M'		
J'	$M-1$	M	$M+1$
$J-1$	$\frac{(J+M)(J+M-1)}{2(2J-1)(2J+1)}$	$\frac{J^2-M^2}{(2J-1)(2J+1)}$	$\frac{(J-M)(J-M-1)}{2(2J-1)(2J+1)}$
$J+1$	$\frac{(J-M+1)(J-M+2)}{2(2J+1)(2J+3)}$	$\frac{(J+1)^2-M^2}{(2J+1)(2J+3)}$	$\frac{(J+M+1)(J+M+2)}{2(2J+1)(2J+3)}$

Each of these expressions corresponds to the radiation of a photon with arbitrary polarization. If we are interested in the radiative rate for the emission of a photon in either of the two possible definite polarization states, these are equal to each other and thus each is one half of the total rate just cited. This fact follows from the observation that, after averaging over projections of the total angular momentum, the rate for emission of a photon with a given polarization does not depend on the direction of this polarization.

Problem 1.22. Compare the rates for radiative transitions between rotational, vibrational, and electronic molecular states.

SOLUTION. Rate differences for radiative transitions are due to differences of transition energies and to differences in the dipole moment operator matrix elements.

The ratio of photon energies for vibrational ($\hbar\omega_{\text{vib}}$) and electron transitions ($\hbar\omega_{\text{el}}$) are of the order of magnitude (see Problem 1.20)

$$\frac{\hbar\omega_{\text{vib}}}{\hbar\omega_{\text{el}}} \sim \left(\frac{m}{\mu}\right)^{1/2},$$

where μ is the reduced molecular mass. The matrix element ratio for the dipole moment operators is of the order of magnitude (see also Problem 1.20)

$$\left|\frac{\langle v|\mathbf{D}|v-1\rangle}{D_{\text{el}}}\right| \sim \left(v\frac{m\omega_{\text{el}}}{\mu\omega_{\text{vib}}}\right)^{1/2} \sim v^{1/2}\left(\frac{m}{\mu}\right)^{1/4},$$

where D_{el} is the matrix of the dipole moment operator for the electron transition, and v is the vibrational quantum number. Using these estimates, we obtain from Eq. (1.13) the ratio of the rates for radiative transitions with a change in the vibrational (w_{vib}) and electron (w_{el}) states

$$\frac{w_{\text{vib}}}{w_{\text{el}}} \sim \left(\frac{\omega_{\text{vib}}}{\omega_{\text{el}}}\right)^3 v\left(\frac{m}{\mu}\right)^{1/2} \sim v\left(\frac{m}{\mu}\right)^2.$$

It is evident that vibrational transition rates are significantly less than electronic transition rates.

Analogous estimates can be made for rotational transition rates (see Problem 1.19). Since the coefficient B in the rotational spectrum

$$\varepsilon_J = BJ(J+1)$$

is of the order of

$$B \sim (m/\mu)\varepsilon_{\text{el}},$$

we obtain for $J \gg 1$ the comparison

$$\frac{w_{\text{rot}}}{w_{\text{el}}} \sim J^3\left(\frac{m}{\mu}\right)^3.$$

We conclude that rotational transition rates are much less than both electronic and vibrational transition rates.

Finally, it follows from the solution of Problem 1.21 that vibrational-rotational transition rates are of the order of the rates for vibrational transitions.

1.5 POLARIZATION OF RADIATION

Every photon is characterized by a direction of propagation that coincides with the wave vector **k** and by a polarization vector **s** that is perpendicular to the vector **k**. The polarization vector can be represented as a superposition of two vectors mutually perpendicular to each other and to the wave vector **k**. These two vectors determine two independent polarizations.

For induced radiation, the ejected photon has the same polarization as the initial photon impacting on the atomic system. However, for spontaneous radiation, the problem of the polarization of the emitted photon remains open for investigation.

Problem 1.23. Obtain the spontaneous emission rate from an excited atom of a photon, expressed in terms of specific polarization and emission directions.

SOLUTION. The rate for the spontaneous transition $k \longrightarrow 0$ is given by Eq. (1.14), where we must put $n_\omega = 0$. The statistical weight g_0 is equal to $(\frac{1}{2})d\Omega/4\pi$, where $d\Omega$ is the solid angle element for the emitted photon, and the factor $\frac{1}{2}$ appears because of the two degrees of freedom for the photon polarization vector. We then obtain the spontaneous emission rate for a photon in a given direction with a given polarization vector **s** as

$$dw_{k0} = \frac{\omega^3}{2\pi\hbar c^3}(\mathbf{D}_{k0} \cdot \mathbf{s})^2 \, d\Omega.$$

We introduce the spherical components s_q and D_q of the vectors **s** and **D** (see Appendix C). Then this expression can be rewritten in the form

$$dw_{k0} = \frac{\omega^3}{2\pi\hbar c^3}\left(\sum_q s_{-q}\langle LM_L\alpha|D_q|L'M_L'\alpha'\rangle\right)^2 d\Omega, \tag{1.44}$$

where L, M_L, α are, respectively, the orbital momentum, its projection on the z axis, and the remaining quantum numbers necessary to define the state. The same quantities with the primes refer to the final state 0. We also introduce the quantity $\Delta M_L = M_L - M_L'$.

Only one term in the sum over q in Eq. (1.44) is nonzero for each of the three possible transitions with $\Delta M_L = 0, \pm 1$. If $\Delta M_L = 0$, then we obtain the rate for emission of a photon polarized in the plane defined by **k** and the z axis. We define θ to be the angle between the z axis and the direction $\mathbf{n} = \mathbf{k}/k$ of the emitted photon. Then the direction of the vector **s** can be chosen to be either in the plane of **n** and the z axis or perpendicular to this plane. In the first case we have $s_z = \sin\theta$, since the vectors **s** and **n** are normal to each other. In the second case $s_z = 0$, since the vector **s** is perpendicular to the z axis. We thus obtain two spontaneous rates: one corresponding to the emitted photon with polarization in the plane of **n** and the z axis; and the second rate corresponding to the emission of a photon with a polarization perpendicular to this plane. These rates are

$$\begin{aligned} dw_1 &= \frac{\omega^3}{2\pi\hbar c^3}|\langle LM_L\alpha|D_z|L'M_L'\alpha'\rangle|^2 \sin^2\theta\, d\Omega, \\ dw_2 &= 0. \end{aligned} \tag{1.45}$$

Summing over the two nonzero terms and integrating over angles, we get the total rate for photon emission polarized in the plane of **n** and the z axis,

$$w(LM_L\alpha \to L'M_L\alpha') = \frac{4\omega^3}{3\hbar c^3}|\langle LM_L\alpha|D_z|L'M_L'\alpha'\rangle|^2.$$

If $M_L = \mp 1$, then only one term is retained in the sum over q in Eq. (1.44), which is

$$s_{\mp 1}\langle LM_L\alpha|D_{\pm 1}|L'\,M_L \pm 1\,\alpha'\rangle = \tfrac{1}{2}(s_x \mp is_y)\langle LM_L\alpha|D_x \pm iD_y|L'\,M_L \pm 1\,\alpha'\rangle.$$

It determines the photon emission in the xy plane with right (left) circular polarization corresponding to the upper (lower) of the ambiguous signs.

If there is no physically imposed reason for selection of a z axis, then an atom will be in each of its M_L substates with equal probability. The transition rate $dw(L\alpha \to L'\alpha')$ is obtained by summing the expression for dw over M_L' and averaging over M_L. There is a simpler alternate way to arrive at the same result. We can elect to align the z axis in Eq. (1.44) with the direction of the polarization vector **s**. We obtain thereby

$$\begin{aligned} dw_{\mathbf{s}}(L\alpha \to L'\alpha') &= \frac{\omega^3}{2\pi\hbar c^3}\frac{1}{2L+1}\sum_{M_L M_L'}|\langle LM_L\alpha|D_z|L'M_L'\alpha'\rangle|^2\, d\Omega \\ &= \frac{\omega^3}{6\pi\hbar c^3}\frac{1}{2L+1}\sum_{M_L M_L'}|\langle LM_L\alpha|\mathbf{D}|L'M_L'\alpha'\rangle|^2\, d\Omega \\ &= \frac{\omega^3}{6\pi\hbar c^3}\frac{1}{2L+1}|\langle L\alpha|D|L'\alpha'\rangle|^2\, d\Omega. \end{aligned}$$

It is seen that the emission is isotropic. That this is so has the simple explanation that all directions in space are equivalent. The quantity $\langle L\alpha|D|L'\alpha'\rangle$ is the radial dipole matrix element. Integration over angles is equivalent to multiplication by 4π because of the isotropy of the emission. That is, we have

$$w_{1,2}(L\alpha \to L'\alpha') = \frac{2\omega^3}{3\hbar c^3}\frac{1}{2L+1}|\langle L\alpha|D|L'\alpha'\rangle|^2.$$

This expression determines the total rate of photon emission with the fixed polarization direction, but it does not depend on this direction. When we sum over the two independent polarizations, we obtain the rate that establishes the lifetime of the particular excited state k to be

$$w(L\alpha \to L'\alpha') = \frac{1}{\tau} = \frac{4\omega^3}{3\hbar c^3}\frac{1}{2L+1}|\langle L\alpha|D|L'\alpha'\rangle|^2.$$

Problem 1.24. The splitting of spectral lines that occurs when an atom is placed in a magnetic field is called the Zeeman effect. Calculate the relative intensity of Zeeman spectral lines both along and perpendicular to the direction of the magnetic field.

SOLUTION. Choose the z axis to be along the direction of the magnetic field vector. We consider first the emission along this axis. That is, the wave vector $\mathbf{k}$ is oriented along the z axis. The polarization vector therefore lies in the xy plane. The photon emission rate, from Eq. (1.14), is proportional to the matrix element

$$|\langle LM_L\alpha|\mathbf{D}|L'M_L'\alpha'\rangle\mathbf{s}|^2,$$

where L, M_L, α and L', M_L', α' are, respectively, the orbital angular momentum, its projection on the z axis, and all remaining quantum numbers for the initial and final states. We choose two independent polarizations, $\mathbf{s}_1 = \mathbf{i}_x$ and $\mathbf{s}_2 = \mathbf{i}_y$, where $\mathbf{i}_x$ and $\mathbf{i}_y$ are unit vectors along the x and y axes. The rate $w_{\|}$, summed over polarization states of the photon, behaves as

$$\begin{aligned} w_{\|} &\sim |\langle LM_L\alpha|D_x|L'M_L'\alpha'\rangle|^2 + |\langle LM_L\alpha|D_y|L'M_L'\alpha'\rangle|^2 \\ &\sim \sum_{q=\pm1} |\langle LM_L\alpha|D_q|L'M_L'\alpha'\rangle|^2. \end{aligned}$$

Recalling the solution of Problem 1.23, we can identify right circularly ($\Delta M_L = 1$) and left circularly ($\Delta M_L = -1$) polarized light propagating along the z axis. Their relative intensities (called σ components) correspond to terms in the above expression with $q = -1$ and $q = +1$, respectively. From Eq. (C.9), these relative intensities have the behavior

$$\begin{aligned} \Delta M_L = 1: &\quad w_{\|}^{+} \sim |\langle L, 1; M_L, -1|L', M_L - 1\rangle|^2, \\ = -1: &\quad w_{\|}^{-} \sim |\langle L, 1; M_L, 1|L', M_L + 1\rangle|^2. \end{aligned}$$

Now we turn our attention to the observation of light in a direction perpendicular to the magnetic field vector. We can select this direction to be the y axis. Then the polarization vector $\mathbf{s}$ of the emitted photon is in the xz plane. With the assignments $\mathbf{s}_1 = \mathbf{i}_z$ and $\mathbf{s}_2 = \mathbf{i}_x$, we find that the photon emission rate, after summation over the polarizations, has the behavior

$$w_{\perp} \sim |\langle LM_L\alpha|D_z|L'M_L'\alpha'\rangle|^2 + |\langle LM_L\alpha|D_x|L'M_L'\alpha'\rangle|^2.$$

By definition, we have $D_x = (D_1 + D_{-1})/\sqrt{2}$, so then

$$
\begin{aligned}
|(D_x)_{k0}|^2 &= \frac{1}{2}\left|\sum_{q=\pm 1}(D_q)_{k0}\right|^2 \\
&= \frac{1}{2}\sum_{q=\pm 1}|(D_q)_{k0}|^2 + \frac{1}{2}\left[(D_1)_{k0}(D_{-1})^*_{k0} + (D_1)^*_{k0}(D_{-1})_{k0}\right].
\end{aligned}
$$

The last term of this expression must be zero, since for fixed states k and 0 both matrix elements $(D_1)_{k0}$ and $(D_{-1})_{k0}$ cannot be nonzero simultaneously. If we have $\Delta M_L = 1$, then the matrix element $(D_{-1})_{k0}$ is zero, and if we have $\Delta M_L = -1$, then the matrix element $(D_1)_{k0}$ is zero. We thus obtain

$$|(D_x)_{k0}|^2 = \frac{1}{2}\sum_{q=\pm 1}|(D_q)_{k0}|^2.$$

The rate $w_\perp$ can thus be written in the form

$$w_\perp \sim |\langle LM_L\alpha|D_z|L'M'_L\alpha'\rangle|^2 + \frac{1}{2}\sum_{q=\pm 1}|\langle LM_L\alpha|D_q|L'M'_L\alpha'\rangle|^2.$$

We again employ Eq. (C.9) in the calculation of the matrix elements. The intensity of the σ components is half that of the intensity along the direction of light propagation, so that we have

$$
\begin{aligned}
\Delta M_L &= 1: \quad w_\perp^+ \sim \tfrac{1}{2}|\langle L, 1; M_L, -1|L', M_L - 1\rangle|^2, \\
&= -1: \quad w_\perp^- \sim \tfrac{1}{2}|\langle L, 1; M_L, +1|L', M_L + 1\rangle|^2.
\end{aligned}
$$

The intensity of the component with $\Delta M_L = 0$ (the so-called π component) is proportional to the quantity

$$w_\perp^0 \sim |\langle L, 1; M_L, 0|L', M_L\rangle|^2.$$

Upon evaluation of the Clebsch–Gordan coefficients (see Table B.2), we find that the relative intensities of the various components can be as given in Table 1.2.

As a supplement to this table, we note that $w_\parallel^0 = 0$, $w_\parallel^+ = 2w_\perp^+$, and $w_\parallel^- = 2w_\perp^-$. Also, there can be no transitions with $L' = L$ for single-electron spectra, due to parity conservation (see Problem 1.17).

TABLE 1.2. Relative Intensities of the Components in Zeeman Splitting

$w_\perp$	$L' = L - 1$	$L' = L$	$L' = L + 1$
$w_\perp^+$	$(L + M_L)(L + M_L - 1)$	$(L + M_L)(L - M_L + 1)$	$(L - M_L + 1)(L - M_L + 2)$
$w_\perp^0$	$4(L^2 - M_L^2)$	$4M_L^2$	$4[(L + 1)^2 - M_L^2]$
$w_\perp^-$	$(L - M_L)(L - M_L - 1)$	$(L - M_L)(L + M_L + 1)$	$(L + M_L + 1)(L + M_L + 2)$

Problem 1.25. The splitting of spectral lines that occurs when an atom is placed in an electric field is called the Stark effect. Find the relative intensity of Stark spectral lines when the light is observed along the direction of the electric field vector, as compared to the direction perpendicular to the field.

SOLUTION. A constant electric field causes both shifting and splitting of atomic levels. If we do not consider the hydrogen atom, then such shifts and splittings are determined by second-order perturbation theory. As in Problem 1.24, the polarization of the radiation is determined by the direction of observation. Let the z axis be along the direction of the electric field vector. When the light is observed along this direction, the emitted photons are polarized in the xy plane and correspond to transitions with $M_L \rightarrow M_L \pm 1$. Such components are called σ components as in the magnetic field case.

In the direction perpendicular to the electric field vector, π components are observed as well as σ components. They correspond to transitions with $M_L \rightarrow M_L$, and are polarized in the xz plane. If $M_L = 0$, then we can use the results of Problem 1.24, putting $M_L = 0$.

If $M_L \neq 0$, then levels are doubly degenerate corresponding to the two signs possible for M_L. Hence the intensities of the π components are double the results found in Problem 1.24. The degeneracy arises from the fact that transitions with $M_L \rightarrow M_L$ and with $-M_L \rightarrow -M_L$ have the same energies.

In the case of σ components, we find that the energies of transitions with $M_L \rightarrow M_L - 1$ and with $-M_L \rightarrow 1 - M_L$ are the same. The first transition is associated with the emission of right-handed circularly polarized light and the second to the emission of left-handed circular polarization of the light. It is seen from Table 1.2 that their intensities are the same, so that the sum of their intensities corresponds to a doubling.

Hence the factor 2 appears in all entries in Table 1.2, so this factor can be omitted altogether for this case. In applying the results of Table 1.2 to the present problem, we should replace $w_\perp^+$ by $w(|M_L| \rightarrow |M_L - 1|)$, and $w_\perp^-$ by $w(|M_L| \rightarrow |M_L + 1|)$.

Problem 1.26. Alkali atoms are excited from the ground state to the resonant excited state ^2P_J by circularly polarized light and then deexcited with the emission of a photon. Neglecting atomic collisions, determine the spin polarization of atoms in the ground state. That is, find the mean value of the atomic spin.

SOLUTION. According to the statement of the problem, an atom in an initial s state is excited to a state with total angular momentum J with projection M onto a given axis. Polarization of an atom appears because the probability of excitation by circularly polarized light depends on its spin polarization. After emission of a photon, the atom returns to the same s state with the projection σ' of its spin on the reference axis. This projection can take the values $+\frac{1}{2}$ and $-\frac{1}{2}$.

We denote by $w(\sigma \rightarrow \sigma')$ the transition rate between states with spin projections σ and σ' in the process of excitation and deexcitation we have described. The number densities of atoms with spin projections $+\frac{1}{2}$ and $-\frac{1}{2}$ are labeled as N_+ and N_-, respectively. The rate equations for populations of these atomic spin states in the initial s state are

$$\frac{dN_+}{dt} = -w(\tfrac{1}{2} \to -\tfrac{1}{2})N_+ + w(-\tfrac{1}{2} \to +\tfrac{1}{2})N_-,$$
$$\frac{dN_-}{dt} = -w(-\tfrac{1}{2} \to +\tfrac{1}{2})N_- + w(\tfrac{1}{2} \to -\tfrac{1}{2})N_+.$$

For a steady-state process, we neglect the left side of these equations, and find the mean spin of the atom to be

$$\overline{S} = \frac{N_+ - N_-}{2(N_+ + N_-)} = \frac{1}{2}\frac{w(-\frac{1}{2} \to +\frac{1}{2}) - w(\frac{1}{2} \to -\frac{1}{2})}{w(-\frac{1}{2} \to +\frac{1}{2}) + w(\frac{1}{2} \to -\frac{1}{2})}.$$

We now establish the relative probabilities of the radiative transitions. We take the light to be right circularly polarized, so that the projection of the orbital angular momentum increases by one upon excitation: $m \to m + 1$. The relative probability for atomic excitation from the state with spin $\frac{1}{2}$ and spin projection σ into the state with total angular momentum J with projection $M = \sigma + 1$ is, from Problem 1.12,

$$|\langle \tfrac{1}{2}, 1; \sigma, 1 | J, \sigma + 1\rangle|^2.$$

The relative probability for photon emission by an atom in this JM state for transition back to an s state with spin projection σ' is proportional to the squared Clebsch–Gordan coefficient

$$|\langle J, 1; \sigma + 1, m | \tfrac{1}{2}, \sigma'\rangle|^2.$$

All of this leads to the transition probabilities

$$w(-\tfrac{1}{2} \to +\tfrac{1}{2}) = |\langle \tfrac{1}{2}, 1; -\tfrac{1}{2}, 1 | J, \tfrac{1}{2}\rangle|^2 |\langle J, 1; \tfrac{1}{2}, m = 0 | \tfrac{1}{2}, \tfrac{1}{2}\rangle|^2,$$
$$w(+\tfrac{1}{2} \to -\tfrac{1}{2}) = 0.$$

The conclusion is that the mean value of the spin, $\overline{S}$, is $\frac{1}{2}$. This signifies that the initially unpolarized atomic system becomes polarized through interaction with the circularly polarized light.

The physical meaning of this result is instructive. Suppose that $J = \frac{1}{2}$. Then only atoms with the initial value $\sigma = -\frac{1}{2}$ are excited. They deexcite to both of the states $\sigma' = \frac{1}{2}$ and $\sigma' = -\frac{1}{2}$. However, reexcitation from the state $\sigma' = -\frac{1}{2}$ can occur, whereas that from $\sigma' = \frac{1}{2}$ cannot. Therefore, the population of the $\sigma = -\frac{1}{2}$ state will diminish with time and all electrons will finally be in the $\sigma = \frac{1}{2}$ state.

If the excited state has the total momentum $J = \frac{3}{2}$, then excitation of both the $\sigma = \frac{1}{2}$ and the $\sigma = -\frac{1}{2}$ initial states will take place. In the latter case, the excited state with $M = \frac{3}{2}$ is produced. The selection rules allow deexcitation of this state only to the state with $\sigma' = \frac{1}{2}$. As time passes, all electrons in the state with $\sigma = -\frac{1}{2}$

will disappear, since transition from $\sigma = \frac{1}{2}$ to the state with $\sigma' = -\frac{1}{2}$ is forbidden. Hence all electrons will concentrate in the $\sigma' = \frac{1}{2}$ state.

Problem 1.27. Establish the time required to achieve the spin polarization discussed in Problem 1.26.

SOLUTION. In contrast to the above problem where the steady-state situation was treated, we now retain the time derivative in the rate equations. We can write these equations as

$$dN_+/dt = wN_-, \qquad dN_-/dt = -wN_-,$$

where w stands for $w(-\frac{1}{2} \to \frac{1}{2})$, since we are taking into account the conclusion from Problem 1.26 that $w(\frac{1}{2} \to -\frac{1}{2}) = 0$. The initial conditions are

$$N_-(0) = \tfrac{1}{2}, \qquad N_+(0) = \tfrac{1}{2},$$

since the initial s state has no polarization before the action of the external field. The solution of the rate equation is, thereby,

$$N_-(t) = \tfrac{1}{2}\exp(-wt),$$

$$N_+(t) = 1 - \tfrac{1}{2}\exp(-wt).$$

The time required to spin polarize the atoms is thus of the order of magnitude of $1/w$.

1.6 RADIATIVE TRANSITIONS WITH TWO PHOTONS

We have presumed in the preceding discussion that typical times for two-photon transitions are large as compared to those for single-photon processes if the radiation field interacts with a nonrelativistic atomic system. The ratio of the perturbation potential $-\mathbf{D} \cdot \mathbf{E}$ to a typical transition energy is the small parameter that explains the negligible rates for two-photon transitions. This ratio is of the order of the ratio of a typical atomic velocity $e^2/\hbar$ to the speed of light c in the case of spontaneous transitions. However, if the single-photon transition is forbidden, then the radiative transition is determined by the less effective two-photon process. A variety of two-photon transitions is presented in Table 1.3.

TABLE 1.3. Two-Photon Processes

Two-Photon Process	Scheme of the Process
Emission of two photons	$A^* \to A + \mathbf{k} + \mathbf{k}'$
Absorption of two photons	$A + \mathbf{k} + \mathbf{k}' \to \mathbf{A}^*$
Rayleigh scattering of light	$A + \mathbf{k} \to A + \mathbf{k}', k' = k$
Raman scattering of light	$A + \mathbf{k} \to A^* + \mathbf{k}', k' \neq k$

In Table 1.3, A and A^* are ground-state and excited atomic electrons, and $\mathbf{k}$ and $\mathbf{k}'$ are the wave vectors of the two photon fields that take part in each of these processes.

Problem 1.28. Find the expression for the rate of a two-photon process.

SOLUTION. We use the same method for the treatment of two-photon processes as was employed earlier for single-photon transitions. (See Section 1.1.) We treat only those transitions that are not in competition with nonelectromagnetic processes. We employ the formalism of perturbation theory and assume that single-photon transitions are forbidden. We take the perturbation potential to be of the form $-\mathbf{D}\cdot\mathbf{E}$.

The transition operator for the interaction of an atomic system with the electromagnetic radiation field has the form

$$V = -\mathbf{E}_1\cdot\mathbf{D}\cos\omega_1 t - \mathbf{E}_2\cdot\mathbf{D}\cos\omega_2 t, \tag{1.46}$$

where $\mathbf{E}_1$ and $\mathbf{E}_2$ are the electric field vectors for the electromagnetic waves with frequencies ω_1 and ω_2, respectively; and $\mathbf{D}$ is the dipole moment operator of the atomic system. The formalism of second-order perturbation theory then defines the transition rate. The system of equations (1.4) gives, for the transition amplitude from the state 0 to the state m,

$$c_m^{(2)}(t) = -\frac{i}{\hbar}\int_0^t \sum_k V_{mk}(t')c_k^{(1)}(t')\exp(i\omega_{mk}t')\,dt'.$$

Here, $\hbar\omega_{mk}$ is the energy difference between states m and k, and the transition amplitude of first-order perturbation theory is given in Eq. (1.5) by the expression

$$\begin{aligned} c_k^{(1)}(t) &= -\frac{i}{\hbar}\int_0^t V_{k0}(t')\exp(i\omega_{k0}t')\,dt' \\ &= -\frac{1}{2\hbar}\mathbf{E}_1\cdot\mathbf{D}_{k0}\left\{\frac{1-\exp[i(\omega_{k0}-\omega_1)t]}{\omega_{k0}-\omega_1} + \frac{1-\exp[i(\omega_{k0}+\omega_1)t]}{\omega_{k0}+\omega_1}\right\} \\ &\quad - \frac{1}{2\hbar}\mathbf{E}_2\cdot\mathbf{D}_{k0}\left\{\frac{1-\exp[i(\omega_{k0}-\omega_2)t]}{\omega_{k0}-\omega_2} + \frac{1-\exp[i(\omega_{k0}+\omega_2)t]}{\omega_{k0}+\omega_2}\right\}. \end{aligned}$$

The two-photon transition is a resonant process, so that the relation

$$\omega_{m0} = \mp(\omega_1+\omega_2)$$

is almost exact, with the upper of the ambiguous signs referring to photon emission and the lower sign corresponding to photon absorption. This relation is the energy conservation law for two-photon transitions and is analogous to the single-photon energy conservation condition employed in Section 1.1.

Because of the resonance behavior, we retain in the two-photon transition amplitude $c_m^{(2)}(t)$ only those terms that contain the quantity $\omega_{m0} = \mp(\omega_1+\omega_2)$ in the denominator. This denominator is zero when the energy conservation condition is fulfilled exactly. We obtain

$$c_m^{(2)}(t) = -\frac{E_1 E_2}{4\hbar^2}\left\{\sum_k \left[\frac{(\mathbf{s}_2\cdot\mathbf{D}_{mk})(\mathbf{s}_1\cdot\mathbf{D}_{k0})}{\omega_{k0}\pm\omega_1} + \frac{(\mathbf{s}_1\cdot\mathbf{D}_{mk})(\mathbf{s}_2\cdot\mathbf{D}_{k0})}{\omega_{k0}\pm\omega_2}\right]\right\} \times \frac{1-\exp[i(\omega_{m0}\pm\omega_1\pm\omega_2)t]}{\omega_{m0}\pm(\omega_1+\omega_2)}, \tag{1.47}$$

where $\mathbf{s}_1$ and $\mathbf{s}_2$ are unit polarization vectors directed along the electric field vectors $\mathbf{E}_1$ and $\mathbf{E}_2$. The upper sign in Eq. (1.47) refers to the emission of a photon, and the lower sign is for the absorption of a photon.

We did not consider the case with $\mathbf{E}_1 = \mathbf{E}_2$ and $\omega_1 = \omega_2$, so that resonances with the absorption or emission of two photons of the same electromagnetic field will not be considered here.

In addition to the resonance terms explicitly exhibited in Eq. (1.47), additional resonance terms exist, arising from terms in the expression for the amplitude $c_k^{(1)}(t)$. The terms we have reference to are those terms that contain $\omega_{mk}\pm\omega_1$ or $\omega_{mk}\pm\omega_2$ in the denominator. They occur because of a sudden turn on of the perturbation. This sudden turn on of a perturbation is a mathematical device used to simplify the calculations and is not a condition achieved in practice. In actuality, the perturbation is turned on sufficiently slowly that energy is conserved, and resonance denominators of the type $\omega_{mk}\pm\omega_1$ or $\omega_{mk}\pm\omega_2$ do not occur. Such terms correspond to Fourier components of a perturbation with sudden onset and are not an attribute of a monochromatic electromagnetic field. For these reasons, we give such terms no further consideration.

We define a transition rate as

$$w_{0m} = \lim_{t\to\infty} |c_m^{(2)}(t)|^2/t,$$

related to the use of the relation

$$\lim_{t\to\infty}\frac{\sin^2\alpha t}{\pi\alpha^2 t} = \delta(\alpha)$$

in Section 1.1. From Eq. (1.47) we thus obtain the two-photon transition rate

$$w_{0m} = \frac{(E_1E_2)^2}{8\hbar^4}\left\{\sum_k \left[\frac{(\mathbf{s}_2\cdot\mathbf{D}_{mk})(\mathbf{s}_1\cdot\mathbf{D}_{k0})}{\omega_{k0}\pm\omega_1} + \frac{(\mathbf{s}_1\cdot\mathbf{D}_{mk})(\mathbf{s}_2\cdot\mathbf{D}_{k0})}{\omega_{k0}\pm\omega_2}\right]\right\}^2 \delta(\omega_{m0}\pm\omega_1\pm\omega_2). \tag{1.48}$$

As previously, the upper sign corresponds to the emission of a photon and the lower sign to the absorption of a photon. The quantity $\hbar\omega_{k0}$ is the transition energy between the two stated levels of an atomic system. Expressions (1.47) and (1.48) are applicable if the energy denominators $\omega_{k0}\pm\omega_1$ or $\omega_{k0}\pm\omega_2$ are large compared to the width of the state k. If this is not true, we have a resonance transition, which is a significantly more complicated problem.

We suppose also that the frequencies ω_1 and ω_2 are sufficiently different from each other that waves from these two modes do not interact with each other (as in the single-photon transitions considered in Problem 1.1). That is, we presume that conditions for phases of the modes or frequency differences of neighboring modes are fulfilled.

Equation (1.48) has been written for the case of classical electromagnetic fields. To obtain the analogous expression when the radiation field is quantized, we use the correspondence principle. This takes the form of expressing the energy density of the electromagnetic field in terms of the number of photons n_ω in a state with energy $\hbar\omega$. This semiclassical substitution is valid when $n_\omega \gg 1$. However, we can generalize this substitution by extending it to the situation where $n_\omega \sim 1$. In this way we can take into account both induced and spontaneous properties.

In the solution of Problem 1.1, we found that

$$\frac{E^2}{8\pi}\frac{d\omega}{\delta\omega} = \hbar\omega n_\omega \frac{d\mathbf{k}}{8\pi^3}.$$

We omit here a factor of 2 on the right-hand side, since we suppose that the photon polarization is fixed. The quantity $\delta\omega$ represents the frequency difference of neighboring modes (see the beginning of Chapter 1). This equation is correct when $n_\omega \gg 1$, as we have employed in its derivation that the field intensity and the number of photons in a given state experience negligible change in a transition. In actuality, the number of photons increases by one in emission and decreases by one in absorption of light. A more exact connection between the radiation field intensity and the photon number can be obtained from the results of Problems 1.1 and 1.2. Specifically, Eqs. (1.9) and (1.13) show that the quantity $E^2(d\omega/\delta\omega)$ is equal to $\hbar\omega(n_\omega + 1)d\mathbf{k}/\pi^2$ for photon emission and is equal to $\hbar\omega n_\omega d\mathbf{k}/\pi^2$ for photon absorption.

When we sum Eq. (1.48) over field modes (as in Problem 1.1) and make the substitution $\sum_k \to d\omega/\delta\omega$, we find

$$dw_{0m} = \left|\sum_k \left[\frac{(\mathbf{s}_2\cdot\mathbf{D}_{mk})(\mathbf{s}_1\cdot\mathbf{D}_{k0})}{\omega_{k0} \pm \omega_1} + \frac{(\mathbf{s}_1\cdot\mathbf{D}_{mk})(\mathbf{s}_2\cdot\mathbf{D}_{k0})}{\omega_{k0} \pm \omega_2}\right]\right|^2 \tag{1.49}$$

$$\times \left(n_{\omega_1} + \frac{1}{2} \pm \frac{1}{2}\right)\left(n_{\omega_2} + \frac{1}{2} \pm \frac{1}{2}\right)\frac{\omega_1\omega_2\, d\mathbf{k}_1\, d\mathbf{k}_2}{8\pi^3\hbar^2}\delta(\omega_{m0} \pm \omega_1 \pm \omega_2).$$

The upper of the ambiguous signs is used for absorption and the lower sign for emission of the corresponding photon. The number of photons in a given state is given by n_ω, and the photon polarizations $\mathbf{s}_1$ and $\mathbf{s}_2$ are taken to be fixed.

Equation (1.49) is the principal formula of the theory of two-photon radiative transitions. It applies to all of the cases listed in Table 1.3. The problems that follow correspond to the specific two-photon processes displayed in Table 1.3.

Problem 1.29. Calculate the lifetime for a two-photon spontaneous decay of an atomic state.

SOLUTION. The scheme of the process, as listed in Table 1.3, is

$$A^* \to A + \mathbf{k}_1 + \mathbf{k}_2.$$

From Eq. (1.49), the rate $1/\tau_2$ for two-photon spontaneous decay is given by

$$\frac{1}{\tau_2} = \frac{1}{8\pi^3\hbar^2c^6}\int \omega_1^3\omega_2^3\, d\omega_1\, d\Omega_{\mathbf{k}_1}\, d\Omega_{\mathbf{k}_2} \tag{1.50}$$
$$\times \left|\sum_j \left[\frac{(\mathbf{D}\cdot\mathbf{s}_1)_{mj}(\mathbf{D}\cdot\mathbf{s}_2)_{j0}}{\omega_{j0}+\omega_2} + \frac{(\mathbf{D}\cdot\mathbf{s}_2)_{mj}(\mathbf{D}\cdot\mathbf{s}_1)_{j0}}{\omega_{j0}+\omega_1}\right]\right|^2.$$

To arrive at this expression we have set $n_{\omega_1} = n_{\omega_2} = 0$, arising from the absence of radiation from the initial state; and, using $\omega = ck$, we have employed the delta function in Eq. (1.49) to accomplish the integration over ω_2. The quantities $d\Omega_{\mathbf{k}_1}$ and $d\Omega_{\mathbf{k}_2}$ in Eq. (1.50) refer to the solid angles for the two emitted photons. We have assigned the index 0 to the initial state and the index m to the final state. The energy conservation law that follows from the delta function of Eq. (1.49) is

$$\omega_{0m} = \omega_1 + \omega_2.$$

It should be kept in mind that the probability for emission of two photons with frequencies ω_1 and ω_2 is very small as compared to the probability of emission of a single photon with the frequency $\omega_1 + \omega_2$. However, if the selection rules prohibit the single-photon transition, then only the two-photon process can occur. From Eq. (1.14), the lifetime for single-photon emission has the order of magnitude

$$\tau_1 \sim (\hbar c/e^2)^3\tau_a,$$

where τ_a is a typical atomic time, which is, in turn, of the order of magnitude

$$\tau_a \sim \hbar^3/(me^4).$$

Equation (1.50) then leads to the estimate

$$\tau_2 \sim (\hbar c/e^2)^6\tau_a$$

for the lifetime with respect to two-photon decay. Therefore the lifetime for two-photon decay is $(\hbar c/e^2)^3 \sim 10^6$ times longer than that for single-photon decay if both processes are allowed by the selection rules. In the following, we shall consider transitions that are forbidden for single-photon decay. Important examples are transitions between two states, both with angular momentum $J = 0$.

Averaging over polarization states and integrating over the directions of propagation for the photons can be done in general. To accomplish this, we employ

$$\overline{s_{1i}s_{1k}} = (\tfrac{1}{3})\delta_{ik} = \overline{s_{2i}s_{2k}},$$

where averaging is carried out both over polarizations and over photon propagation directions. Summing over these quantities gives a result greater by a factor $2 \times 4\pi$

than averaging. We find for the inverse lifetime the result

$$\frac{1}{\tau_2} = \frac{8}{9\pi\hbar^2 c^6} \int \omega_1^3 \omega_2^3 \, d\omega_1$$
$$\times \sum_{i,k} \left\{ \sum_j \left[\frac{(\mathbf{D}_i)_{mj}(\mathbf{D}_k)_{j0}}{\omega_{j0} + \omega_2} + \frac{(\mathbf{D}_k)_{mj}(\mathbf{D}_i)_{j0}}{\omega_{j0} + \omega_1} \right] \right\}^2 .$$

The factor $\omega_1^3\omega_2^3$ in the integrand has a sharp maximum at the point where $\omega_1 = \omega_2 = \omega_{0m}/2$, since $\omega_1 + \omega_2 = \omega_{0m}$. If the state 0 is the first excited state, then we have $\omega_{j0} \geq 0$ for all other states j. This implies that $\omega_{j0} \neq -\omega_1, -\omega_2$. In this case there can be no resonance denominators in the above expression, and we can set $\omega_1 = \omega_2 = \omega_{0m}/2$ in these energy denominators in accordance with the mean value theorem. Since we now have

$$\int_0^{\omega_{0m}} \omega_1^3 \omega_2^3 \, d\omega_1 = \omega_{0m}^7/140,$$

we obtain

$$\frac{1}{\tau_2} = \frac{2\omega_{0m}^7}{315\pi\hbar^2 c^6} \sum_{i,k} \left[\sum_j \frac{(\mathbf{D}_i)_{mj}(\mathbf{D}_k)_{j0} + (\mathbf{D}_k)_{mj}(\mathbf{D}_i)_{j0}}{\omega_{j0} + \omega_{0m}/2} \right]^2 ,$$

where i, k can take on the values -1, 0, $+1$ corresponding to vector spherical coordinates.

In summary, we see that two-photon emission will occur with both photons having about the same frequency and that this frequency corresponds to about half the total transition energy.

The initial and final states can be degenerate with respect to the magnetic quantum number. If this is the case, we should average the transition rate over the magnetic quantum numbers of the initial state and sum over the magnetic quantum numbers of the final state. A typical example is the two-photon radiation from the $2s_{1/2}$ metastable excited state of hydrogen. The transition is to the $1s_{1/2}$ ground state of hydrogen. The wave functions of the initial, final, and intermediate states are of the form

$$\Psi_m = (4\pi)^{-1/2} R_{10}(r), \qquad \Psi_0 = (4\pi)^{-1/2} R_{20}(r),$$
$$\Psi_j = R_{n1}(r) Y_{1m}(\Omega); \qquad M = -1, 0 - +1; \qquad n = 2, 3, 4, \ldots,$$

where R_{nl} is the radial wave function of the hydrogen atom. We introduce the matrix element of the radial dipole moment,

$$D_{nn'} = \int_0^\infty R_{n'0}(r) R_{n1}(r) r^3 \, dr,$$

in terms of which we can write the two-photon transition rate from the $2s_{1/2}$ state of hydrogen in the form

$$\frac{1}{\tau_2} = \frac{2\omega_{21}^7}{315\pi\hbar^2 c^6} \sum_{M=0,\pm1} \left[\sum_{n\geq 2} \frac{D_{n2}D_{n1}(\frac{1}{3}+\frac{1}{3})}{\omega_{n2}+\omega_{21}/2} \right]^2$$

$$= \frac{8\omega_{21}^7}{945\pi\hbar^2 c^6} \left[\sum_{n\geq 2} \frac{D_{n2}D_{n1}}{\omega_{n2}+\omega_{21}/2} \right]^2 .$$

The sum over the principal quantum numbers can be evaluated only by numerical means. The sum contains both bound and continuum states. The primary term in the sum is the one corresponding to the $2p_{1/2}$ state ($n = 2$). If we retain only this term, we obtain the estimate $\tau_2 = 0.045\, s$ for the lifetime, which is of the order of the exact value $\tau_2 = 0.15\, s$.

Problem 1.30. Calculate, within the constant energy denominator approximation, the lifetime for an excited atomic state to decay by two-photon spontaneous emission.

SOLUTION. Label an initial state with the subscript 0 and a final state with m. The general expression for the two-photon decay rate was given by Eq. (1.50) in the preceding problem as

$$\frac{1}{\tau_2} = \frac{1}{8\pi^3\hbar^2 c^6} \int \omega_1^3 \omega_2^3\, d\omega_1\, d\Omega_{\mathbf{k}_1}\, d\Omega_{\mathbf{k}_2}$$

$$\times \left| \sum_j \left[\frac{(\mathbf{D}\cdot\mathbf{s}_1)_{mj}(\mathbf{D}\cdot\mathbf{s}_2)_{j0}}{\omega_{j0}+\omega_2} + \frac{(\mathbf{D}\cdot\mathbf{s}_2)_{mj}(\mathbf{D}\cdot\mathbf{s}_1)_{j0}}{\omega_{j0}+\omega_1} \right] \right|^2 .$$

We wish to carry out the summations in this expression within the constant energy denominator approximation. We accomplish this by the introduction of a fitting parameter representing the mean value of the energy differences in the denominators in the summation considered. By this means, we obtain

$$\sum_j \frac{(\mathbf{D}\cdot\mathbf{s}_2)_{mj}(\mathbf{D}\cdot\mathbf{s}_1)_{j0}}{\omega_{j0}+\omega_1} = \frac{1}{\omega_{k0}+\omega_1} \sum_j (\mathbf{D}\cdot\mathbf{s}_2)_{mj}(\mathbf{D}\cdot\mathbf{s}_1)_{j0}$$

$$= \frac{1}{\omega_{k0}+\omega_1} \left[(\mathbf{D}\cdot\mathbf{s}_2)(\mathbf{D}\cdot\mathbf{s}_1)\right]_{m0},$$

where we have used the completeness property of the eigenfunctions. The quantity ω_{k0} is the fitting parameter for this sum. The approximation will be accurate if the most important terms in the sum are those that have nearly equal energies.

In this fashion, we obtain the two-photon transition rate in the form

$$\frac{1}{\tau_2} = \frac{1}{8\pi^3\hbar^2 c^6} \int \omega_1^3 \omega_2^3\, d\omega_1\, d\Omega_{\mathbf{k}_1}\, d\Omega_{\mathbf{k}_2}$$

$$\left(\frac{1}{\omega_{k0}+\omega_1} + \frac{1}{\omega_{k0}+\omega_2} \right)^2 \left| \left[(\mathbf{D}\cdot\mathbf{s}_1)(\mathbf{D}\cdot\mathbf{s}_2)\right]_{m0} \right|^2 .$$

We wish now to carry out the integrations over the directions of photon emission. Since both photons are unpolarized and are emitted isotropically, we can replace the solid angle of the ejected photon $d\Omega_{\mathbf{k}}$ by the solid angle $d\Omega_{\mathbf{s}}$, where $\mathbf{s}$ is the unit vector for the photon polarization. Consequently, the integrals over the angles take the form

$$\int d\Omega_{\mathbf{s}_1}\, d\Omega_{\mathbf{s}_2} \left| \left[(\mathbf{D}\cdot\mathbf{s}_1)(\mathbf{D}\cdot\mathbf{s}_2) \right]_{m0} \right|^2 .$$

In the interests of simplicity, we now specialize to a single-electron atom, so that $\mathbf{D} = e\mathbf{r}$. The wave functions of the initial and final states are Ψ_0 and Ψ_m, and we suppose that these wave functions are real. The integral just written is equal to

$$e^4 \int \Psi_m(\mathbf{r}')\Psi_m(\mathbf{r})\Psi_0(\mathbf{r}')\Psi_0(\mathbf{r})\, d\Omega_{\mathbf{s}_1}\, d\Omega_{\mathbf{s}_2} (\mathbf{r}\cdot\mathbf{s}_1)(\mathbf{r}'\cdot\mathbf{s}_1)(\mathbf{r}\cdot\mathbf{s}_2)(\mathbf{r}'\cdot\mathbf{s}_2)\, d\mathbf{r}\, d\mathbf{r}'.$$

From the relation

$$\cos\theta_{\mathbf{r}'\mathbf{s}} = \cos\theta_{\mathbf{r}\mathbf{r}'}\cos\theta_{\mathbf{r}\mathbf{s}} + \sin\theta_{\mathbf{r}\mathbf{r}'}\sin\theta_{\mathbf{r}\mathbf{s}}\cos\varphi,$$

we see that

$$\int d\Omega_{\mathbf{s}} (\mathbf{r}\cdot\mathbf{s})(\mathbf{r}'\cdot\mathbf{s}) = \frac{4\pi}{3}(\mathbf{r}\cdot\mathbf{r}'),$$

where the indices on the θ angles identify the vectors between which the angle is taken, and φ is an azimuthal angle for the vectors $\mathbf{r}$ and $\mathbf{s}$. With this result taken into account, we find

$$\begin{aligned}
&\int d\Omega_{\mathbf{s}_1}\, d\Omega_{\mathbf{s}_2} |[(\mathbf{D}\cdot\mathbf{s}_1)(\mathbf{D}\cdot\mathbf{s}_2)]_{m0}|^2 \\
&\quad = \frac{16\pi^2 e^4}{9} \int \Psi_m(\mathbf{r}')\Psi_m(\mathbf{r})\Psi_0(\mathbf{r}')\Psi_0(\mathbf{r})(\mathbf{r}\cdot\mathbf{r}')^2\, d\mathbf{r}\, d\mathbf{r}' \\
&\quad = \frac{16\pi^2 e^4}{9} \sum_{i,k} |(r_i r_k)_{m0}|^2 = \frac{16\pi^2}{9} \sum_{i,k} |(D_i D_k)_{m0}|^2 .
\end{aligned}$$

Altogether, we find that the lifetime is given by

$$\frac{1}{\tau_2} = \frac{2}{9\pi\hbar^2 c^6} \sum_{i,k} |(D_i D_k)_{m0}|^2 \int \omega_1^3 \omega_2^3 \left(\frac{1}{\omega_{k0}+\omega_1} + \frac{1}{\omega_{k0}+\omega_2} \right)^2 d\omega_1 .$$

This expression should be multiplied by 4 to take into account the two polarization states of each of the two photons. We should also keep in mind that the energy conservation condition $\omega_{0m} = \omega_1 + \omega_2$ is fulfilled in this result.

We can achieve notational economies by introducing the dimensionless variable of integration x, defined by

$$\omega_1 = (\omega_{0m}/2)(1+x), \quad \text{or} \quad \omega_2 = (\omega_{0m}/2)(1-x); \quad |x| < 1,$$

and representing the extent to which each of the photon frequencies deviates from the mean frequency. In these terms, the two-photon transition rate can be written in the form

$$\frac{1}{\tau_2} = \frac{\omega_{0m}^5}{9\pi\hbar^2 c^6} \sum_{i,k} |(D_i D_k)_{m0}|^2 f\left(\frac{2\omega_{k0}}{\omega_{0m}}\right),$$

where the function f is

$$f(t) = (t+1)^2 \int_0^1 \frac{(1-x^2)^3}{[(t+1)^2 - x^2]^2}\, dx, \qquad t \geq 0.$$

An important property of this function is that it is a smoothly varying function of its argument, and varies only slightly over the entire range of the argument. For $t \to \infty$, we have $f(t) = \frac{16}{35} \approx 0.457$, while at the opposite limit, when $t = 0$, we have $f(t) = \frac{2}{3} \approx 0.667$. For intermediate values of the argument t, the function $f(t)$ takes on values between the two limits shown.

In particular, for two-photon emission from the $2s_{1/2}$ excited state of hydrogen to the $1s_{1/2}$ ground state, upon substitution of the explicit radial wave functions for hydrogen, the result is that

$$\frac{1}{\tau_2} = \frac{\omega_{21}^5}{27\pi\hbar^2 c^6} \left| \int_0^\infty R_{10}(r) R_{20}(r) r^4\, dr \right|^2 f\left(\frac{\omega_{k2}}{\omega_{21}}\right),$$

where the indices 1 and 2 denote the ground and first excited s states, respectively. When we substitute $f = 0.56$, the mean of the extreme values of f, we obtain $\tau_2 = 0.12\, s$, which is close to the known exact result $0.15\, s$.

Problem 1.31. Find the selection rules for vibrational-rotational transitions in the Raman scattering of light by molecules.

SOLUTION. Raman scattering is a process in which light is scattered by a quantum system such that the initial and final quantum states of the system are different.

We use selection rules for vibrational and rotational transitions that were found previously for single-photon transitions. To apply these rules to Raman scattering, we use them first for the transition from the initial to an intermediate state, and then for the subsequent transition from this intermediate state to the final state.

According to the solution of Problem 1.20, the vibrational quantum number v changes by one in a single-photon transition. Therefore, the change of the vibrational quantum number v between initial and final states in Raman scattering is

$$\Delta v = 0, \pm 2.$$

In similar fashion, we found in Problem 1.19 the single-photon radiative transition selection rule for a linear molecule that ($\Lambda = 0$)

$$\Delta J = \pm 1, \qquad \Delta M = 0, \pm 1,$$

where ΔJ and ΔM are the change in the rotational angular momentum and in its projection on the molecular axis, respectively. When we apply these rules to the radiative transition between initial and intermediate states, and then from the intermediate state to the final state, we find

$$\Delta J = 0, \pm 2, \qquad \Delta M = 0, \pm 1, \pm 2.$$

According to the solution of Problem 1.21, these same selection rules are also valid for vibrational-rotational transitions.

1.7 POLARIZABILITY OF MOLECULAR GAS

We define the tensor for the dynamic polarizability of a molecule in the scattering of light by using Eq. (1.48). We obtain thereby

$$\alpha_{qq'} = \sum_k \left[\frac{(D_q)_{mk}(D_{q'})_{k0}}{\hbar(\omega_{k0} - \omega_1)} + \frac{(D_{q'})_{mk}(D_q)_{k0}}{\hbar(\omega_{k0} + \omega_2)} \right], \tag{1.51}$$

where D_q is the component of the dipole moment in terms of projections q defined either in the spherical vector coordinate system ($q = -1, 0, +1$) or in the rectangular Cartesian coordinate system ($q = x, y, z$). The quantities ω_1 and ω_2 are the frequencies of the initial photon and the scattered photon. The polarizability tensor defines the polarization properties of light upon being scattered by the molecules.

When we take the frequencies ω_1 and ω_2 to be small as compared to typical electron transition frequencies, we can neglect ω_1 and ω_2 in Eq. (1.51). The concomitant consequence of this condition is that light of such low frequency cannot excite electronic transitions. Therefore, the state m in Eq. (1.51) can be different from the initial state 0 only in terms of vibrational or rotational quantum numbers.

Because of the small energy intervals in vibrational and rotational spectra, we consider such scattering to be quasi-elastic, which allows us to denote both initial and final molecular states by the index 0. Furthermore, in the calculation of the quasi-elastic scattering rate, we should average this rate over final vibrational and rotational substates with nearby energies. The small energy intervals in the vibrational and rotational spectra permits such averaging to be done classically, that is to say, over orientations of the molecule in space.

From Eq. (1.48), and recalling the condition that photon energies are small as compared to electron transition energies, we find that the scattering rate for a photon of fixed polarization has the behavior

$$w \sim \overline{\left| \sum_k 2 \frac{(\mathbf{D}_{0k} \cdot \mathbf{s}_1)(\mathbf{D}_{0k} \cdot \mathbf{s}_2)}{\hbar \omega_{k0}} \right|^2}. \tag{1.52}$$

The averaging indicated in Eq. (1.52) by the superior bar refers to an averaging over final vibrational and rotational states at a fixed electron state 0. The sum over the index k is taken over all intermediate molecular states, including electron states. The polarization state of the incident light is labeled by $\mathbf{s}_1$, while $\mathbf{s}_2$ is the polarization of the final scattered light.

From Eq. (1.51), the static polarizability tensor is of the form

$$\alpha_{qq'} = \sum_k 2\frac{(D_q)_{0k}(D_{q'})_{k0}}{\omega_{0k}}.$$

Equation (1.52) then shows that we can express the transition rate in terms of the polarization tensor as

$$w \sim \overline{\left|\sum_{q,q'} \alpha_{qq'} s_q s_{q'}\right|^2}. \tag{1.53}$$

We shall define the degree of polarization P. First, we select space axes so that the z axis lies along the polarization direction $\mathbf{s}_1$ of the incident light, and the x axis is oriented such that the wave vector $\mathbf{k}_2$ of the scattered light lies in the xz plane. The polarization of light is normal to the direction of its propagation, which is the direction of the wave vector. Therefore the polarization vector $\mathbf{s}_2$ of the scattered light can be either in the xz plane or normal to this plane. In the latter case, it is along the direction of the y axis. Then the degree of polarization is defined as

$$P = \frac{w(\mathbf{s}_2 \perp \mathbf{i}_y) - w(\mathbf{s}_2 \| \mathbf{i}_y)}{w(\mathbf{s}_2 \perp \mathbf{i}_y) + w(\mathbf{s}_2 \| \mathbf{i}_y)}. \tag{1.54}$$

The quantity P as defined by this relation is such that $P = 1$ corresponds to total polarization of the radiation (in the xz plane), while the value $P = 0$ corresponds to a lack of any polarization.

Problem 1.32. Calculate the degree of polarization of linearly polarized light scattered by a gas of linear molecules through the angle θ. This scattering angle is the angle between the wave vector of the incident light and the wave vector of the scattered light. The frequency of the scattered light is presumed to be much less than typical frequencies for atomic electron transitions. The result should be expressed in terms of the eigenvalues of the polarizability tensor.

SOLUTION. We take the polarization vector $\mathbf{s}_2$ of the scattered light to be in the xz plane. From Eq. (1.53), we find that the intensity of the light with polarization in this plane is of the form

$$w(\mathbf{s}_2 \perp \mathbf{i}_y) \sim \overline{(\alpha_{zz}\sin\theta + \alpha_{xz}\cos\theta)^2}.$$

We now align the y axis with $\mathbf{s}_2$. The intensity of the light polarized in the direction normal to its incident polarization has the behavior

$$w(\mathbf{s}_2 \| \mathbf{i}_y) \sim \overline{\alpha_{zy}^2}.$$

Hence we obtain the degree of polarization of the scattered light in the form

$$P = \frac{\overline{(\alpha_{zz} \sin \theta + \alpha_{xz} \cos \theta)^2} - \overline{\alpha_{zy}^2}}{\overline{(\alpha_{zz} \sin \theta + \alpha_{xz} \cos \theta)^2} + \overline{\alpha_{zy}^2}}.$$

Now we take into account the symmetry properties that arise when averaging over the rotational states is done. From the classical point of view, this averaging is over the rotation of the molecule around the axis normal to the molecular symmetry axis. It follows from these considerations that

$$\overline{\alpha_{xy}^2} = \overline{\alpha_{xz}^2} = \overline{\alpha_{zy}^2} \quad \text{and} \quad \overline{\alpha_{zz}\alpha_{xz}} = 0.$$

We are thus led to

$$P = \frac{(\alpha_0^2 - \alpha_1^2) \sin^2 \theta}{\alpha_0^2 \sin^2 \theta + \alpha_1^2(1 + \cos^2 \theta)},$$

where we define

$$\alpha_0^2 = \overline{\alpha_{zz}^2}, \qquad \alpha_1^2 = \overline{\alpha_{xy}^2} = \overline{\alpha_{yz}^2}.$$

When $\theta = 0$, then we have $P = 0$, while P has its maximum value at $\theta = \pi/2$, where

$$P_{\max} = \frac{\alpha_0^2 - \alpha_1^2}{\alpha_0^2 + \alpha_1^2}.$$

We denote by $\alpha_{\parallel}$ the eigenvalue of the polarizability tensor of the linear molecule along its axis, so that

$$\alpha_{\parallel} = \sum_k \frac{2(D_\xi)_{0k}(D_\xi)_{k0}}{\hbar\omega_{k0}},$$

where ξ is the axis along the moving molecular axis. We also define $\alpha_{\perp}$ as the eigenvalue in the direction η or in the direction ζ, the mutually perpendicular axes normal to the molecular axis as well as to each other. This eigenvalue has the expression

$$\alpha_{\perp} = \sum_k \frac{2(D_\eta)_{0k}(D_\eta)_{k0}}{\hbar\omega_{k0}} = \sum_k \frac{2(D_\zeta)_{0k}(D_\zeta)_{k0}}{\hbar\omega_{k0}},$$

and we note that both the η and ζ axes are moving axes. From the definition of the eigenvalues, we find that

$$\sum_k \frac{2(D_\xi)_{0k}(D_\eta)_{k0}}{\hbar\omega_{k0}} = 0.$$

The next step is to express the degree of polarization P via these eigenvalues. The angle between the molecular axis and the fixed axis z is denoted by θ. The z axis defines the direction of polarization $\mathbf{s}_1$ of the incident light. We elect to place the molecular axis in the xz plane. (This does not affect the end result, which would be the same were we to place the axis in the yz plane.) We then find

$$D_z = D_\zeta \cos\theta + D_\eta \sin\theta, \qquad D_x = -D_\zeta \sin\theta + D_\eta \cos\theta.$$

It follows that

$$\alpha_{zz} = \alpha_{\|} \cos^2\theta + \alpha_{\perp} \sin^2\theta,$$
$$\alpha_{xz} = (\alpha_{\perp} - \alpha_{\|}) \sin\theta \cos\theta.$$

When these expressions are squared and averaged over the angle θ, we obtain

$$\begin{aligned}\alpha_0^2 &= \alpha_{\|}^2\overline{\cos^4\theta} + \alpha_{\perp}^2\overline{\sin^4\theta} + 2\alpha_{\perp}\alpha_{\|}\overline{\sin^2\theta\cos^2\theta}\\ &= \tfrac{1}{15}(8\alpha_{\|}^2 + 4\alpha_{\|}\alpha_{\perp} + 3\alpha_{\perp}^2),\\ \alpha_1^2 &= (\alpha_{\perp} - \alpha_{\|})^2\overline{\sin^2\theta\cos^2\theta} = \tfrac{2}{15}(\alpha_{\perp} - \alpha_{\|})^2.\end{aligned}$$

In particular, the maximum degree of polarization is

$$P_{\max} = \frac{6\alpha_{\|}^2 + 8\alpha_{\|}\alpha_{\perp} + \alpha_{\perp}^2}{5(\alpha_{\perp}^2 + 2\alpha_{\|}^2)}.$$

This can achieve the value $P_{\max} = 1$ when $\alpha_{\|} = \alpha_{\perp}$.

Problem 1.33. Express the polarizability tensor of a rotating linear molecule in terms of its eigenvalues in the moving coordinate system associated with the molecular axis. The photon frequencies are presumed to be small as compared to typical electron transition frequencies.

SOLUTION. The polarizability tensor [see Eq. (1.48)] for Raman scattering can, according to Eq. (1.51), be written as

$$\alpha_{q'q} = \sum_k \left[\frac{(D_{q'})_{mk}(D_q)_{k0}}{\hbar(\omega_{k0} - \omega_1)} + \frac{(D_q)_{mk}(D_{q'})_{k0}}{\hbar(\omega_{k0} + \omega_2)}\right].$$

Here, D_q is the component of the dipole moment expressed in spherical vector coordinates, and ω_1 and ω_2 are the frequencies of the incident and scattered light, respectively. Our goal is to separate out the rotational degrees of freedom and to express the polarizability tensor via its eigenvalues in the coordinate system moving with the molecule.

The general formulas (C.11) relate the matrix element of the dipole moment operator in the rest system to the values in the moving coordinate system. We have

$$(D_q)_{k0} = \langle J'M'\Lambda'k|D_q|JM\Lambda 0\rangle$$
$$= \sum_{\mu}\left(\frac{2J'+1}{2J+1}\right)^{1/2}\langle J'1, Mq|JM\rangle\langle J'1, \Lambda'\mu|J\Lambda\rangle\langle\Lambda'k|D_\mu|\Lambda 0\rangle,$$

where J, M are the initial state values of the rotational momentum and its projection on the fixed direction, J', M' are the values of the same quantum numbers in the intermediate state, and Λ and $\Lambda' = \Lambda - \mu$ are projections of the electron angular momentum on the molecular axis in the initial and intermediate states, respectively. The indices 0 and k represent all other quantum numbers of the initial and intermediate states, respectively.

When we account for the fact that the energy spacing of neighboring rotational molecular states is small as compared to energy differences for electronic and vibrational states, we find the polarizability tensor to be

$$\alpha_{q'q} = \sum_{J',\mu'}\frac{2J'+1}{[(2J+1)(2J_1+1)]^{1/2}}\left\{\sum_k\frac{|(D_\mu)_{0k}|^2}{\hbar(\omega_{k0}-\omega_1)}\langle J'1, M'q|JM\rangle\right. \quad (1.55)$$
$$\times\langle J'1, \Lambda'\mu|J\Lambda\rangle\langle J'1, M'q'|J_1M_1\rangle\langle J'1, \Lambda'\mu|J_1\Lambda\rangle$$
$$+\langle J'1, M'q|J_1M_1\rangle\langle J'1, \Lambda'\mu|J_1\Lambda\rangle\langle J'1, M'q'|JM\rangle$$
$$\left.\times\langle J'1, \Lambda'\mu|J\Lambda\rangle\sum_k\frac{|(D_\mu)_{0k}|^2}{\hbar(\omega_{k0}+\omega_2)}\right\}.$$

In this expression, J_1 and M_1 are the final state rotational momentum and projection onto the fixed direction, and D_μ is the projection of the rotational-coordinate-system dipole moment onto the molecular axis. In arriving at Eq. (1.55), we have taken into account that the projection of the electron angular momentum onto the molecular axis is unchanged from initial to final state, since we have that only the rotational state of the molecule changes in Raman scattering.

Equation (1.55) gives the connection between the polarizability component and its values in the coordinate system where the symmetry axis coincides with the molecular axis. In particular, if the photon frequency is small compared to the typical frequency for electron transitions, this equation allows us to express the polarizability tensor component as a function of its eigenvalues as

$$\alpha_{\|} = \sum_k\frac{2|(D_\xi)_{0k}|^2}{\hbar\omega_{k0}} = \sum_k\frac{2|(D_{\mu=0})_{0k}|^2}{\hbar\omega_{k0}},$$

where D_ξ and $D_{\mu=0}$ are projections of the dipole moment operator onto the ξ axis directed along the molecular axis. In analogous fashion, we can express the other component of the polarizability tensor as

$$\alpha_{\perp} = \sum_k \frac{2|(D_\eta)_{0k}|^2}{\hbar\omega_{k0}} = \sum_k \frac{2|(D_\zeta)_{0k}|^2}{\hbar\omega_{k0}} = \sum_k \frac{2|(D_{\mu=\pm 1})_{0k}|^2}{\hbar\omega_{k0}},$$

where D_η and D_ζ are projections of the dipole moment onto the η and ζ axes, which are perpendicular to the molecular axis. The indices 0 and k refer to states of the nonrotating molecule.

In the limit of large (i.e., classical) rotational momenta of the molecule, where $J \approx J' \approx J_1 \gg 1$, and recalling the condition that the photon frequencies are to be small, Eq. (1.55) gives

$$\alpha_{qq'} = \frac{1}{2}\sum_{J'} [\langle J'1, M'q|JM\rangle\langle J'1, M'q'|J_1M_1\rangle + \langle J'1, M'q'|JM\rangle\langle J'1, M'q|J_1M_1\rangle]$$

$$\{\alpha_{\|}\langle J'1, \Lambda 0|J\Lambda\rangle\langle J'1, \Lambda 0|J_1\Lambda\rangle + \alpha_{\perp}[\langle J'1, \Lambda' 1|J\Lambda\rangle\langle J'1, \Lambda' 1|J_1\Lambda\rangle$$
$$+ \langle J'1, \Lambda' - 1|J\Lambda\rangle\langle J'1, \Lambda' - 1|J_1\Lambda\rangle]\}.$$

In deriving this result, we have made use of the approximation that

$$\frac{2J' + 1}{[(2J + 1)(2J_1 + 1)]^{1/2}} \approx 1,$$

which is justified by the large rotational momenta premise, $J \approx J' \approx J_1 \gg 1$. We have also used $D_\xi = 0$, and $D_\zeta = D_\eta = D_{\pm 1}$.

2

RADIATIVE TRANSITIONS BETWEEN DISCRETE STATES IN ATOMIC SYSTEMS

The formalism for radiative transitions developed in Chapter 1 will now be applied to transitions between discrete states in a bound system, which is the simplest application for this formalism. Transition energies as determined by energy conservation laws are uniquely specified and can take on only discrete values. In application to real physical systems, however, it is necessary to take into account the broadening of spectral lines arising from various interaction mechanisms.

Equation (1.7) states that the probability per unit time for a radiative transition from a state labeled 0 to another state labeled k has the form

$$w \sim \delta(\omega_{k0} - \omega),$$

where $\hbar\omega_{k0}$ is the energy difference between the states. We introduce the distribution function a_ω for the absorption or emission of a photon of frequency ω. That is, $a_\omega d\omega$ is the probability that the photon has a frequency in the interval from ω to $\omega + d\omega$. In accordance with the behavior of the transition probability given above, this distribution function has the form

$$a_\omega \, d\omega = \delta(\omega_{k0} - \omega) \, d\omega, \tag{2.1}$$

and it obeys the normalization condition

$$\int a_\omega \, d\omega = 1. \tag{2.2}$$

It follows from Eq. (2.1) that, on a scale of the order of ω_{k0}, the photon frequency distribution function has the form of a delta function. This means that the frequency difference between most photons and ω_{k0} is small compared to ω_{k0}. However, on smaller scales, there will be a departure from the delta function form for the distribution function. In other words, the distribution function does possess some structure. It is this function that determines the photon absorption and emission cross sections for the atom. We shall now study the structure of the distribution function a_ω and its dependence on the mechanisms for the broadening of spectral lines.

2.1 RADIATIVE AND DOPPLER BROADENING OF SPECTRAL LINES

The spontaneous emission of a photon by an atomic electron in an initial excited state causes this initial state to be nonstationary. Therefore the state is not sharply defined in energy, and photons emitted from such states will have some distribution of frequencies. This broadening of spectral lines is called radiative broadening.

Doppler broadening of spectral lines arises from the motion of the radiating atomic system. The frequency of the radiation emitted depends upon the velocity of the source of emission relative to the observer, and the distribution of frequencies is determined by the distribution of atomic velocities.

In contrast to the radiative broadening of spectral lines, Doppler broadening will occur even if all photons have a single, well-defined frequency at the time of emission. A Doppler-broadened spectral line can be regarded as a sum of a large number of narrow lines. For this reason Doppler broadening is called nonuniform broadening, whereas radiative broadening is a property of any single emission, and is a uniform broadening of the spectral lines.

Problem 2.1. Establish the shape of a spectral line if the broadening arises from a finite lifetime τ_k of the initial state in the transition.

SOLUTION. A finite lifetime for an initial state k means that the state is nonstationary. We shall consider the single-photon transition between the initial excited state k and the ground state 0. If w is the transition rate as established by Eq. (1.13), then for times t such that $t \gg 1/w$, the atom will be in its ground state with a probability approaching unity. We wish to determine the distribution function a_ω for the emitted photon frequencies under these circumstances. Equations (1.4) for the coefficients in a perturbation expansion have the form

$$i\hbar\,\dot{c}_k = \sum_\omega V_{k,0\omega} c_{0\omega} \exp\left[i(\omega_{k0} - \omega)t\right], \tag{2.3}$$

$$i\hbar\,\dot{c}_{0\omega} = V_{0\omega,k} c_k \exp\left[-i(\omega_{k0} - \omega)t\right],$$

where $c_{0\omega}$ is the probability amplitude for the atomic electron to be in the state 0 after the emission of a photon of frequency ω. The initial conditions for the system of equations in (2.3) are $c_k(0) = 1$ and $c_{0\omega}(0) = 0$.

We express the solution of (2.3) in the form $c_k(t) = \exp(-wt/2)$, where w is a quantity to be determined. When we substitute this exponential decay behavior into the second equation of (2.3), we obtain

$$c_{0\omega}(t) = V_{0\omega,k} \frac{\exp[-i(\omega_{k0} - \omega)t - wt/2] - 1}{\hbar(\omega_{k0} - \omega - iw/2)}. \tag{2.4}$$

The resonant character of this equation justifies the resonance approach we have employed in the interaction of an atomic electron with the electromagnetic field. Also, from Eq. (1.2), we have $V_{0\omega,k} = -\mathbf{D}_{0k}\cdot\mathbf{E}/2$. Substitution of Eq. (2.4) into the first equation of (2.3) leads to the expression (1.13) for w with $n_\omega = 0$. Hence w is $1/\tau_k$, where τ_k is the lifetime of state k. Correspondingly, from Eq. (2.4), the probability for the emission of a photon with frequency ω is

$$a_\omega = |c_{0\omega}(\infty)|^2 = \frac{1}{2\pi\tau_k} \frac{1}{(\omega - \omega_{k0})^2 + [1/(2\tau_k)]^2}. \tag{2.5}$$

Equation (2.5) satisfies the normalization condition (2.2). It is valid for $\omega_{k0}\tau_k \gg 1$. For frequencies such that $\omega \sim \omega_{k0}$, Eq. (2.5) exhibits the behavior expressed in Eq. (2.1).

We remark that we neglect a very small shift in the spectral line, which is known as the Lamb shift. It is due to the interaction of an atomic electron with the "vacuum" electromagnetic field. This has reference to the effect of virtual electromagnetic processes that can occur even in the absence of an externally applied field.

Equation (2.5) expresses the radiative broadening of spectral lines. The line shape exhibited in Eq. (2.5) is of the Lorentz form.

Problem 2.2. Two sources of line broadening both lead to line shapes of Lorentz form, with widths of $1/\tau_1$ and $1/\tau_2$. Show that the sum of the two processes also leads to a Lorentz line shape, and determine the width of the line.

SOLUTION. If we have two independent broadening mechanisms with distribution functions $a_1(\omega)$ and $a_2(\omega)$, the total distribution function has the form

$$a_\omega = \int_{-\infty}^{\infty} a_1(\omega')a_2(\omega + \omega_{k0} - \omega')\,d\omega', \tag{2.6}$$

where ω_{k0} is the central frequency of the line being examined.

For the circumstances defined in the statement of this problem, Eqs. (2.5) and (2.6) give

$$a_\omega = \frac{1}{4\pi^2\tau_1\tau_2} \int_{-\infty}^{\infty} \frac{d\omega'}{\{(\omega' - \omega_{k0})^2 + [1/(2\tau_1)]^2\}\{(\omega - \omega')^2 + [1/(2\tau_2)]^2\}}.$$

The integrand in this expression can be rewritten as

$$\left[\frac{1}{\omega' - \omega_{k0} + i/(2\tau_1)} - \frac{1}{\omega' - \omega - i/(2\tau_2)}\right]$$
$$\times \left[\frac{1}{\omega' - \omega_{k0} - i/(2\tau_1)} - \frac{1}{\omega' - \omega + i/(2\tau_2)}\right]$$
$$\times \left[\frac{1}{\omega_{k0} - \omega - i/(2\tau_1) - i/(2\tau_2)}\right]\left[\frac{1}{\omega_{k0} - \omega + i/(2\tau_1) + i/(2\tau_2)}\right].$$

The integral can then be accomplished in the complex plane by employing a path that encloses the two poles in the upper half plane to give the result

$$a_\omega = \frac{1}{2\pi\tau_{12}} \frac{1}{(\omega - \omega_{k0})^2 + [1/(2\tau_{12})]^2}, \tag{2.7}$$

where the notation is used such that

$$\frac{1}{\tau_{12}} = \frac{1}{\tau_1} + \frac{1}{\tau_1}. \tag{2.8}$$

Thus the combined spectral line is Lorentzian, and its width is the sum of the widths of the individual lines. This result can be generalized to the sum of any number of independent sources of the Lorentz broadening. That is, the total spectral line will be of Lorentz form, and its width will be the sum of the widths of all the individual lines.

We can generalize Eq. (2.5) to the case where the initial state k can decay to other states in addition to the ground state. The expression $c_k(t) = \exp(-wt/2)$ can be retained, but now the quantity w characterizes the total width of the state k. To accommodate this extension of the theory, the right-hand side of the first element of Eq. (2.3) must have terms added to it to describe transitions to other states. Equation (2.5) will be obtained again, but with τ_k now representing the total lifetime of the state, which can be composed of the partial lifetimes by means of Eq. (2.8).

Another generalization that can be made is if the lower state 0 is not the ground state but can itself decay to other states, with the result that it has a lifetime τ_0. Then Eq. (2.7) is replaced by

$$a_\omega = \frac{1}{2\pi\tau_{k0}} \frac{1}{(\omega - \omega_{k0})^2 + [1/(2\tau_{k0})]^2}, \tag{2.9}$$

where τ_{k0} determines the so-called reduced lifetime, and the spectral width is given by

$$\frac{1}{\tau_{k0}} = \frac{1}{\tau_k} + \frac{1}{\tau_0}, \tag{2.10}$$

replacing Eq. (2.8).

Problem 2.3. Find the spectral line shape that arises from the thermal motion of atoms.

SOLUTION. We label as v the relative velocity of the emitter of radiation with respect to the system that receives the radiation. If the emitted frequency is ω', the Doppler law gives ω for the apparent frequency of the received radiation, where

$$\omega = \omega'(1 + v_x/c). \tag{2.11}$$

In this expression, c is the velocity of light, v_x is the projection of the relative velocity of the radiator and receiver onto the line joining them, and $v_x > 0$ corresponds to the radiator and receiver approaching each other.

We shall determine the frequency distribution of the emitted photons in the laboratory frame of reference. The probability that an atom has a velocity in the interval from v_x to $v_x + dx$ is $f(v_x)dv_x$, which must be normalized to unity, as in

$$\int_{-\infty}^{\infty} f(v_x)dv_x = 1.$$

When we employ the connection (2.11) between the frequency of the radiation and the velocity of the atom, we obtain the frequency distribution

$$a_\omega \, d\omega = f(v_x)\, dv_x = \frac{c}{\omega'} f\left(c\frac{\omega - \omega'}{\omega'}\right) d\omega. \tag{2.12}$$

Specifically, the Maxwell velocity distribution function is

$$f(v_x) = \left(\frac{M}{2\pi T}\right)^{1/2} \exp\left(-\frac{Mv_x^2}{2T}\right),$$

where M is the mass of the emitting atom and T is the temperature in energy units. Then Eq. (2.12) gives

$$a_\omega \, d\omega = \frac{1}{\omega'}\left(\frac{Mc^2}{2\pi T}\right)^{1/2} \exp\left[-\frac{Mc^2(\omega - \omega')^2}{2T\omega'^2}\right]. \tag{2.13}$$

The Doppler function thus has a Gaussian form. From this expression, a typical Doppler width for a spectral line can be estimated to be $\Delta\omega_D \sim \omega'[T/(Mc^2)]^{1/2}$. The expression (2.13) describes a symmetrical spectral distribution centered at the transition frequency ω_{k0}, with the width $\Delta\omega_D$. For example, for atoms with mass number $A \sim 100$ at room temperature, we have $\Delta\omega_D \sim 10^{-6}\omega'$.

Problem 2.4. Find the spectral line shape resulting from the simultaneous effects of Doppler broadening and limited lifetime of the state.

SOLUTION. The spectral line shape in this problem is established by Eq. (2.9), but with the central frequency shifted as a result of the atomic motion that gives rise to the Doppler effect. The frequency distribution function found from Eq. (2.6) with the help of Eqs. (2.9) and (2.12), is

$$a_\omega = \frac{1}{2\pi\tau_{k0}} \int_{-\infty}^{\infty} \frac{f[c(\omega' - \omega_{k0})/\omega_{k0}]}{(\omega - \omega')^2 + [1/(2\tau_{k0})]^2} \frac{c}{\omega_{k0}} d\omega'.$$

We take the $f(v)$ in this equation to be the Maxwell velocity distribution. Rather than attempt the full analytical evaluation of the integral, we shall analyze certain limiting cases. First we assume that the Doppler width is small compared with the Lorentz width, which is to say that

$$\Delta\omega_D \equiv \omega_{k0}\left(\frac{2T}{Mc^2}\right)^{1/2} \ll \frac{1}{\tau_{k0}}.$$

Then if we make the replacement

$$\frac{c}{\omega_{k0}} f\left(c\frac{\omega' - \omega_{k0}}{\omega_{k0}}\right) \to \delta(\omega' - \omega_{k0}),$$

we obtain Eq. (2.9), which describes the Lorentz mechanism for the broadening of spectral lines. The Doppler broadening can be neglected in this case.

The opposite limiting case is expressed by the relation

$$\Delta\omega_D \tau_{k0} \gg 1.$$

Then for $|\omega - \omega_{k0}| \leq \Delta\omega_D$, we can replace the δ function by the Lorentz function in the above formula. This step gives Eq. (2.13) for the distribution function a_ω, which corresponds to Doppler broadening of the spectral lines. A more complicated case arises from $|\omega - \omega_{k0}| \gg \Delta\omega_D$. In this case, it is convenient to separate the integral into the two regions $\omega' - \omega_{k0} \sim \Delta\omega_D$ and $\omega' - \omega \sim 1/\tau_{k0}$. The contribution from the first region gives

$$(2\pi\tau_{k0})^{-1}(\omega - \omega_{k0})^{-2},$$

while the integral from the second region is equal to

$$\frac{1}{\Delta\omega_D \pi^{1/2}} \exp\left[-\frac{(\omega - \omega_{k0})^2}{(\Delta\omega_D)^2}\right].$$

Thus the Doppler form of the spectral line is preserved in the wings of the distribution if the condition

$$\exp\left[-\frac{(\omega - \omega_{k0})^2}{(\Delta\omega_D)^2}\right] \gg \frac{\Delta\omega_D}{\tau_{k0}(\omega - \omega_{k0})^2}$$

is fulfilled. If, on the other hand, the opposite relation

$$\exp\left[-\frac{(\omega - \omega_{k0})^2}{(\Delta\omega_D)^2}\right] \ll \frac{\Delta\omega_D}{\tau_{k0}(\omega - \omega_{k0})^2}$$

holds true, then the distribution function has the form

$$a_\omega = \frac{1}{2\pi\tau_{k0}(\omega - \omega_{k0})^2}.$$

Thus, in the case of competition between Lorentz and Doppler broadening of spectral lines, the wings of the spectral lines are always determined by the Lorentz mechanism, whereas the central part of the spectral line is determined by whichever mechanism is the stronger. In the intermediate region one can use the sum of Lorentz and Doppler contributions, because under the condition $|\omega - \omega_{k0}| \gg \Delta\omega_D$ their contributions to a_ω are independent.

In the general case, the spectral line is said to have the Voigt form. It is intermediate between the Lorentz and Doppler shapes.

Problem 2.5. Two sources of broadening lead to spectral lines of Gaussian shape with widths of Δ_1 and Δ_2. Prove that the total spectral line has a Gaussian form, and determine its width.

SOLUTION. From Eqs. (2.6) and (2.13), we have

$$\begin{aligned} a_\omega &= \frac{1}{\pi\Delta_1\Delta_2}\int_{-\infty}^{\infty}\exp\left[-\frac{(\omega'-\omega_{k0})^2}{\Delta_1^2}-\frac{(\omega'-\omega)^2}{\Delta_2^2}\right]d\omega' \\ &= \frac{1}{\pi^{1/2}\Delta}\exp\left[-\left(\frac{\omega-\omega_{k0}}{\Delta}\right)^2\right], \end{aligned}$$

where $\Delta = (\Delta_1^2+\Delta_2^2)^{1/2}$. Thus the combined line is Gaussian, and its width is the root mean square of the partial widths.

In addition to the Doppler effect, another broadening mechanism that leads to a Gaussian form of the spectral line is the broadening that occurs with atoms embedded in a crystal. Then fluctuations in local stresses cause frequency shifts in emitted photons that lead to a Gaussian form of the line.

Problem 2.6. Determine the time-of-flight broadening of an absorption line arising from the interaction of an atomic beam with an intersecting laser beam.

SOLUTION. In this case, the width of the absorption line is determined by the interaction time of the electrons with the electromagnetic field. This time is that required for an atom to cross the photon beam. Broadening takes place if the time of flight is short compared to the lifetime of the excited atoms. We assume the spatial distribution of the atoms to be Gaussian, in which case the electric field for a mode with frequency ω has the form

$$E(\rho,t) = E_0\exp(-\rho^2/a^2)\cos\omega t$$

for a beam of Gaussian width a. The coordinate ρ is the distance from the axis of the laser beam. We take the atomic beam to be narrow as compared to the laser beam, and we assume the two beams intersect perpendicularly. If the atomic velocity is v, a characteristic interaction time of an atom with the laser beam is a/v. That is, the width of the absorption line goes as v/a. We shall find its form below.

The transition amplitude for single-photon absorption is, from the perturbation theory expression (1.4),

$$c_k^{(1)} = \frac{i}{\hbar}\int_{-\infty}^{\infty}\mathbf{E}\cdot\mathbf{D}_{k0}\exp(-i\omega_{k0}t)\,dt.$$

In this expression, $\mathbf{D}_{k0}$ is the matrix element of the dipole interaction operator, ω_{k0} is the interaction frequency, and $\mathbf{E}$ is the electric field vector. We take the origin of the time coordinate to be that moment at which an atom crosses the axis of the laser beam, so that $\rho = vt$. The transition amplitude is then

$$c_k^{(1)} = \frac{i}{\hbar}\mathbf{E}_0\cdot\mathbf{D}_{k0}\int_{-\infty}^{\infty}\cos\omega t\exp\left(-i\omega_{k0}t-\frac{v^2t^2}{a^2}\right)dt.$$

The distribution function is proportional to $|c_k^{(1)}|^2$, and assuming that $|\omega - \omega_{k0}| \ll \omega$, we find

$$a_\omega \sim \exp\left[-(\omega - \omega_{k0})^2 a^2/(2v)^2\right].$$

When we apply the normalization condition of Eq. (2.2) to this expression, we obtain

$$a_\omega = \frac{2v}{a\sqrt{\pi}} \exp\left[\frac{-(\omega - \omega_{k0})^2 a^2}{(2v)^2}\right].$$

Thus the time-of-flight broadening caused by a laser beam of Gaussian spatial distribution is itself Gaussian.

Now consider the laser beam to have a rectangular profile, and take the atoms to be in the laser field for a time T, where $T = a/v$. With the same cosine time dependence of the electric field as employed above, we find that

$$a_\omega = \frac{2}{\pi} \frac{\sin^2[(\omega - \omega_{k0})/2]}{(\omega - \omega_{k0})^2 T}.$$

Problem 2.7. An excited atom undergoes diffusive motion in a gas and experiences many collisions during its radiative lifetime. Determine the form of the spectral line caused by the Doppler mechanism.

SOLUTION. Collisions in the gas cause the excited atom to change its direction many times before radiating. This problem is the reverse of Problem 2.3, where it is assumed that the direction of motion of the atom is unchanged during radiation. We fix the origin of coordinates on an atom at time $t = 0$, and then consider the probability $W(x_0, t)$ that the atom is to be found at x_0 at time t. This probability satisfies the diffusion equation

$$\frac{\partial W}{\partial t} = \mathcal{D} \frac{\partial^2 W}{\partial t^2}, \tag{2.14}$$

where $\mathcal{D}$ is the diffusion coefficient of an atom in the gas. The probability W is normalized by the condition

$$\int_{-\infty}^{\infty} W(x_0, t)\, dx_0 = 1.$$

The electric field strength for the electromagnetic wave emitted by the atom is

$$\mathbf{E} = \mathbf{E}_0 \exp\left[ik(x - x_0) - i\omega_{k0} t\right],$$

where $k = \omega_{k0}/c$. Averaging this field strength over the positions of the radiating atom gives

$$\overline{\mathbf{E}} = \int_{-\infty}^{\infty} \mathbf{E}\, W(x_0, t)\, dx_0 = \mathbf{E}_0 J(t) \exp(ikx - i\omega_{k0} t),$$

where

$$J(t) = \int_{-\infty}^{\infty} \exp(-ikx_0) W(x_0, t)\, dx_0.$$

When we multiply Eq. (2.14) by $\exp(-ikx_0)$ and integrate it over the positions x_0, we obtain

$$\frac{dJ}{dt} = \mathcal{D} \int_{-\infty}^{\infty} \exp(-ikx_0) \frac{\partial^2 W}{\partial t^2}\, dx_0.$$

Integration by parts, with application of the boundary condition $W(x_0 = \pm\infty) = 0$, gives

$$\frac{dJ}{dt} = -\mathcal{D}k^2 J.$$

The solution of this equation with the initial condition $J(0) = 1$ is

$$J(t) = \exp(-\mathcal{D}k^2 t).$$

From this it follows that the average strength of the electric field is

$$\overline{\mathbf{E}} = \mathbf{E}_0 \exp(ikx - i\omega_{k0}t - \mathcal{D}k^2 t).$$

In a classical treatment, this field is generated by a dipole moment that oscillates with a frequency ω_{k0}, that is, the dipole moment behaves as

$$\mathbf{D} \sim \exp(-i\omega_{k0}t - \mathcal{D}k^2 t).$$

According to Eq. (1.13), the photon frequency distribution function, proportional in the classical case to the square of a Fourier component of a dipole moment, is

$$a_\omega = \frac{k^2\mathcal{D}}{\pi[(\omega - \omega_{k0})^2 + (k^2\mathcal{D})^2]}. \tag{2.15}$$

Thus the distribution function has a Lorentz form with the width $2k^2\mathcal{D}$. Then, in accordance with Problem 2.2, taking account of the limited lifetime τ_{k0} of the initial state in the transition leads to a Lorentz spectral line shape with the width $2k^2\mathcal{D} + 1/\tau_{k0}$.

We remark that only the motion of the atom in the direction of propagation is significant, even though the diffusion is three dimensional. There is an analogy with Doppler broadening, where the transverse motion of the radiating atom is also nonessential. The validity criterion for Eq. (2.15) is that the time τ_{k0} must be large compared to the time for transit of a mean free path in the gas. Such a situation exists in masers. The opposite limit leads to the Doppler broadening considered in Problem 2.3.

2.2 COLLISION BROADENING OF SPECTRAL LINES

The collision mechanism (or impact mechanism) for the broadening of spectral lines follows from the fact that an excited atom will sometimes experience collisional interactions with neighboring gas molecules. This interaction is strong, but it acts only during a small portion of the excited atom's lifetime for photon emission. The result is the collision broadening of the emitted photon spectrum.

An excited atom will emit a monochromatic electromagnetic wave until such time as it experiences interaction with other atoms in the gas. The result of collisions with other atoms produces strong changes in the frequency emitted, but only for a time short as compared to the radiative lifetime. One can take this effect into account by viewing each collision as causing a major shift in the phase of the emitted electromagnetic wave. The emitted field will then consist of a succession of separate monochromatic segments, with the phase of each segment shifted with respect to adjacent segments. The limited duration of each of these independent time segments in the emission determines the width of the resulting spectral line.

We decompose the wave functions of the initial state 0 and the final states k into stationary states with no broadening interaction present. For simplicity, we set $\hbar = 1$ in the equations that follow. The expansions of the wave functions are

$$\Psi_0 = \sum_n a_{0n}(t)\psi_n \exp(-i\varepsilon_n t), \qquad \Psi_k = \sum_m a_{km}(t)\psi_m \exp(-i\varepsilon_m t), \tag{2.16}$$

where ψ_n, ψ_m are the spatial parts of the stationary state wave functions of the atom, and ε_n, ε_m are the energies of these states. The coefficients a_{0n}, a_{km} are determined by interactions with other perturbing atoms. Initial conditions for these coefficients are $a_{0n}(0) = \delta_{0n}$, $a_{km}(0) = \delta_{km}$.

We now introduce the quantities

$$f_{nm}(t) = a_{0n}(t)a^*_{km}(t),$$

and denote the special function $f_{0k}(t)$ by $f(t)$. In terms of the quantities $\varphi(t)$ defined by

$$\varphi(t) = f(t)\exp\left[i(\varepsilon_k - \varepsilon_0)t\right], \tag{2.17}$$

we can express the correlation function as

$$\Phi(\tau) = \lim_{T\to\infty}\frac{1}{T}\int_0^T \varphi(t)\varphi^*(t+\tau)\,dt.$$

Because the time dependences of a_{0n} and a_{km} are determined by phase factors, the amplitude of φ is simply $|\varphi(t)| = 1$. With $\varphi(-t) = \varphi^*(t)$, and a change in the interval of integration from 0 to T in the correlation function to the interval $-T/2$ to $T/2$, the correlation function can be expressed as

$$\Phi(\tau) = \lim_{T\to\infty}\frac{1}{T}\int_{-T/2}^{T/2} \varphi(t)\varphi^*(t+\tau)\,dt. \tag{2.18}$$

Problem 2.8. Express the photon frequency distribution function $a(\omega)$ in terms of the correlation function $\Phi(\tau)$.

SOLUTION. The interaction (1.1) between an atomic electron and the radiation field is of dipole character, and so the transition amplitude is proportional to the matrix element of the dipole moment operator between the initial and final states in the transition, that is,

$$\langle \Psi_k|\mathbf{D}|\Psi_0\rangle = \sum_{m,n} \mathbf{D}_{nm} f_{nm}(t)\exp(i\omega_{nm}t), \tag{2.19}$$

where $\mathbf{D}_{nm} = \langle \psi_n|\mathbf{D}|\psi_m\rangle$. The Fourier transform of Eq. (2.19) gives the amplitude for the emission of a photon of a given frequency. The probability for radiation of a photon of frequency ω is therefore proportional to

$$a_\omega \sim \left|\int e^{-i\omega t}\langle \Psi_k|\mathbf{D}|\Psi_0\rangle\, dt\right|^2. \tag{2.20}$$

We are interested in the production of photons with frequencies near ω_{k0}. We restrict Eq. (2.19) to terms with $m = k$ and $n = 0$, and with the assistance of Eqs. (2.16) and (2.17), we obtain

$$a_\omega \sim \left|\int e^{-i\omega t}\varphi(t)\, dt\right|^2. \tag{2.21}$$

This integral can be written in the form

$$a_\omega \sim \int dt_1 \int dt_2 \varphi(t_1)\varphi^*(t_2)\exp\left[i\omega(t_2 - t_1)\right].$$

In terms of new variables $\tau = t_1 - t_2$ and $t = t_2$, we have

$$a_\omega \sim \int d\tau\, e^{-i\omega\tau} \int dt\, \varphi^*(t)\varphi(t+\tau).$$

On the basis of the definition of the correlation function in Eq. (2.18), we obtain

$$a_\omega = \frac{1}{2\pi}\int_{-\infty}^{\infty} \Phi(\tau)e^{-i\omega\tau}\, d\tau, \tag{2.22}$$

where we have employed the normalization condition (2.2) for the distribution function a_ω. Equation (2.22) connects the emitted photon distribution function a_ω with the Fourier transform of the correlation function for the corresponding frequency of the radiated photon.

Problem 2.9. Determine the shape of a collision-broadened spectral line when it arises from collisions of an excited atom with atoms of a surrounding gas. Assume the motion of the particles to be classical.

SOLUTION. The wave function of the atomic electron in the excited state k satisfies the Schrödinger equation

$$i\hbar \frac{d\Psi_k}{dt} = \varepsilon_k \Psi_k + V_{kk}(t)\Psi_k, \tag{2.23}$$

where V_{kk} is the diagonal matrix element of the interaction between the excited electron and the perturbing particles in the gas. We neglect inelastic collisions, since elastic collisions have the dominant line-broadening effect for a wide variety of physical conditions. On the basis of Eq. (2.23), we have the simple solution for the a_{kk} amplitude of Eq. (2.16),

$$a_{kk}(t) = \exp\left[-\frac{i}{\hbar}\int^t V_{kk}(t')\,dt'\right]. \tag{2.24}$$

An interaction between the excited atom and the perturbing gas atoms takes place during a very short interval of time. Times of collision t_i are distributed randomly. Assume the interaction potential between colliding particles to be determined by the spatial locations of these particles, which is to say that we use a classical description of their motion. We introduce the coordinates $\mathbf{R}_l$ of the colliding particles l, and we postulate a certain time dependence $\mathbf{R}_l(t)$. As a result of a collision with a particle, the wave function of the initial state acquires the additional phase

$$\varkappa_{lk} = \frac{1}{\hbar}\int_{-\infty}^{\infty} V_{kk}(\mathbf{R}_l)\,dt. \tag{2.25}$$

Thus Eq. (2.24) has the form

$$a_{kk}(t) = \exp\left[-i\sum_l \varkappa_{lk}\,\eta(t-t_l)\right], \tag{2.26}$$

where the step function $\eta(t)$ is defined by

$$\eta(t) = \begin{cases} 0, & t<0 \\ 1, & t>0 \end{cases}.$$

This function expresses the fact that the duration of a collision is small as compared to a radiative lifetime.

Using an expression $a_{00}(t)$ similar to that for $a_{kk}(t)$, we can write the function $\varphi(t)$ of Eq. (2.17) in the form

$$\varphi(t) = \exp\left[i\omega_{k0}t + i\sum_l \chi_l\,\eta(t-t_l)\right], \tag{2.27}$$

in which we have used the notation

$$\chi_l = \varkappa_{lk} - \varkappa_{l0} = \frac{1}{\hbar}\int_{-\infty}^{\infty}\left[V_{kk}(\mathbf{R}_l) - V_{00}(\mathbf{R}_l)\right]dt. \tag{2.28}$$

The quantity χ_l is the phase shift introduced by the difference between the interaction potentials for the upper and lower states in the transition.

The correlation function in Eq. (2.18) can be written as

$$\Phi(\tau) = \overline{\varphi^*(t)\varphi(t+\tau)},$$

where the superior bar indicates a time average. To find $\Phi(\tau)$, we use the evolution expression

$$\begin{aligned}\Delta\Phi(\tau) &= \Phi(\tau) - \exp(-i\omega_{k0}\Delta\tau)\Phi(\tau+\Delta\tau) \\ &= \overline{\varphi^*(t)\left[\varphi(t+\tau) - \exp(-i\omega_{k0}\Delta\tau)\varphi(t+\tau+\Delta\tau)\right]}.\end{aligned} \tag{2.29}$$

The time interval $\Delta\tau$ is assumed to be small compared to the time τ characterizing the collision broadening, which is given by the free-path flight time of the excited atom. However, $\Delta\tau$ is large compared to a collision time. Since we have the inequality $\Delta\tau \ll \tau$, there can be only one collision during $\Delta\tau$, and that collision has small probability. With the insertion of Eq. (2.27), we have

$$\Delta\Phi(\tau) = \overline{\varphi^*(t)\varphi(t+\tau)\left[1 - \exp\left\{i\sum_l \chi_l[\eta(t+\tau+\Delta\tau-t_l) - \eta(t+\tau-t_l)]\right\}\right]}. \tag{2.30}$$

The sum in this expression is taken over times in the interval between $t+\tau$ and $t+\tau+\Delta\tau$. We note that simultaneous collisions with two particles are neglected because such collisions are rare.

Because of the random character of the collisions, we can do the averaging as if no previous collisions had occurred. The average value in Eq. (2.30) can thus be obtained as a product of individual averages, which gives

$$\Delta\Phi(\tau) = \Phi(\tau)\overline{[1 - e^{i\chi(\rho)}]}, \qquad t+\tau < t_l < t+\tau+\Delta\tau. \tag{2.31}$$

The quantity ρ is the impact parameter of the collision associated with the value χ from Eq. (2.28).

Averaging this expression over time amounts to averaging the collision times t_l contained within some larger time interval. From a classical point of view the impact parameter of the collision is unambiguously connected with t_l, so it is convenient to carry out the averaging in terms of ρ. The collision probability per unit time in an interval of impact parameters between ρ and $\rho + d\rho$ is equal to $2\pi\rho\, d\rho\, Nv$, so the average required for Eq. (2.31) is

$$\overline{[1 - e^{i\chi(\rho)}]} = \Delta\tau v N \int_0^\infty 2\pi\rho\, d\rho\, \left[1 - \exp(i\chi)\right], \tag{2.32}$$

where N is the number density of the perturbing particles and v is the relative velocity of the collision.

When we go to the limit $\Delta\tau \to 0$ in Eq. (2.29) and take Eqs. (2.31) and (2.32) into account, we obtain

$$\begin{aligned}\Delta\Phi(\tau) &= \Phi(\tau) - (1 - i\omega_{k0}\Delta\tau)\left[\Phi(\tau) + \frac{d\Phi}{d\tau}\Delta\tau\right] \\ &= -\Delta\tau\left(\frac{d\Phi}{d\tau} - i\omega_{k0}\Phi\right) \\ &= \Phi(\tau)\Delta\tau N v\int_0^\infty 2\pi\rho\, d\rho\, \left[1 - \exp(i\chi)\right].\end{aligned} \tag{2.33}$$

This is equivalent to

$$\frac{d\Phi}{d\tau} - i\omega_{k0}\Phi = -\Phi N v(\sigma' + i\sigma''), \tag{2.34}$$

with the notation that

$$\sigma' = \int_0^\infty 2\pi\rho\, d\rho\, (1 - \cos\chi), \tag{2.35}$$

$$\sigma'' = -\int_0^\infty 2\pi\rho\, d\rho\, \sin\chi. \tag{2.36}$$

The solution of Eq. (2.34) is

$$\Phi(\tau) \underset{\tau>0}{=} \exp\left[i\omega_{k0}\tau - N v(\sigma' + i\sigma'')\tau\right]. \tag{2.37}$$

When we substitute Eq. (2.37) into Eq. (2.22), and employ the property $\Phi(\tau) = \Phi^*(-\tau)$, we find the photon frequency distribution function to be

$$a_\omega = \frac{\nu}{2\pi[(\omega - \omega_{k0} + \Delta\nu)^2 + (\nu/2)^2]}, \tag{2.38}$$

where we employ the notation

$$\nu = 2Nv\sigma', \qquad \Delta\nu = Nv\sigma''. \tag{2.39}$$

We have found that the collision broadening of spectral lines exhibits the Lorentzian form as in radiative broadening [see Eq. (2.9)]. In contrast to the broadening due to a nonzero lifetime for radiative transitions, we now have both a broadening of the spectral line and a significant shift in its central frequency. The shift in the radiative broadening case, the so-called Lamb shift, is small as compared to the line width, while for collision broadening the shift and width are comparable.

At room temperature and atmospheric gas pressure, a typical value for the collision broadened width is $\nu \sim 10^{11}\,\mathrm{s}^{-1}$. For radiative transitions in the visible range of the spectrum, we have $\omega_{k0} \sim 10^{15}\,\mathrm{s}^{-1}$, so the ratio of width to frequency is $\nu/\omega_{k0} \sim 10^{-4}$. This is consistent with the approximations employed. We observe that the radiative width $1/\tau_{k0} \sim 10^8 - 10^9\,\mathrm{s}^{-1}$ is small compared with the above values. Also, the kinetic theory of gases identifies ν as the time of flight for a mean free path.

Problem 2.10. Connect the parameters of the collision broadening of a spectral line with the properties of elastic collisions of an excited atom with the surrounding atoms. Assume the interaction energy to be small as compared to the thermal energy of the atoms.

SOLUTION. Parameters describing collision broadening are connected with the phase shift $\chi(\rho)$. Consider the case in which the interaction potential is small compared with the thermal energy of the gas atoms, that is, when the inequality holds that $U(\mathbf{R}) \ll \mu v^2$, where μ is the reduced mass of the colliding particles and v is their relative velocity. From Eq. (2.28), we have

$$\chi(\rho) = \frac{1}{\hbar}\int_{-\infty}^{\infty} U(\mathbf{R})\,dt,$$

where $\mathbf{R}$ is the vector distance between the colliding particles and $U(\mathbf{R}) = V_{kk}(\mathbf{R}) - V_{00}(\mathbf{R})$. The quantum theory of scattering gives the scattering phase shift for elastic collisions in the form

$$\delta(\rho) = -\frac{1}{2\hbar}\int_{-\infty}^{\infty} U(\mathbf{R})\,dt, \tag{2.40}$$

where $U(\mathbf{R})$ is the interaction potential of the particles. Thus we have $\chi = -2\delta$. The total cross section for elastic scattering of atoms is

$$\sigma_t = 8\pi\int_0^{\infty} \sin^2\delta(\rho)\rho\,d\rho. \tag{2.41}$$

Using the relation $\chi = -2\delta$, Eqs. (2.35) and (2.41) yield

$$\sigma' = 4\pi\int_0^{\infty} \sin^2\delta(\rho)\rho\,d\rho = \frac{\sigma_t}{2}, \tag{2.42}$$

where the total cross section for elastic scattering refers to the interaction potential $U(\mathbf{R}) = V_{kk} - V_{00}$. For the cross section related to the spectral line shift, we have

$$\sigma'' = 2\pi\int_0^{\infty} \sin[2\delta(\rho)]\rho\,d\rho. \tag{2.43}$$

We can estimate the cross sections σ' and σ''. We have

$$\delta(\rho) = \frac{1}{2\hbar}\int_{-\infty}^{\infty} U\left(\sqrt{\rho^2 + v^2t^2}\right)dt \sim \frac{Ut}{\hbar} \sim \frac{U\rho_0}{\hbar v}.$$

This gives $\sigma' \sim \sigma'' \sim \sigma_t \sim \rho_0^2$, where the parameter ρ_0 is determined by

$$\delta(\rho_0) \sim \frac{U(\rho_0)\rho_0}{\hbar v} \sim 1. \tag{2.44}$$

The quantity ρ_0 is called the Weisskopf radius. It is the impact parameter of the scattering if the scattering phase shift is of order unity.

We now analyze the validity of the classical description of collisions for the physical problem being treated. We introduce the collisional angular momentum $l_0 = \mu\rho_0 v/\hbar$, and rewrite Eq. (2.44) in terms of l_0, which gives $[l_0 U(\rho_0)/\mu v^2] \sim 1$. The classical description is valid if the collisional angular momentum is large, that is, if $l_0 \gg 1$. The criterion for validity of a classical treatment is thus

$$\frac{U(\rho_0)}{\mu v^2} \ll 1.$$

This establishes the correctness of the conditions used in obtaining Eq. (2.40). In the classical sense, we can consider the motions of the atoms to be free.

We now give expressions for classical cross sections if the difference of the interaction potentials is approximated by the inverse power law dependence $U(R) = CR^{-n}$. We then have

$$\begin{aligned}\delta(\rho) &= -\frac{C}{2\hbar}\int_{-\infty}^{\infty}\frac{dt}{(\rho^2+v^2t^2)^n}\\ &= -\frac{C\pi^{1/2}}{2\hbar v\rho^{n-1}}\frac{\Gamma[(n-1)/2]}{\Gamma(n/2)},\end{aligned}$$

where Γ is the gamma function. It is seen that the total cross section for elastic scattering does not depend on the sign of the interaction potential, while the cross section σ'' has the same sign as the interaction potential.

Inserting values of $\delta(\rho)$ into the expressions for the cross sections, we obtain

$$\sigma_t = 8\pi\int_0^{\infty}\rho\,d\rho\,\sin^2\left(\frac{A}{\rho^{n-1}}\right) = \frac{8\pi A^{2/(n-1)}}{n-1}\int_0^{\infty} y^{-(n+1)/(n-1)}\sin^2 y\,dy,$$

$$\sigma'' = -2\pi\int_0^{\infty}\rho\,d\rho\,\sin\left(\frac{2A}{\rho^{n-1}}\right) = -\frac{2\pi A^{2/(n-1)}}{n-1}\int_0^{\infty} y^{-(n+1)/(n-1)}\sin 2y\,dy,$$

with the definition

$$A = \frac{C\pi^{1/2}}{2\hbar v}\frac{\Gamma[(n-1)/2]}{\Gamma(n/2)}.$$

With the integrals evaluated as

$$\int_0^{\infty} y^{\mu-1}\sin^2 y\,dy = \frac{\Gamma(\mu)\cos(\mu\pi/2)}{2^{\mu+1}},$$

$$\int_0^{\infty} y^{\mu-1}\sin 2y\,dy = \frac{\Gamma(\mu)\sin(\mu\pi/2)}{2^{\mu}},$$

where $\mu = -2/(n-1)$, we obtain

$$\sigma_t = a_n \left(\frac{|C|}{\hbar v} \right)^{2/(n-1)}, \qquad |\sigma''| = b_n \left(\frac{|C|}{\hbar v} \right)^{2/(n-1)}, \tag{2.45}$$

in terms of the quantities

$$\begin{aligned} a_n &= \frac{8\pi}{n-1} \left\{ \frac{\pi^{1/2}}{2} \frac{\Gamma[(n-1)/2]}{\Gamma(n/2)} \right\}^{2/(n-1)} \frac{|\Gamma[-2/(n-1)]|}{2^{(n-3)/(n-1)}} \cos\left(\frac{\pi}{n-1} \right), \\ b_n &= \frac{2\pi}{n-1} \left\{ \frac{\pi^{1/2}}{2} \frac{\Gamma[(n-1)/2]}{\Gamma(n/2)} \right\}^{2/(n-1)} \frac{|\Gamma[-2/(n-1)]|}{2^{-2/(n-1)}} \sin\left(\frac{\pi}{n-1} \right). \end{aligned} \tag{2.46}$$

In these expressions, we restrict n to $n > 3$. For $n = 3$, we have $b_n = \infty$. The parameters a_n and b_n are connected by the relation

$$a_n = 2b_n \cot\left(\frac{\pi}{n-1} \right).$$

Table 2.1 lists values of a_n and b_n for a variety of values of n.

Problem 2.11. Find the cross section for broadening of a spectral line if the interaction potential is a sharply peaked function of the distance between atomic nuclei. Assume the particle motion to be classical.

SOLUTION. We employ the results found above for broadening cross sections, and take into account the sharply peaked character of the interaction potential of the colliding particles, using a classical context. The broadening cross section is given in Eq. (2.42) as

$$\sigma' = \frac{\sigma_t}{2} = 4\pi \int_0^\infty \sin^2 \delta(\rho) \rho \, d\rho,$$

where the scattering phase shift is given by the expression

$$\delta(\rho) = -\frac{1}{2\hbar} \int_{-\infty}^{\infty} U\left(\sqrt{\rho^2 + v^2 t^2} \right) dt = -\frac{1}{\hbar v} \int_0^\infty U\left(\sqrt{\rho^2 + z^2} \right) dz,$$

where $z = vt$. We now wish to evaluate the integral containing the scattering phase shift when the interaction potential has a sharp peak. The main contribution to the integral comes from the vicinity of $z = 0$. We shall evaluate the integral by the

TABLE 2.1. Representative Values of the a_n, b_n Quantities

n	4	6	8	12	16
a_n	11.4	8.1	7.2	6.6	6.4
b_n	9.8	2.9	1.7	0.96	0.68

method of stationary phase. We write the integral in the form

$$\delta(\rho) = -\frac{1}{\hbar v}\int_0^\infty \exp\left[\ln U(\sqrt{\rho^2 + z^2})\right] dz,$$

and use a series expansion of the exponent near the stationary phase point at $z = 0$. We obtain

$$\ln U\left(\sqrt{\rho^2 + z^2}\right) \approx \ln U(\rho) + \left(\sqrt{\rho^2 + z^2} - \rho\right)\frac{d \ln U(\rho)}{d\rho}$$
$$\approx \ln U(\rho) + \frac{U'(\rho)}{U(\rho)}\frac{z^2}{2\rho}.$$

We now introduce the parameter

$$\alpha = -\frac{U'(\rho)}{2\rho U(\rho)} > 0,$$

which is proportional to the logarithmic derivative of the interaction potential. In terms of this parameter, the scattering phase shift is

$$\delta(\rho) = -\frac{U(\rho)}{\hbar v}\int_0^\infty \exp(-\alpha z^2)\, dz = -\frac{U(\rho)}{\hbar v}\frac{1}{2}\left(\frac{\pi}{\alpha}\right)^{1/2}$$
$$= -\frac{U(\rho)}{\hbar v}\left[-\frac{\pi\rho}{2}\frac{U(\rho)}{U'(\rho)}\right]^{1/2}.$$

We assume the integral to be convergent in a small region near $z = 0$ of extent $z \sim \alpha^{-1/2} \ll \rho$, where ρ is a typical size that characterizes the behavior of the interaction potential. From this it follows that the condition corresponding to the assumption of a sharply peaked interaction potential is

$$\left|\frac{\rho U'(\rho)}{U(\rho)}\right| \gg 1.$$

This is the criterion for the validity of these expressions.

For example, for the power law interaction potential $U(R) = CR^{-n}$, we obtain $\alpha = n/(2R)$, and the validity condition has the form $n \gg 2$. For the power law interaction potential, the scattering phase shift is

$$\delta(\rho) = -\frac{C}{\hbar v \rho^{n-1}}\sqrt{\frac{\pi}{2n}}.$$

We wish to compare this result with a precise expression from the preceding problem. When we employ the previous expression in the limit $n \to \infty$, we obtain

$$\delta(\rho) = -\frac{C\sqrt{\pi}}{2\hbar v\rho^{n-1}}\frac{\Gamma[(n-1)/2]}{\Gamma(n/2)} \underset{n\gg 1}{\approx} -\frac{C}{\hbar v\rho^{n-1}}\sqrt{\frac{\pi}{2n}},$$

where we have used Stirling's formula for the gamma function. Thus we see that the two results coincide.

The parameters a_n and b_n of the preceding problem approach the limiting values

$$\lim_{n\to\infty} a_n = 2\pi, \qquad \lim_{n\to\infty} b_n = \pi^2/n.$$

These values are consistent with the entries in Table 2.1.

We wish to evaluate the integral that gives the broadening cross section. The total cross section for elastic scattering is

$$\sigma_t = 8\pi \int_0^\infty \rho\, d\rho\, \sin^2\left[\frac{U(\rho)}{\hbar v}\sqrt{-\frac{\pi\rho U(\rho)}{2U'(\rho)}}\right].$$

For small impact parameters, the square of the sin function can be replaced by its average value of $\frac{1}{2}$ because of the rapid oscillations of this function. This gives the simple result

$$\sigma_t = 8\pi \int_0^{\rho_0} \rho\, d\rho\, \frac{1}{2} = 2\pi\rho_0^2,$$

where the end point value ρ_0 is given by that ρ for which rapid oscillations of the argument of the sin function cease. This is determined by

$$\beta \equiv \frac{U(\rho)}{\hbar v}\sqrt{-\frac{\pi\rho U(\rho)}{2U'(\rho)}} \sim 1.$$

We check this general result using the example of the power law interaction potential $U(R) = CR^{-n}$, with $n \gg 1$. This leads to the Weisskopf radius

$$\rho_0 = \left(\frac{|C|\sqrt{\pi}}{\hbar v\beta\sqrt{2n}}\right)^{1/(n-1)},$$

and hence to the total cross section

$$\sigma_t = 2\pi\left(\frac{|C|^2\pi}{2n\hbar^2 v^2\beta^2}\right)^{1/(n-1)}.$$

It follows from this that

$$a_n = 2\pi\left(\frac{\pi}{2n}\right)^{1/(n-1)},$$

and, because of the stated properties that $n \gg 1$ and $\beta \sim 1$, then $\beta^{-2/(n-1)} \to 1$, and the dependence on n disappears. In the limit of large n, the quantity $(\pi/2n)^{1/(n-1)}$ can be replaced by unity, and we return to the asymptotic result $a_n = 2\pi$.

Problem 2.12. Express the parameters for the collision broadening of spectral lines in terms of the characteristics of the collisions of excited atoms with the perturbing atoms of the gas. Take the collisions to be described in quantum terms, and assume the excited atomic state to be nondegenerate.

SOLUTION. We assume for simplicity that the broadening arises only from the upper level of the radiative transition in the excited atom. Collisions with surrounding gas atoms lead to an abrupt change in the phase of the wave function for the upper state in the transition, and this leads to the broadening of the radiated spectral line. However, in contrast to Problem 2.10, the collision process is not now described by classical laws, although in both cases the process corresponds to a strong interaction between the colliding bodies, acting during a short time. For the elastic scattering cross section it is now necessary to use the quantum expression

$$\sigma_t = \frac{4\pi}{K^2}\sum_{l=0}^{\infty}(2l+1)\sin^2\delta_l^{(k)}, \tag{2.47}$$

where $\delta_l^{(k)}$ is the elastic scattering partial-wave phase shift for the collisional angular momentum l. The quantity $K = \mu v/\hbar$ is the wave number for the colliding particles. When we repeat the considerations of Problem 2.10 up to Eq. (2.38), we obtain the same connection between the broadening cross section and the total cross section for elastic scattering as in the classical case: $\sigma' = \sigma_t/2$. This means that the total cross sections for elastic scattering and for collision broadening of spectral lines in the problem under consideration are determined by the same interaction.

We now take into consideration broadening involving both the upper and lower states in the transition. The Lorentz character of the broadening means that the overall width of the line will be a sum of the partial widths due to each of the interactions. The broadening cross section is then the average of the elastic scattering cross sections of these states, or

$$\sigma' = \frac{1}{2}\left[\sigma_t^{(0)} + \sigma_t^{(k)}\right] = \frac{2\pi}{K^2}\sum_{l=0}^{\infty}(2l+1)\left[\sin^2\delta_l^{(0)} + \sin^2\delta_l^{(k)}\right], \tag{2.48}$$

where the index 0 refers to the lower state in the transition. This result differs from the classical Eq. (2.41), which obtains for large collisional momenta. This difference is the consequence of a quantum mechanical scattering phase interference.

Based on the analogy between quantum and classical results, we obtain the quantum cross section for the shift in the spectral line associated with collision broadening, when the broadening arises only from the upper state in the transition, as

$$\sigma'' = \frac{\pi}{K^2}\sum_{l=0}^{\infty}(2l+1)\sin\left[2\delta_l^{(k)}\right].$$

If both upper and lower states participate in the shifting of the spectral line, this expression becomes

$$\sigma'' = \frac{\pi}{K^2} \sum_{l=0}^{\infty} (2l+1) \left\{ \sin\left[2\delta_l^{(0)}\right] + \sin\left[2\delta_l^{(k)}\right] \right\}. \tag{2.49}$$

Note that we have implicitly used asymptotic expressions for the wave functions. This leads to final expressions that involve elastic scattering cross sections and requires that the collision velocity must be large. The validity criteria for this step will be considered in Problem 2.14.

Problem 2.13. An excited state of an atom is degenerate with respect to the projections m of the orbital angular momentum. Determine the parameters for the collision broadening of spectral lines in this case, using the assumption that the broadening is created by collisions where only the excited states contribute.

SOLUTION. In this case the line broadening is a result of both the elastic collisions of the excited atom and of transitions with a change in the magnetic quantum number m. The wave function of the excited state is labeled ψ_m. Assume that at $t = 0$ the atom is in the nth angular projection sublevel, and transitions to other states occur subsequently. If $a_{nm}(t)$ is the probability amplitude that the system is in the state n, the atomic wave function is

$$\Psi(\mathbf{r}, t) = \sum_m a_{nm}(t)\psi_m \exp(-i\varepsilon t/\hbar).$$

When we expand the wave function in the Legendre polynomials $P_l(\cos\theta)$, the general expression for the scattering amplitude f is

$$f(\theta) = \frac{1}{2iK} \sum_{l=0}^{\infty} (2l+1) \left(\sum_m S_{nm}^l - 1 \right) P_l(\cos\theta), \tag{2.50}$$

where S_{nm}^l is an element of the scattering matrix. Because of the isotropy of the collisions, the expression (2.50) must be averaged over all directions. Equation (C.1) gives this averaged scattering matrix element as

$$\overline{S_{nm}^l} = \sum_{ij} D_{nl}^{*L}(\vartheta) S_{ij}^l D_{jm}^L(\vartheta),$$

where L is the angular momentum of an atomic electron in a given state with projections labeled by n and m. The quantity $D_{jm}^L(\vartheta)$ is the generalized spherical function, or rotation function, for the angle ϑ between the collision direction and a fixed quantization axis. Using the relation (C.2),

$$\overline{D_{nl}^{*L}(\vartheta) D_{jm}^L(\vartheta)} = \frac{1}{2L+1} \delta_{ij}\delta_{nm},$$

we obtain

$$\overline{S^l_{nm}} = \frac{\delta_{nm}}{2L+1} \sum_{l=-L}^{L} S^l_{ii}, \tag{2.51}$$

where n, m, i are magnetic quantum numbers of a degenerate excited state. In the absence of degeneracy, we would obtain from this the usual S-matrix element of scattering. When we substitute Eq. (2.50) into (2.51) and use the optical theorem connecting the total cross section with the scattering amplitude for the forward direction, we obtain the broadening cross section

$$\sigma' = \frac{\pi}{K^2} \sum_{l=0}^{\infty} \sum_{m=-L}^{L} \frac{2l+1}{2L+1} (1 - \operatorname{Re} S^l_{mm}). \tag{2.52}$$

We now wish to calculate σ''. Equation (2.37) shows that σ'' appears in the correlation function in the combination $\sigma' + i\sigma''$. Hence one can obtain σ'' by taking the real part of Eq. (2.50) and repeating the steps that led to Eq. (2.52). The result is

$$\sigma'' = -\frac{\pi}{K^2} \sum_{l=0}^{\infty} \sum_{m=-L}^{L} \frac{2l+1}{2L+1} \operatorname{Im} S^l_{mm}. \tag{2.53}$$

Problem 2.14. Determine the parameters for broadening of a spectral line arising from transitions from P states to S states in an atom immersed in a gas of like atoms. Use the framework of the collision broadening theory and the quasi-classical theory of elastic collisions. The number density of atoms in the ground S state is N.

SOLUTION. To use quasi-classical methods we must have $U \ll E$, where U is the interaction potential and E is the kinetic energy of the relative motion between colliding particles. Projections of the orbital angular momentum on a fixed axis can be used as good quantum numbers for an excited atom as long as inelastic transitions between sublevels of an excited atom can be neglected. The scattering S matrix is $S^l_{nn} = \exp(2i\delta^l_n)$, where δ^l_n is the scattering phase shift for a sublevel n and collisional orbital angular momentum l. The quasi-classical expression for this phase shift is

$$\delta^l_n = \frac{1}{2\hbar} \int_{-\infty}^{\infty} U_n \, dt,$$

where U_n is the interaction potential for a P-state atom with an angular projection n.

Cross sections for broadening and shifting are given by Eqs. (2.52) and (2.53). Since the principal contributions to these sums come from large values of the collision momenta, the sums can be replaced by integrals over the impact parameter ρ, related to the angular momentum l by the relation $l = K\rho$, where K is the magnitude of the wave vector of the relative motion of the colliding atoms. The index n has six values, arising from the three projections $m = -1, 0, +1$ for a P-state atom combined with the parity of the state. Equations (2.52) and (2.53) give the broadening and shifting cross sections

$$\sigma' = \int_0^\infty 4\pi\rho\, d\rho\, \frac{1}{6} \sum_{n=1}^{6} \sin^2\left[\frac{1}{2\hbar}\int_{-\infty}^{\infty} U_n(R)\, dt\right],$$

$$\sigma'' = \int_0^\infty 4\pi\rho\, d\rho\, \frac{1}{6} \sum_{n=1}^{6} \sin\left[-\frac{1}{\hbar}\int_{-\infty}^{\infty} U_n(R)\, dt\right].$$

Interaction potentials for atoms in ground and excited states have the same values for gerade and ungerade states of interacting atoms, but different signs. For this reason, terms in the σ'' expression mutually cancel, leading to $\sigma'' = 0$. That is, there is no shifting of spectral lines in this case.

For calculation of σ', we employ a fixed rectangular coordinate system whose axes are the impact parameter of collision ρ, the relative velocity of the atoms $\mathbf{v}$, and the angular momentum $\mathbf{M}$ of the relative motion of the atoms. Take the wave functions of the excited states such that the origin of x corresponds to zero projection of momentum on the axis ρ, the origin of y corresponds to zero projection of the atom momentum onto the axis $\mathbf{v}$, and the origin of z corresponds to zero projection of the angular momentum onto the axis $\mathbf{M}$ that is perpendicular to the plane of motion. We take into consideration that one can regard the classical trajectories of the relative motion of the atoms to be rectilinear.

The wave function of the ground state is designated by φ and that of the excited state by ψ. Then the wave function of the quasi molecule—a system of colliding atoms—has the form $(\psi_1\varphi_2 \pm \psi_2\varphi_1)/\sqrt{2}$. Here the subscript labels the atom, the upper sign corresponds to the gerade state of the quasi molecule, and the lower sign refers to the ungerade state. The interaction operator of the atoms has the form (see Problem 1.10)

$$V = \frac{1}{R^3}\left[\mathbf{D}_1 \cdot \mathbf{D}_2 - 3(\mathbf{D}_1 \cdot \mathbf{n})(\mathbf{D}_2 \cdot \mathbf{n})\right],$$

where $\mathbf{D}_1$, $\mathbf{D}_2$ are the dipole moment operators of the atoms, and $\mathbf{n}$ is the unit vector directed along the quasi molecule axis. Since the atoms are neutral, this interaction is most important for large distances between the atoms. In first-order perturbation theory we have

$$U(R) = \frac{1}{2R^3}\langle(\psi_1\varphi_2 \pm \psi_2\varphi_1)|\mathbf{D}_1 \cdot \mathbf{D}_2 - 3(\mathbf{D}_1 \cdot \mathbf{n})(\mathbf{D}_2 \cdot \mathbf{n})|(\psi_1\varphi_2 \pm \psi_2\varphi_1)\rangle.$$

Using the selection rule for matrix elements of the dipole momentum operator, we obtain

$$U(R) = \pm\frac{1}{2R^3}\langle\psi_1\varphi_2|\mathbf{D}_1 \cdot \mathbf{D}_2 - 3(\mathbf{D}_1 \cdot \mathbf{n})(\mathbf{D}_2 \cdot \mathbf{n})|\psi_2\varphi_1\rangle,$$

where the plus and the minus refer to gerade and ungerade states, respectively, of the quasi molecule. Matrix elements of different projections of the dipole moment operator are expressed by

$$D = \int \psi(r)\varphi(r)r^3\, dr,$$

where $\psi(r)$ and $\varphi(r)$ are radial wave functions for excited and ground atomic states. This gives the value of the matrix element as

$$\langle\varphi|\mathbf{D}\cdot\mathbf{k}|\psi\rangle = D\cos\theta/\sqrt{3},$$

where **k** is some unit vector and θ is the angle between this vector and the quantization axis, which is the line joining the two atomic nuclei.

In terms of the oscillator strength f_{0k} of the $S - P$ transition, given in Eq. (1.18), the matrix element D is expressed by

$$D^2 = \frac{\hbar e^2}{2m\omega_{0k}} f_{0k},$$

where $\hbar\,\omega_{0k}$ is the energy difference of the states. Table 2.2 gives D^2 values in atomic units ($e^2a_0^2$).

TABLE 2.2. Values of the Squared Dipole Moment Matrix Element

Atom (transition)	D^2
H ($1^2S \longrightarrow 2^2P$)	1.665
Li ($2^2S \longrightarrow 2^2P$)	11.0
Na ($3^2S \longrightarrow 2^2P$)	18.8
K ($4^2S_{1/2} \longrightarrow 4^2P_{1/2}$)	26.4
Rb ($5^2S_{1/2} \longrightarrow 5^2P_{1/2}$)	28.4
Rb ($5^2S_{1/2} \longrightarrow 5^2P_{3/2}$)	28.0
Cs ($6^2S_{1/2} \longrightarrow 6^2P_{1/2}$)	34.8
Cs ($6^2S_{1/2} \longrightarrow 6^2P_{3/2}$)	34.8
He ($2^1P \longrightarrow 2^1P$)	0.53
Mg ($3^1P \longrightarrow 3^1P$)	11.3
Ca ($4^1P \longrightarrow 4^1P$)	20.7
Sr ($5^1P \longrightarrow 5^1P$)	23.4
Ba ($5^1P \longrightarrow 5^1P$)	25.5
Zn ($4^1P \longrightarrow 4^1P$)	8.5
Hg ($6^1P \longrightarrow 6^1P$)	7.23
Hg ($6^1P \longrightarrow 6^3P$)	0.25

On the basis of the above expressions, we have

$$U_{xx}(R) = \pm\frac{1}{3R^3}D^2(1 - 3\cos^2\theta),$$

$$U_{yy}(R) = \pm\frac{1}{3R^3}D^2(1 - 3\sin^2\theta),$$

$$U_{zz}(R) = \pm\frac{1}{3R^3}D^2.$$

Other interaction operator matrix elements are zero. The indices x, y, z correspond to states with zero projection of the angular momentum vector of the P-state atom onto the ρ, v, M axes, respectively; θ is the angle between the ρ axis and the R (quasi-molecular) axis; the plus sign corresponds to the gerade state of the quasi molecule, and the minus sign refers to the ungerade state.

The character of the atomic collision process must be analyzed. Simple elastic scattering is accompanied by a process in which the excitation energy of one atom is exchanged with that of the other. Atomic depolarization also occurs. This refers to the possibility that states with different angular momentum projections m become mixed as a result of the scattering. This will influence the broadening cross section.

Transitions between states with different projections of the angular momentum can be viewed as rotations of the interatomic axis of the quasi molecule. However, states with zero projection of the P-state atom on the axis perpendicular to the plane defined by the motion of the atoms are not mixed with other states. We represent the wave function of the system of colliding atoms with this angular momentum projection as

$$\Psi_z = a_{+z}\psi_{+z} + a_{-z}\psi_{-z},$$

where $\psi_{+z} = (\psi_{1z}\varphi_{2z} + \psi_{2z}\varphi_{1z})/\sqrt{2}$, $\psi_{-z} = (\psi_{1z}\varphi_{2z} - \psi_{2z}\varphi_{1z})/\sqrt{2}$, with wave functions as defined above, and where the index z characterizes the angular momentum projection of an excited atom.

The Schrödinger equation $i\hbar\partial\Psi/\partial t = H\Psi$ has the form

$$i\hbar\,\dot{a}_{+z} = (D^2/3R^3)a_{+z}, \qquad i\hbar\,\dot{a}_{-z} = -(D^2/3R^3)a_{-z},$$

where $R^2 = \rho^2 + v^2t^2$. Solving these equations for the free relative motion of the atoms gives the scattering matrix elements, or probability amplitude for survival of the corresponding state,

$$S_{+z}(\rho) = a_{+z}(\infty) = \exp\left[-i2D^2/(3\hbar\rho^2 v)\right],$$

$$S_{-z}(\rho) = a_{-z}(\infty) = \exp\left[i2D^2/(3\hbar\rho^2 v)\right].$$

In evaluation of these scattering matrix elements, we impose the initial condition $a_{\pm z}(-\infty) = 1$. We then obtain the results for the scattering matrix elements for excitation exchange S_{ex} and for elastic scattering S_{el}:

$$S_{ex} = \frac{(S_{+z} - S_{-z})}{2} = i \sin\left(\frac{2D^2}{3\hbar\rho^2}\right),$$

$$S_{el} = \frac{(S_{+z} + S_{-z})}{2} = \cos\left(\frac{2D^2}{3\hbar\rho^2}\right).$$

The cross sections for excitation exchange, σ_{ex}, and for elastic scattering, σ_{el}, are

$$\sigma_{ex} = \int_0^\infty 2\pi\rho\, d\rho\, |S_{ex}|^2 = \frac{\pi^2 D^2}{3\hbar v},$$

$$\sigma_{el} = \int_0^\infty 2\pi\rho\, d\rho\, |1 - S_{el}|^2 = \frac{\pi^2 D^2}{3\hbar v}.$$

The total cross section is the sum of these cross sections

$$\sigma_{tot} = \sigma_{ex} + \sigma_{el} = 2\pi^2 D^2/(3\hbar v),$$

and the broadening cross section is

$$\sigma'_z = \sigma_{tot}/2 = \pi^2 D^2/(3\hbar v).$$

Scattering matrices for other projections of the angular momentum cannot be found analytically. To simplify an approximate treatment of these matrix elements, we assume that there are no transitions between states with different angular momentum projections in the excited state during collisions. In analogy with the case just treated, we find the equations for the coefficients of the wave functions to be

$$i\hbar\, \dot{a}_{+x} = (1 - 3\cos^2\theta)(D^2/3R^3)a_{+x}, \qquad i\hbar\, \dot{a}_{-x} = -(1 - 3\cos^2\theta)(D^2/3R^3)a_{-x},$$

$$i\hbar\, \dot{a}_{+y} = (1 - 3\sin^2\theta)(D^2/3R^3)a_{+y}, \qquad i\hbar\, \dot{a}_{-y} = -(1 - 3\sin^2\theta)(D^2/3R^3)a_{-y}.$$

The solutions of these equations give the scattering matrices

$$S_{+x}(\rho) = S^*_{-x}(\rho) = \exp\left(i\frac{2D^2}{3\hbar\rho^2 v}\right),$$

$$S_{+y}(\rho) = S^*_{-y}(\rho) = 0.$$

These results give the average cross sections for excitation exchange and for elastic scattering,

$$\overline{\sigma}_{ex} = \overline{\sigma}_{el} = 2\pi^2 D^2/(9\hbar v),$$

and the average broadening cross section,

$$\overline{\sigma}' = 2\pi^2 D^2/(9\hbar v). \tag{2.54}$$

The width of the spectral line, a quantity that appears in Eq. (2.38), with the definition given in Eq. (2.39), is

$$\nu = 2Nv\sigma' = 4\pi^2 D^2 N/(9\hbar). \tag{2.55}$$

We see that the spectral line width does not depend on the velocities of the atoms. The error associated with the approximation employed here can be appraised by comparing with the outcome of a more accurate procedure. The accurate solution gives the excitation exchange cross section $\overline{\sigma}_{ex} = 2.26\pi D^2/(3\hbar v) = 2.37D^2/(\hbar v)$, while the approximation used in this problem gives $\overline{\sigma}_{ex} = 2\pi^2 D^2/(9\hbar v) = 2.19D^2/(\hbar v)$. The precise value for the elastic scattering cross section is $\overline{\sigma}_{el} = 2.58\pi D^2/(3\hbar v) = 2.70D^2/(\hbar v)$, as compared to the outcome of the currently applied approximations of $\overline{\sigma}_{el} = 2.19D^2/(\hbar v)$. Finally, the accurate broadening cross section is $2.53D^2/(\hbar v)$, versus the present approximation of $2.19D^2/(\hbar v)$.

We wish now to assess the range of variables within which the current approximation is useful. The validity criterion for the collision broadening of spectral lines found in the following in Eq. (2.66) is that the Lorentz line form is appropriate in the frequency range $|\omega - \omega_{k0}|$ when the condition $v/\sqrt{\sigma'} \gg |\omega - \omega_{k0}|$ is fulfilled. This implies that

$$\hbar v^3/D^2 \gg |\omega - \omega_{k0}|^2.$$

That is, the velocity of the atoms is constrained by the condition

$$v \gg \frac{D^{2/3}|\omega - \omega_{k0}|^{2/3}}{\hbar^{1/3}}. \tag{2.56}$$

This means that the broadening of the spectral line arises from separate and relatively high velocity collisions of an excited atom with the surrounding ground-state atoms. In particular, for the main part of the spectrum where $|\omega - \omega_{k0}| \sim \nu$, this criterion is

$$\frac{\hbar v^3}{D^2} \gg \frac{D^4 N^2}{\hbar^2}, \quad \text{or} \quad v \gg \frac{N^{2/3} D^2}{\hbar}. \tag{2.57}$$

Collision broadening occurs, therefore, when the density of atoms is low.

2.3 QUASI-STATIC THEORY OF SPECTRAL LINE BROADENING

The quasi-static theory of spectral line broadening corresponds to physical conditions opposite to those associated with the collision mechanism of line broadening. In the quasi-static case, broadening is created during times that are small compared to times associated with motion of the atoms. One can therefore consider the perturbing atoms to be motionless, and the spectral frequency shift resulting from the interaction of the radiating atom with the surrounding atoms is a sum of the shifts due to a pairwise interaction with each perturbing atom, regarding the spatial configuration of these atoms to be fixed. This interaction energy is assumed to be small compared to typical atomic energies. Perturbation theory is thus appropriate for the treatment of this broadening.

A general goal of the quasi-static theory of spectral line broadening is the determination of the frequency distribution function of the emitted photons. It is connected with a spatial distribution of the perturbing atoms that depends on their interaction with the radiating atom. To characterize the influence of this interaction on the distribution of the perturbing atoms in space, we introduce a function $w(R)$, which is the ratio of the probability for this atom to be at a distance R from the radiating atom to the probability that the atom is at infinity.

Problem 2.15. Express the frequency distribution function $a(\omega)$ of the emitted photons in terms of $w(R)$ within the framework of the quasi-static line-broadening theory.

SOLUTION. In lowest order perturbation theory, the shift in the frequency of an emitted photon due to interaction of the excited atom with neighboring atoms is

$$\omega_{k0} - \omega = \frac{1}{\hbar} \sum_m U(R_m). \tag{2.58}$$

Here, m is the index identifying the perturbing atom, R_m is the distance between the perturbing atom and the radiating atom, and $U(R_m)$ is the difference between the interaction potentials for the upper and the lower states in the radiative transition. Equation (2.58) is assumed to be averaged over the quantum states of both particles, that is, it is the diagonal matrix element of the pairwise interaction operator between the atoms.

Equation (2.58) gives the dependence of the shift of the emitted frequency on the spatial locations of the perturbing atoms. We introduce the notation $U_m = U(R_m)$ and the probability $p(U_m)\,dU_m$ that the pairwise interaction potential of the atoms lies in the interval between U_m and $U_m + dU_m$. Then, according to the definition of a_ω, we have

$$a_\omega\, d\omega = \prod_m p(U_m)\, dU_m \quad \text{at} \quad \omega_{k0} - \omega = \frac{1}{\hbar} \sum_m U(R_m). \tag{2.59}$$

It is assumed that the spatial location of any of the perturbing atoms is independent of that of the others.

When we introduce the spatial probability function $w(R)$ described above, we have

$$p(U_m)\, dU_m = w(R_m) \frac{d\mathbf{R}_m}{\Omega}, \tag{2.60}$$

where Ω is a normalization volume which includes the perturbing atoms and the radiating atom. If $w = 1$, the ratio $d\mathbf{R}_m/\Omega$ is the probability for the mth perturbing atom to be located in the element of space $d\mathbf{R}_m$.

We calculate first the Fourier transform of the frequency distribution function a_ω. From Eqs. (2.59) and (2.60), we have

$$\mu(t) \equiv \int_{-\infty}^{\infty} \exp[it(\omega - \omega_{k0})] a_\omega \, d\omega \tag{2.61}$$
$$= \prod \int \exp\left(\frac{it}{\hbar} U_m\right) p(U_m) \, dU_m.$$

We next define the quantity $\Phi(t)$ by

$$\Phi(t) \equiv \int \exp\left(\frac{it}{\hbar} U\right) p(U) \, dU,$$

which, according to Eq. (2.60), can be expressed as

$$\Phi(t) = \frac{1}{\Omega} \int \exp\left[\frac{it}{\hbar} U(R)\right] w(R) \, d\mathbf{R}. \tag{2.62}$$

Equations (2.61) and (2.62) combine to give

$$\mu(t) = \prod_m \Phi(t).$$

If N is the number density of perturbing atoms, then the number of these atoms in the normalization volume is $N\Omega$, and hence we have

$$\mu(t) = \Phi^{N\Omega}(t) = \lim_{N\Omega \to \infty} \left\{1 + \frac{N}{N\Omega} \int \left[\exp\left(\frac{itU}{\hbar}\right) - 1\right] w(R) \, d\mathbf{R}\right\}^{N\Omega}$$
$$= \exp\left\{N \int \left[\exp\left(\frac{itU(R)}{\hbar}\right) - 1\right] w(R) \, d\mathbf{R}\right\}.$$

To obtain the frequency distribution function a_ω, we perform the inverse Fourier transform and find the result

$$a_\omega = \frac{1}{2\pi} \int_{-\infty}^{\infty} \exp[-it(\omega - \omega_{k0})] \mu(t) \, dt$$
$$= \frac{1}{2\pi} \int_{-\infty}^{\infty} \exp\left\{-it(\omega - \omega_{k0}) + N \int \left[\exp\left(\frac{itU(R)}{\hbar}\right) - 1\right] w(R) \, d\mathbf{R}\right\} dt. \tag{2.63}$$

The function $U(R)$ in Eq. (2.63) corresponds to the $U(R)$ in Problem 2.10 for the collision broadening of spectral lines.

In general, the distribution function a_ω has a complicated character and does not correspond to a Lorentzian. When the exponent $\exp(i\nu t/\hbar)$ is expanded in a series, we obtain the Lorentz line shape (2.38) with zero width and the Stark shift

$$\Delta\nu = -\frac{1}{\hbar} N \int U(R) w(R) \, d\mathbf{R}. \tag{2.64}$$

In particular, if we assume in Eq. (2.64) that $w = 1$ and $R = vt$ (corresponding to a straight-line classical trajectory, since $U \ll E$), Eq. (2.64) leads directly to Eq. (2.36) with $\sin\chi$ replaced by χ. Thus, within the context of perturbation theory, the collision theory and the quasi-static theory of spectral line broadening lead to the same result.

Second-order perturbation theory in Eq. (2.63) gives a nonzero addition to the broadening cross section, but it is not of Lorentz form. The additional term is proportional to t^2, whereas it must be proportional to t if it is to give rise to a Lorentzian line shape.

The validity criterion for perturbation theory, where the collision and quasi-static theories of line broadening coincide, is $\chi \ll 1$. In this case the broadening cross section is small compared to the line shift cross section because it corresponds to a second-order perturbative effect. Equation (2.40) gives the validity criterion in the form

$$|U(R)R/(\hbar v)| \ll 1,$$

where R is a typical interatomic distance associated with the broadening cross section, and v is the relative velocity of the atoms.

Problem 2.16. Establish the validity criteria for the collision theory and for the quasi-static theory of spectral line broadening.

SOLUTION. We must analyze the physical conditions in which the collision or the quasi-static theory of spectral line broadening is valid. The basis of the collision theory is that one can find only one perturbing particle in the region of significant interaction near the radiating atom, and the probability of the participation of two perturbing atoms with the excited atom is small. The collision theory of broadening corresponds to a low-pressure gas situation where the effective radius of interaction is small compared to the average interparticle distance.

The collision time is of the order of ρ/v, where v is the velocity of the collision and ρ is a typical impact parameter. The impact parameter can be estimated to be given by the Weisskopf radius $\sqrt{\sigma_t}$, where σ_t is the total elastic scattering cross section for the interacting particles. A typical collision time is thus of the order of $\sqrt{\sigma_t}/v$. The rationale for collision broadening requires that this time be small as compared to the flight time involved in a mean free path for the colliding particles, which is $(Nv\sigma_t)^{-1}$. The collision time should also be short compared to the detection time for the frequency shift of the spectral line, which is given by the Heisenberg uncertainty principle as $(\omega - \omega_{k0})^{-1}$. Thus the collision theory of spectral line broadening requires for its validity that

$$\frac{v}{\sqrt{\sigma_t}} \gg \mathrm{Max}\left[Nv\sigma_t, (\omega - \omega_{k0})\right]. \tag{2.65}$$

We also deduce from the foregoing that the collision theory of broadening is valid for the wings of the spectral line when

$$|\omega - \omega_{k0}| \ll \frac{v}{\sqrt{\sigma_t}}. \tag{2.66}$$

For the central part of the spectral line, we have $|\omega - \omega_{k0}| \sim Nv\sigma_t$, and so the collision theory of line broadening is valid when

$$N\sigma_t^{3/2} \ll 1. \tag{2.67}$$

We can also conclude from Eq. (2.66) that the collision theory of broadening is not valid in the far wings of the line.

We now turn to the validity criteria for the quasi-static theory of spectral line broadening. This theory will work if a typical time $(\omega - \omega_{k0})^{-1}$, during which the frequency shift $\omega - \omega_{k0}$ is detected, is small compared to a characteristic time ρ/v for the motion of a perturbing atom. If this constraint is satisfied, the perturbing atoms will not change their positions significantly during a broadening time. If the Weisskopf radius is used for ρ, the validity criterion takes the form

$$|\omega - \omega_{k0}| \gg \frac{v}{\sqrt{\sigma_t}}. \tag{2.68}$$

This requirement is the direct opposite of that found in Eq. (2.66) for the collision theory. For the central part of the spectral line, our new criterion is

$$N\sigma_t^{3/2} \gg 1, \tag{2.69}$$

which is also opposite to that found in Eq. (2.67) for the collision theory. Thus the collision theory and the quasi-static theory are converse to each other. This conclusion does not contradict the earlier finding that the two theories coincide when perturbation theory is valid. We remark in closing that according to the criteria (2.66) and (2.68), the far wings of spectral lines are described better by the quasi-static broadening theory.

Problem 2.17. Obtain an expression for the distribution function in the wings of a spectral line within the framework of the quasi-static theory of broadening.

SOLUTION. We start with the general expression (2.63) for the frequency distribution function of the emitted photons. If $|\omega - \omega_{k0}| \to \infty$, the integral over t is determined by very small times. In the opposite case the integrand would be a strongly oscillating function, since typical times t are of the order of $(\omega - \omega_{k0})^{-1}$. We consider that portion of the integrand given by

$$\exp\left(N \int \left\{\exp\left[\frac{itU(R)}{\hbar}\right] - 1\right\} w(R)\, d\mathbf{R}\right), \tag{2.70}$$

and estimate the exponent in this expression. This estimate can be represented in the form Nr_0^3, where r_0 is a typical radius in the integral over R. The value of r_0 is determined by the relation

$$\frac{U(r_0)}{\hbar(\omega - \omega_{k0})} \sim 1.$$

The function $U(R)$ decreases for large values of R when $R > r_0$, and

$$\exp\left[itU(R)/\hbar\right] - 1 \to 0,$$

so that the contribution of such values of R is small. The value of r_0 can thus be found from the relation

$$U(r_0) \sim \hbar(\omega - \omega_{k0}).$$

If $(\omega - \omega_{k0}) \to \infty$, then $r_0 \to 0$. Hence we have $Nr_0^3 \ll 1$ for sufficiently remote parts of the wings of the spectral line. The exponent (2.70) can therefore be expanded in a Taylor series, and Eq. (2.63) leads to

$$a_\omega = \frac{N}{2\pi} \int w(R)\, d\mathbf{R} \int_{-\infty}^{\infty} \exp\left[-it(\omega - \omega_{k0})\right] \left[\exp\left(\frac{itU}{\hbar}\right) - 1\right] dt.$$

Since $(\omega - \omega_{k0}) \neq 0$, the second term in this expression is zero. The first term gives a delta function, and so we obtain

$$\begin{aligned} a_\omega &= N \int \delta\left[\omega - \omega_{k0} - \frac{U(R)}{\hbar}\right] w(R)\, d\mathbf{R} \\ &= 4\pi\hbar N R^2 w(R) \left[\frac{dU(R)}{dR}\right]^{-1}_{U(R)=\hbar(\omega-\omega_{k0})}. \end{aligned} \tag{2.71}$$

This equation may be obtained by a simpler alternative. We have seen that the wing of the spectral line is produced by interactions at small distances, $r_0 \to 0$. The probability of having two or more perturbing particles at small distances is vanishingly small, so we can neglect this possibility. The probability of finding one perturbing particle in the range between $\mathbf{R}$ and $\mathbf{R} + d\mathbf{R}$ is $Nw(R)\, d\mathbf{R}$. Since the distance between the perturbing atom and the radiating atom determines the frequency ω of the emitted photon, we have

$$a_\omega\, d\omega = Nw(R)\, d\mathbf{R}.$$

Equation (2.58) allows us to write

$$d\omega = dU(R)/\hbar,$$

and so we arrive again at Eq. (2.71).

In the special case where $U(R) = C/r^n$ and $w(R) = 1$, we find the explicit expression for the frequency distribution function to be

$$a_\omega = \frac{4\pi\hbar N R^{n+3}}{Cn} = \frac{4\pi N}{n}\left(\frac{C}{\hbar}\right)^{3/n}\frac{1}{|\omega - \omega_{k0}|^{1+3/n}}. \quad (2.72)$$

We see that the wing of the spectral line is of Lorentzian form only if $n = 3$. If $n > 3$, then a_ω decreases more slowly than for a Lorentz profile. We can neglect the shift of the spectral line in this limit.

Problem 2.18. The center of the spectral line is described by the collision theory of broadening. Examine how the transition from collision broadening to quasi-static broadening takes place as the frequency departs from the peak of the spectrum.

SOLUTION. Since the principal part of the spectral line is described by the collision theory of broadening, the criterion of applicability, Eq. (2.67), is $N\sigma_t^{3/2} \ll 1$. Now we consider the wing of the spectral line, where $\hbar|\omega - \omega_{k0}|$ exceeds the typical width $Nv\sigma_t$ of the line. Equations (2.66) and (2.68) indicate that the transition from collision broadening to quasi-static broadening takes place at frequencies such that

$$|\omega - \omega_{k0}| \sim v/\sqrt{\sigma_t}.$$

Since it is true that

$$v/\sqrt{\sigma_t} \gg Nv\sigma_t,$$

then the transition from collision broadening to quasi-static broadening occurs in the far wings of the spectral line.

We now examine this region of the line. The quasi-static broadening theory, according to Eq. (2.71), gives the frequency distribution function of the photons in the wing region of the line as

$$a_\omega = 4\pi\hbar N R^2 w(R)(dU/dR)^{-1}. \quad (2.73)$$

To estimate this quantity, we note that we have the behavior $|\omega - \omega_{k0}| \sim v/\sqrt{\sigma_t}$, or $U(R) \sim \hbar v/\sqrt{\sigma_t}$, in the transition region. We thus find that $R \sim \sqrt{\sigma_t}$. Hence, the distribution function (2.73) is of order

$$a_\omega \sim \frac{\hbar N R^3}{U(R)} \sim \frac{N\sigma_t^{3/2}}{|\omega - \omega_{k0}|} \sim \frac{N\sigma_t^2}{v}.$$

From another point of view, we can obtain from Eqs. (2.38) and (2.39) for the far wings of the spectral line within the context of the collision theory, the orders of magnitude

$$a_\omega \sim \frac{\nu}{(\omega - \omega_{k0})^2} \sim \frac{Nv\sigma_t}{(v^2/\sigma_t)} \sim \frac{N\sigma_t^2}{v}.$$

Thus, in the transition region where $|\omega - \omega_{k0}| \sim v/\sqrt{\sigma_t}$, the results of the collision and quasi-static theories are of the same order of magnitude, as we would expect. Outside this region, the dependence of a_ω on the frequency ω, and the order of magnitude of a_ω are fundamentally different within the two theories of broadening.

Problem 2.19. Estimate the dependence on physical parameters of the width and of the frequency distribution function in the wings of a spectral line arising from a resonantly excited atom. This excited atom is in a gas of like atoms, with all others being in the ground state. The excited atom interacts with the other atoms of the gas through dipole–dipole interactions.

SOLUTION. The dipole–dipole interaction is given in Problem 2.14. From Eq. (2.55), the width ν of the spectral line is of the order of

$$\nu \sim ND^2/\hbar \tag{2.74}$$

in the collision-broadening approximation. The applicability criterion is given by the inequality in Eq. (2.56). From Eq. (2.38), the collision approximation frequency distribution function in the wings of the line is

$$a_\omega = \frac{\nu}{2\pi|\omega - \omega_{k0}|^2}, \tag{2.75}$$

where we have assumed that $|\omega - \omega_{k0}| \gg \nu$.

We now make estimates in the quasi-static broadening theory. From Problem 2.17, we have $a_\omega \sim NR^2\, dR/d\omega$. Within the dipole–dipole approximation, Eq. (2.58) gives

$$\hbar(\omega - \omega_{k0}) \sim D^2/R^3,$$

so the quasi-static approximation distribution function behaves as

$$a_\omega \sim N\frac{d(R^3)}{d\omega} \sim \frac{ND^2/\hbar}{|\omega - \omega_{k0}|^2}. \tag{2.76}$$

When we compare Eqs. (2.75) and (2.76), we see that both theories of broadening give the same dependence on physical parameters for the far wings of the spectral line. For collision broadening, according to Eq. (2.56), the condition

$$ND^2/\hbar \ll |\omega - \omega_{k0}| \ll \hbar^{1/2}v^{3/2}/D \tag{2.77}$$

must be fulfilled. For the quasi-static theory, the condition given in Eq. (2.68) can be rewritten with the help of Eq. (2.54) to give the requirement

$$|\omega - \omega_{k0}| \gg \hbar^{1/2} v^{3/2}/D. \tag{2.78}$$

Thus, with respect to the condition (2.57) we have two theories for the wings of the spectral line with qualitatively the same form and the same dependence on parameters. Under conditions (2.77) the collision-broadening theory is valid, that is, Eq. (2.75) holds. In the more distant wings of the spectral line when Eq. (2.78) holds true, the quasi-static theory is valid; in other words, Eq. (2.76) is applicable to the photon frequency distribution function.

The case opposite to Eq. (2.57) is when Eq. (2.69) is satisfied instead. This last condition can be written

$$N(D^2/\hbar v)^{3/2} \gg 1.$$

Then both the central part of the spectral line as well as its wings are correctly described by the quasi-static broadening theory, and, in particular, Eq. (2.76) is valid for the wings of the line.

We conclude that the qualitative form (2.76) is always correct for the wings of the spectral line irrespective of the choice of collision or quasi-static broadening mechanism. We point out that the region of the line wings is given by $|\omega - \omega_{k0}| \gg ND^2/\hbar$, since the line width in the quasi-static limit is of the same order as the value (2.74) for collision broadening. This follows from Eq. (2.58), in that

$$|\omega - \omega_{k0}| \sim U(R)/\hbar \sim D^2/(\hbar R^3) \sim ND^2/\hbar.$$

Problem 2.20. Estimate the dependence on $(\omega - \omega_{k0})$ of the antistatic wing of the spectral line when the signs of the difference $(\omega - \omega_{k0})$ and of the interaction potential $U(R)$ are opposite. Employ the quasi-static theory of broadening, and consider the case of the inverse power law potential $U(R) = C/R^n$.

SOLUTION. The expression (2.71) for the static wing of the spectral line was obtained under the condition that the quantities $(\omega - \omega_{k0})$ and $U(R)$ have the same signs. Here we suppose that $(\omega - \omega_{k0})$ and $U(R)$ have opposite signs, that is, we consider the opposite wing of the line to that investigated in Problem 2.17. Since we consider the far wing of the line, the expansion in power series of the exponent (2.70) remains appropriate. Equation (2.63) thus leads to

$$a_\omega \sim \int d\mathbf{R} \int_{-\infty}^{\infty} \exp\left[-it(\omega - \omega_{k0}) + itU(R)/\hbar\right] dt.$$

However, we can no longer extract a delta function from this expression since we now postulate that $(\omega - \omega_{k0})$ and $U(R)$ have opposite signs. In this case the values of R^* that supply the principal contribution to the integral can be found by the condition of phase stationarity. The condition for stationary phase is $\omega - \omega_{k0} = U(R^*)$, but this can be satisfied only for complex values of R^*.

In particular, for the potential $U(R) = C/R^n$ (e.g., with $\dot{C} > 0$), we find

$$R^* = \left[C/(\omega_{k0} - \omega)\right]^{1/n} \exp(\pi i/n).$$

The essential times t^* that determine the integral over t in the expression for a_ω are of the order of $t^* \sim R^*/v$. Using the method of steepest descent (the saddle-point method), we obtain for the exponentially small wing of the spectral line, the frequency distribution function

$$a_\omega \sim \exp\left[-\frac{\text{const}}{v} C^{1/n} (\omega_{k0} - \omega)^{1-1/n}\right],$$

where v is the relative velocity of the colliding particles.

Problem 2.21. Find the line shape for the transition $nn_1n_2 \to n'n_1'n_2'$ in a hydrogen atom in an ideal hydrogen plasma (where n_1 and n_2 are the parabolic quantum numbers). Broadening of the line is caused by the ions in the vicinity of the excited atom. This is the so-called Holtzmark broadening.

SOLUTION. Let **E** be the total electric field produced by the ions around the atom being examined. This field can be regarded as a quasi-static potential since the ions are of relatively large mass and their velocities can be regarded as low. As is well known, the static electric field produces a linear Stark shift in an excited hydrogen atom. Thus the upper and lower states of the radiative transition in the hydrogen atom are split into Stark components. These components are characterized by parabolic quantum numbers. The quantum numbers for the initial and final states will be denoted by nn_1n_2 and $n'n_1'n_2'$, respectively. The value of the linear Stark shift is

$$\Delta \mathcal{E}_{nn_1n_2} = \left(\tfrac{3}{2}\right) n(n_1 - n_2)E.$$

Hence the transition frequency $\omega_{nn'}$ for the given radiative transition is shifted by the amount

$$\Delta\omega_{nn'} = \left(\tfrac{3}{2}\right)\left[n(n_1 - n_2) - n'(n_1' - n_2')\right] E \equiv \varkappa E.$$

Here and in the following we use the atomic system of units where $e = \hbar = m = 1$.

We are interested in the distribution function a_ω for the photon frequencies ω that are absorbed or emitted in the transition. We write this function in the form

$$a_\omega \, d\omega = w(E)\, dE,$$

where the function $w(E)$ is the probability that the field strength is E. The values of ω and E are connected by the relation [see also Eq. (2.58)]

$$\omega = \omega_{nn'} + \varkappa E,$$

so that

$$dE/d\omega = \varkappa^{-1}.$$

We must therefore calculate the function $w(E)$. The value of E is the sum of all the field strengths of the surrounding ions. From Coulomb's law we have

$$\mathbf{E} = \sum_m \frac{\mathbf{R}_m}{R_m^3},$$

where $\mathbf{R}_m$ is the radius vector of the mth ion in a coordinate system with the radiating atom at the origin. For a fixed distribution of ions in space we obviously have

$$w(E) = \delta\left(\mathbf{E} - \sum_m \mathbf{R}_m/R_m^3\right),$$

where $\delta(x)$ is the Dirac delta function. A change in the $\mathbf{R}_m$ vectors produces a change in the electric field strength $\mathbf{E}$, and hence a shift in the transition frequency $\omega_{nn'}$. This produces a broadening of the spectral line. To calculate the broadening, we average the function $w(E)$ over the positions $\mathbf{R}_m$ of the ions.

The probability that the mth ion is in the volume $d\mathbf{R}_m$ is $d\mathbf{R}_m/\Omega$ [see also Eq. (2.60)], where Ω is the normalized volume of the plasma. Such probabilities for the individual ions must be multiplied by each other, on the assumption that the ions do not interact with each other. The absence of such interaction is the underlying principle of the ideal plasma.

We obtain in this fashion

$$\begin{aligned} w(E) &= \left\langle \delta\left(\mathbf{E} - \sum_m \frac{\mathbf{R}_m}{R_m^3}\right)\right\rangle \\ &= \frac{1}{\Omega^{N\Omega}} \int \cdots \int d\mathbf{R}_1 d\mathbf{R}_2 \ldots d\mathbf{R}_{N\Omega} \delta\left(\mathbf{E} - \sum_m \frac{\mathbf{R}_m}{R_m^3}\right), \end{aligned}$$

where N is the density of the ions and $N\Omega$ is their total number. To calculate this integral, we employ the integral representation for the Dirac delta function given by

$$\delta\left(\mathbf{E} - \sum_m \frac{\mathbf{R}_m}{R_m^3}\right) = \frac{1}{(2\pi)^3} \int d\mathbf{r} \exp\left[i\left(\mathbf{E} - \sum_m \frac{\mathbf{R}_m}{R_m^3}\right)\mathbf{r}\right].$$

From the fact that each of the integrals over $\mathbf{R}_m$ is the same as all the others, the probability $w(E)$ can be written as

$$w(E) = \int \frac{d\mathbf{r}}{(2\pi)^3} \exp(i\mathbf{E}\cdot\mathbf{r}) \left[\frac{1}{\Omega}\int d\mathbf{R} \exp\left(\frac{-i\mathbf{R}\cdot\mathbf{r}}{R^3}\right)\right]^{N\Omega},$$

where $\mathbf{R}$ is any one of the $\mathbf{R}_m$ coordinates of the ions.

We perform the limit $N\Omega \to \infty$ in this expression by following the same procedure as that employed in arriving at Eq. (2.63). That is, we add and subtract unity to

$\exp(-i\mathbf{R}\cdot\mathbf{r}/R^3)$, and make use of $\int d\mathbf{R} = \mathbf{\Omega}$ to arrive at the result

$$w(E) = \int \exp\left\{N\int d\mathbf{R}\left[\exp\left(\frac{-i\mathbf{R}\cdot\mathbf{r}}{R^3}\right) - 1\right]\right\}\exp(i\mathbf{E}\cdot\mathbf{r})\frac{d\mathbf{r}}{(2\pi)^3}.$$

The normalization volume Ω vanishes, as it should. Evaluation of the integral over $\mathbf{R}$ is done directly, leading to the result

$$w(E) = \int \exp\left(-\frac{8\pi N}{15}r^{3/2} + i\mathbf{E}\cdot\mathbf{r}\right)\frac{d\mathbf{r}}{(2\pi)^3}.$$

We can evaluate the angular part of $\int d\mathbf{r}$ and find that

$$w(E) = \frac{1}{E_0}\mathcal{H}\left(\frac{E}{E_0}\right),$$

where

$$E_0 = 2\pi(4N/15)^{2/3},$$

and

$$\mathcal{H}(x) = \frac{2}{\pi x}\int_0^\infty t\sin t\exp\left[-\left(\frac{t}{x}\right)^{3/2}\right]dt.$$

This is the so-called Holtzmark distribution.

The function $\mathcal{H}(x)$ can be evaluated explicitly in the limit of small or of large argument. If $x \ll 1$, then

$$\mathcal{H}(x) = 4x^2/3.$$

If $x \gg 1$, then we obtain

$$\mathcal{H}(x) = \left(\tfrac{3}{2}\right)/x^{5/2}.$$

The function $\mathcal{H}(x)$ has a single maximum at $x = 1.7$, with a value of approximately 0.35.

The frequency distribution function a_ω which determines the profile of the spectral line is of the form

$$a_\omega = \frac{1}{\varkappa E_0}\mathcal{H}\left(\frac{\omega - \omega_{nn'}}{\varkappa E_0}\right).$$

The Holtzmark width of this spectral line is $\varkappa E_0 \sim N^{2/3}$. The quantity E_0 is the mean strength of the ion field that acts on the excited atom under consideration. We see that the distribution function a_ω decreases as $(\omega - \omega_{nn'})^{-5/2}$. This agrees with the general expression (2.72) for the quasi-static wing of the spectral line at $n = 2$. Indeed, for the case we consider, Eq. (2.58) gives

$$\omega - \omega_{nn'} \sim E \sim R^{-2}$$

due to the Coulomb law. That is, we have $n = 2$. In this limit, broadening is produced only by the nearest ion.

The frequency distribution function for the entire transition line for $n \to n'$ is obtained by simply summing the above partial distribution functions over the parabolic quantum numbers $n_1 n_2$ and $n_1' n_2'$ of the initial and final states in the transition. This follows from the independence of the broadening of the separate Stark components of the levels of the hydrogen atom.

2.4 CROSS SECTIONS FOR ABSORPTION AND INDUCED EMISSION OF PHOTONS: ABSORPTION COEFFICIENT

Our goal is the calculation of the parameters relevant to the fundamental interaction processes between radiation and atomic electrons. If the atomic transitions consist of the absorption or emission of a single photon, a basic parameter is the absorption or emission cross section. For sufficiently small intensities, the cross section does not depend on the radiation intensity or on the density of atoms.

We want to define the photoabsorption cross section for transitions between the bound states 0 and k of the atomic electron. This process is, schematically,

$$A_0 + n\hbar\omega \to A_k + (n - 1)\hbar\omega,$$

where A is the atom, the index labels its state, and $\hbar\omega$ is the energy of a photon of frequency ω. The photon absorption cross section is the ratio of the rate of photon absorption w_a to the flux of incoming photons j_ω in the given frequency range. That is, the cross section is

$$\sigma_a = w_a / j_\omega. \tag{2.79}$$

For induced emission, corresponding to the scheme

$$A_k + n\hbar\omega \to A_0 + (n + 1)\hbar\omega,$$

the cross section is the ratio of the induced emission rate w_r to the photon flux density j_ω, or

$$\sigma_r = w_r / j_\omega. \tag{2.80}$$

In terms of the absorption and induced emission cross sections, we introduce the quantity that determines the propagation of the radiation in a gas. We denote by I_ω the intensity of radiation of frequency ω moving through a gas. This intensity varies during propagation of the radiation as a result of absorption and induced emission. Since the number of absorption or induced emission events is proportional to the number of photons that take part in these processes, the change of the photon intensity with distance is proportional to this intensity, so that

$$dI_\omega/dx = -k_\omega I_\omega, \tag{2.81}$$

where x is the coordinate in the direction of propagation. The proportionality coefficient k_ω is called the absorption coefficient.

Problem 2.22. Relate the absorption and induced emission cross sections to the photon frequency distribution function. Obtain also the general expression for the absorption coefficient.

SOLUTION. Consider first the photon absorption process. We take into account that this process does not really occur at a fixed value ω_{k0} of the frequency, but rather it involves a range of frequencies in the vicinity of ω_{k0}. Designate by a_ω the frequency distribution function for the absorbed photons in the range $[\omega, \omega + d\omega]$. Previous sections of this chapter were devoted to the calculation of this function for a variety of cases.

From Eq. (1.9), the photon absorption rate is

$$w_{0k} = \frac{4\omega_{k0}^3}{3\hbar c^3}|\mathbf{D}_{k0}|^2 g_k n_\omega,$$

where g_k is the statistical weight of the final state k. We can relate the discrete absorption spectrum to the continuous frequency spectrum by means of the frequency distribution function a_ω by setting

$$dw_a = w_{0k} a_\omega\, d\omega. \tag{2.82}$$

This rate is an average over the polarization states of the incoming photons.

As in Eq. (1.13), we introduce the lifetime τ of the state k with respect to spontaneous transition to the state 0 (setting $n_\omega = 0$), and obtain

$$\tau^{-1} = 4\omega_{k0}^3|\mathbf{D}_{k0}|^2 g_0/3\hbar c^3,$$

where g_0 is the statistical weight of the state 0. Then we can rewrite Eq. (2.82) in the form

$$dw_a = \frac{1}{\tau}\frac{g_k}{g_0} n_\omega a_\omega\, d\omega. \tag{2.83}$$

We must now find the photon flux impinging on a single atom, given n_ω photons in a given state. In a unit volume, the number of photons is

$$dN_\omega = n_\omega 2\, d\mathbf{k}/(2\pi)^3 = n_\omega \omega^2\, d\omega/(\pi^2 c^3).$$

The factor 2 takes into account the two possible polarization states of the photons. Hence the photon flux for the frequency range $[\omega, \omega + d\omega]$ is

$$dj_\omega = c\, dN_\omega = n_\omega \omega^2\, d\omega/(\pi c)^2. \tag{2.84}$$

Equations (2.79), (2.83), and (2.84) give

$$\sigma_a = \frac{dw_a}{dj_\omega} = \left(\frac{\pi c}{\omega}\right)^2 \frac{a_\omega}{\tau} \frac{g_k}{g_0}. \tag{2.85}$$

Now we consider the process of induced emission. With Eq. (1.13) and the definition of the frequency distribution function a_ω for the emitted photons, we can express the induced photon emission rate in the form

$$dw_r = (1/\tau) n_\omega a_\omega \, d\omega.$$

When we divide by the photon flux (2.84), we obtain according to Eq. (2.80), the cross section for induced emission

$$\sigma_r = \left(\frac{\pi c}{\omega}\right)^2 \frac{a_\omega}{\tau}. \tag{2.86}$$

Now we calculate the absorption coefficient for a gas, associated with the radiative transition between the states 0 and k of the atomic electron. We designate by N_0 and N_k the densities of the atoms in states 0 and k, respectively. From Eq. (2.81), the quantity k_ω is that portion of the photon absorption that takes place in a unit length of the photon beam. The decrease of the photon population in the beam is given by $N_0 \sigma_a$, while the increase in the photon population in the beam is $N_k \sigma_r$. Consequently, k_ω is

$$k_\omega = N_0 \sigma_a - N_k \sigma_r. \tag{2.87}$$

Substituting Eqs. (2.85) and (2.86) into (2.87) gives the absorption coefficient

$$k_\omega = N_k \left(\frac{\pi c}{\omega}\right)^2 \frac{a_\omega}{\tau} \left(\frac{N_0}{N_k} \frac{g_k}{g_0} - 1\right). \tag{2.88}$$

Problem 2.23. Calculate the resonance fluorescence cross section. Resonance fluorescence is the process in which a photon incident on an atom is absorbed, followed by photon emission with a return to the initial state. Take the photon energy to be equal to the excitation energy from one atomic energy level to another. Broadening arises from spontaneous decay.

SOLUTION. Resonance fluorescence stems from the absorption by an atomic electron of a photon from the distribution function a_ω determined by the spontaneous width $1/\tau_k$ of the excited state k. The resonance fluorescence cross section is found from Eq. (2.85), in which is incorporated the statistical weight $g_k = (2J_k + 1)$ associated with the degeneracy of the excited state k with respect to projections of the angular momentum J_k of this state. In like fashion, $g_0 = (2J_0 + 1)$, where J_0 is the angular momentum of the state 0.

The quantity τ in Eq. (2.85) represents the lifetime τ_k^0 of the state k with respect to spontaneous decay into the state 0. In the case when this is the only decay channel open, we have the obvious relation $\tau_k^0 = \tau_k$. The function a_ω in Eq. (2.85) is determined by Eq. (2.9), with the width established by the total lifetime τ_k of the state k. From Eq. (2.85) we then obtain the photon absorption cross section

$$\sigma_a = \frac{2J_k + 1}{2(2J_0 + 1)} \left(\frac{\pi c^2}{\omega^2} \right) \frac{(\tau_k \tau_k^0)^{-1}}{(\omega_{k0} - \omega)^2 + [1/(2\tau_k)]^2}. \tag{2.89}$$

The resonance fluorescence cross section as obtained from Eq. (2.89) is that part of the total cross section arising from the channel that returns the atomic electron back to the state 0 instead of to some other final state. Hence, we can write

$$\sigma_f = \frac{\tau_k}{\tau_k^0} \sigma_a = \frac{2J_k + 1}{2(2J_0 + 1)} \frac{\pi c^2 (\omega \tau_k^0)^{-2}}{(\omega_{k0} - \omega)^2 + [1/(2\tau_k)]^2}. \tag{2.90}$$

As an alternative, the factor τ_k/τ_k^0 could be inserted into the statistical weight g_k of the final state corresponding to the channel with return to the initial state 0.

The order of magnitude of the value of Eq. (2.90) is

$$\sigma_f \sim \pi c^2/\omega^2 \sim \pi(\lambda/2\pi)^2,$$

where $\lambda = 2\pi c/\omega$ is the photon wavelength. We see that the cross section does not depend on the fine structure constant $\alpha = e^2/\hbar c$, in contrast to analogous cross sections in nonresonant cases. The resonance fluorescence cross section thus exceeds usual nonresonant cross sections for light scattering by a factor $(\omega_{k0}\tau_k)^2$.

Problem 2.24. Calculate the photoabsorption cross section integrated over all frequencies of the absorbed photons. Assume that the atomic electron is initially in its ground state.

SOLUTION. We first integrate the photoabsorption cross section in Eq. (2.85) over those frequencies in the neighborhood of the fixed absorption line for the transition $0 \to k$. Using the normalization condition (2.2), we obtain

$$\int \sigma_a(0 \to k)\, d\omega = \frac{4\pi^2 \omega_{k0} |\mathbf{D}_{k0}|^2}{3\hbar c}, \tag{2.91}$$

where we have set $g_k = g_0 = 1$. Integration over all frequencies of the absorbed photons corresponds to performing a sum in Eq. (2.91) over all possible final states of the atomic electron, including continuum states. When we use the sum rule (1.21) and take into account that

$$|\mathbf{D}_{k0}|^2 = 3|(D_z)_{0k}|^2,$$

we obtain

$$\int \sigma_a\, d\omega = \frac{2\pi^2 e^2 n}{mc}, \tag{2.92}$$

where m is the electron mass and n is the number of electrons in the atom.

Problem 2.25. Calculate the maximum photoabsorption cross section in the atomic transition from state 0 to state k.

SOLUTION. The photoabsorption cross section in Eq. (2.85) is proportional to the distribution function a_ω for the frequency ω of the absorbed photons. It follows from the definition of a_ω that the minimum width of the spectral line corresponds to the maximum of a_ω, that is, to the maximum of σ_a. We have seen above that the minimum width occurs with spontaneous broadening. The function a_ω is then determined by Eq. (2.5). If we set $\omega_{k0} - \omega = 0$ in this equation, we can rewrite Eq. (2.85) in the form

$$\sigma_a = \left(\frac{\pi c}{\omega}\right)^2 \frac{1}{\tau_k^0} \frac{2\tau_k}{\pi} \frac{g_k}{g_0} = 2\pi \left(\frac{c}{\omega}\right)^2 \frac{\tau_k}{\tau_k^0} \frac{g_k}{g_0}, \tag{2.93}$$

where we recall the meaning of g_k and g_0 as the statistical weights of the k and 0 states of the atomic electron.

Although the cross section (2.93) is greater than the resonance fluorescence cross section σ_f in Eq. (2.90), it is nevertheless of the same order of magnitude. Fluorescence is one of the channels for deexcitation of an atomic electron.

2.5 CROSS SECTIONS FOR SCATTERING AND RAMAN SCATTERING OF PHOTONS

We here regard photon scattering as a two-photon process in which the first photon of frequency ω_1 is absorbed and a second photon of frequency ω_2 is emitted, with the concomitant transition of an atomic electron from the initial state 0 to the final state m. This process differs from fluorescence in that the final state m can differ from the initial state 0.

Problem 2.26. Calculate the cross sections for resonant and nonresonant Raman scattering.

SOLUTION. Consider first the case of resonant Raman scattering, where the photon frequency ω_1 is nearly the same as the excitation frequency ω_{k0} of the state k of the atomic electron. The photoabsorption cross section σ_a is given by Eq. (2.85). Subsequent steps in the solution of this problem parallel the solution of Problem 2.22 about resonance fluorescence, except that Eq. (2.89) for the photoabsorption cross section should be multiplied by the factor τ_k/τ_k^m, where τ_k^m is the lifetime of the state k with respect to the transition into the final state. As a result, instead of Eq. (2.90) we find

$$\sigma_c = \frac{2J_k + 1}{2(2J_0 + 1)} \frac{\pi c^2}{\omega^2} \frac{(\tau_k^m \tau_k^0)^{-1}}{(\omega_{k0} - \omega)^2 + [1/(2\tau_k)]^2}. \tag{2.94}$$

In particular, the result at exact resonance is

$$\sigma_c^{\max} = \frac{2J_k + 1}{2(2J_0 + 1)} \frac{4\pi c^2}{\omega^2} \frac{(\tau_k)^2}{\tau_k^m \tau_k^0}.$$

As should be the case, this quantity is less than Eq. (2.93), which is the maximum value for the photoabsorption cross section. We observe that Eq. (2.94) is of the same order of magnitude as the resonance fluorescence cross section in Eq. (2.90).

We now treat nonresonant Raman scattering. We write the cross section as the ratio of the two-photon transition rate (1.46) to the incident photon flux. Both absorbed and emitted photons in Eq. (1.46) have definite polarizations ($\mathbf{s}_1$ and $\mathbf{s}_2$, respectively), so the appropriate flux is one half of Eq. (2.84). We can presume that there are no scattered photons in the incident beam, that is, $n_{\omega_2} = 0$, and we then obtain the cross section for nonresonant Raman scattering,

$$d\sigma_c = \frac{\omega_1 \omega_2^3}{\hbar^2 c^4} \left| \sum_k \left[\frac{(\mathbf{s}_2 \cdot \mathbf{D}_{mk})(\mathbf{s}_1 \cdot \mathbf{D}_{k0})}{\omega_{k0} - \omega_1} + \frac{(\mathbf{s}_1 \cdot \mathbf{D}_{mk})(\mathbf{s}_2 \cdot \mathbf{D}_{k0})}{\omega_{k0} + \omega_2} \right] \right|^2 d\Omega_2. \tag{2.95}$$

In obtaining this expression from Eq. (1.46), we integrated over the frequency ω_2 of the emitted photon, using the energy-conserving delta function that appears in Eq. (1.46). The sign conventions in Eq. (1.46) for ω_1 and ω_2 are chosen so as to describe the absorption of a photon with frequency ω_1 and the emission of a photon with frequency ω_2. The quantity Ω_2 is the solid angle of the scattered photon, and m is the index labeling the final state of the atomic electron. Energy conservation gives $\omega_{m0} = \omega_1 - \omega_2$. In particular, when the initial and final states are the same, that is, when $m = 0$ and $\omega_1 = \omega_2$, Eq. (2.95) gives the nonresonant fluorescence cross section.

Problem 2.27. Calculate the cross section for photon scattering by a free electron, using the general expression (2.95).

SOLUTION. As everywhere above, we suppose that the energy of the incident photon $\hbar\omega$ is small as compared to the electron rest energy, that is, $\hbar\omega \ll mc^2$. The photon momentum is $\hbar\omega/c$. The change of photon momentum in the scattering process and the electron momentum after scattering are of the same order of magnitude: $\hbar\omega/c$ (excluding scattering through very small angles). The energy gained by the electron in the collision is of the order of $(\hbar\omega)^2/(mc^2)$. It is seen that the energy gained is small as compared to the rest energy, which is equivalent to the statement that the velocity increment of the electron from the scattering is small as compared to c. The electron motion is thus nonrelativistic. Since the change of the energy of the photon, also $\sim (\hbar\omega)^2/(mc^2)$, is small compared to the initial photon energy $\hbar\omega$, the photon-electron scattering is quasi-elastic, which means $\omega_1 \approx \omega_2$.

We shall use Eq. (2.95) for the calculation of the cross section supposing that $\omega_1 = \omega_2 = \omega$, and using the semiclassical approximation for the free electron states due to the semiclassical character of the initial continuum electron state 0. The dipole operator of an electron is $\mathbf{D} = -e\mathbf{r}$, where $\mathbf{r}$ is the electron coordinate. Equation (2.95) leads to

$$d\sigma = \left(\frac{2\omega^2 e^2}{\hbar c^2}\right)^2 \left|\sum_k (\mathbf{s}_2 \cdot \mathbf{r})_{0k} (\mathbf{s}_1 \cdot \mathbf{r})_{k0} \frac{\omega_{k0}}{\omega_{k0}^2 - \omega^2}\right|^2 d\Omega, \tag{2.96}$$

where the index k identifies the free electron continuum quantum states.

The principal contribution to the sum over k comes from states that have energies near the energy of the initial state 0. Indeed, matrix elements $\mathbf{r}_{0k}$ are semiclassically small for large differences in energy between the states k and 0. Consequently, we have $\omega \gg \omega_{k0}$, that is, the frequency of the scattered photon is large compared to typical electron frequencies. When ω_{k0} is neglected in the denominator of Eq. (2.96), we obtain the expression, independent of ω,

$$d\sigma = \left(\frac{2e^2}{\hbar c^2}\right)^2 \left|\sum_k (\mathbf{s}_2 \cdot \mathbf{r})_{0k} (\mathbf{s}_1 \cdot \mathbf{r})_{k0} \omega_{k0}\right|^2 d\Omega. \tag{2.97}$$

We wish to evaluate the sum in Eq. (2.97). The coordinate axes are selected so that $\mathbf{s}_1$ is along the z axis and $\mathbf{s}_2$ is in the xz plane. The sum is then of the form

$$s_{2x} \sum_k \omega_{k0} x_{0k} z_{k0} + s_{2z} \sum_k \omega_{k0} |z_{k0}|^2.$$

The first of the sums in this expression is zero because of the odd parity of the product $x_{0k} z_{k0}$. It changes sign when $z \to -z$. The second sum can be calculated using the sum rule (1.21) for dipole transitions. The above expression then yields

$$\hbar s_{2z}/(2m) = \hbar(\mathbf{s}_1 \cdot \mathbf{s}_2)/(2m).$$

Substituting this expression into Eq. (2.97) gives the cross section in the form

$$d\sigma = r_e^2 (\mathbf{s}_1 \cdot \mathbf{s}_2)^2 \, d\Omega. \tag{2.98}$$

The quantity $r_e = e^2/(mc^2)$ is the classical electron radius. Equation (2.98) is called the Thomson formula. It is a purely classical result, since the Planck constant does not appear.

To calculate the total cross section, we integrate Eq. (2.98) over the solid angle. We select the polar axis of a system of spherical coordinates to lie along the polarization vector $\mathbf{s}_1$ of the incident photon. The vector $\mathbf{s}_2$ lies in the plane determined by $\mathbf{s}_1$ and the wave vector $\mathbf{k}_2$ of the scattered photon. For the direction normal to this plane, the scattering cross section is zero. (In this direction, $\mathbf{s}_1$ and $\mathbf{s}_2$ are perpendicular to each other.) Let θ be the angle between $\mathbf{s}_1$ and $\mathbf{k}_2$. Since $\mathbf{s}_2$ and $\mathbf{k}_2$ are perpendicular, we obtain $\mathbf{s}_1 \cdot \mathbf{s}_2 = \sin\theta$. The total cross section is thus

$$\sigma = r_e^2 \int \sin^2\theta \, d\Omega = \frac{8\pi}{3} r_e^2. \tag{2.99}$$

Equations (2.98) and (2.99) are also attainable in classical radiation theory by solving the Newtonian equations of motion for induced electron oscillations and considering the emission of secondary waves with the same frequency. The classical results fail when the photon energy $\hbar\omega$ is of the order of the electron rest energy mc^2 or greater. Then most of the incident photon energy is transferred to the electron, and the scattering is therefore inelastic. In this case, relativistic and quantum effects will be important simultaneously, and the electron spin will be an essential element in the description of the scattering.

Problem 2.28. Calculate the low-energy scattering cross section for a photon scattering from an atom with zero angular momentum. Evaluate the result in closed form for the ground state of hydrogen.

SOLUTION. The frequency of the photon is taken to be small compared to typical atomic frequencies, or $\omega \lll \omega_{k0}$. This limit is thus opposite to that considered in Problem 2.27 for a photon scattering off a free electron with $\omega \ggg \omega_{k0}$. The small frequency condition allows us to simplify Eq. (2.95) to

$$d\sigma = \left(\frac{\omega^2}{\hbar c^2}\right)^2 \left|\sum_k 2\frac{(\mathbf{s}_2\cdot\mathbf{D}_{0k})(\mathbf{s}_1\cdot\mathbf{D}_{k0})}{\omega_{k0}}\right|^2 d\Omega. \tag{2.100}$$

The initial state is specified to be an S state, so that its magnetic quantum number is $M_0 = 0$. The state k is therefore a P state in accordance with the dipole selection rule, and so $M_k = 0, \pm 1$. We take the axis of quantization z to lie along $\mathbf{s}_1$. The vector $\mathbf{D}$ is along the z direction, so it is along $\mathbf{s}_1$. In the opposite case, the quantity $\mathbf{D}\cdot\mathbf{s}_1$ vanishes. Hence we have

$$\mathbf{s}_2\cdot\mathbf{D}_{0k} = (\mathbf{s}_1\cdot\mathbf{s}_2)(D_z)_{0k}.$$

We now define the polarizability tensor

$$\alpha_{ij} = 2\sum_k \frac{(D_i)_{0k}(D_j)_{k0}}{\hbar\omega_{k0}}$$

(see Section 1.7). It follows from the above considerations that α_{ij} is a diagonal tensor, so that $\alpha_{ij} = \alpha\delta_{ij}$, and

$$\alpha = \frac{2e^2}{\hbar}\sum_k \frac{|z_{k0}|^2}{\omega_{k0}}.$$

When this result is substituted into Eq. (2.100), we find the scattering cross section

$$d\sigma = \left(\frac{\omega}{c}\right)^4 \alpha^2(\mathbf{s}_1\cdot\mathbf{s}_2)^2\, d\Omega = \frac{\omega^4\alpha^2}{c^4}\sin^2\theta\, d\Omega, \tag{2.101}$$

where θ is the angle between the polarization direction $\mathbf{s}_1$ of the incident photon and the direction $\mathbf{k}_2$ of the wave vector of the scattered photon. If $\mathbf{s}_2$ is normal to the plane containing the vectors $\mathbf{s}_1$ and $\mathbf{k}_2$, then $\mathbf{s}_1\cdot\mathbf{s}_2 = 0$, since the directions $\mathbf{s}_1$ and

D coincide. In the opposite case, if the vector $\mathbf{s}_2$ is in the plane defined by $\mathbf{s}_1$ and $\mathbf{k}_2$, then $\mathbf{s}_1 \cdot \mathbf{s}_2 = \sin\theta$, since $\mathbf{s}_2$ is perpendicular to $\mathbf{k}_2$. After integration over the angular coordinates, we find the total cross section for photon scattering by an atom in the low photon frequency limit to be

$$\sigma = \frac{8\pi\omega^4}{3c^4}\alpha^2. \tag{2.102}$$

We now wish to solve the same problem by classical methods. From Eq. (1.17), the intensity of scattered light is

$$I = 2\ddot{\mathbf{D}}^2/3c^3,$$

where **D** is the induced dipole moment produced by the field of an electromagnetic wave with an electric field given by $\mathbf{E}\cos\omega t$. By the definition of atomic polarizability α, we have $\mathbf{D} = -\alpha\mathbf{E}\cos\omega t$. We are thus led to the intensity of scattered light expressed as

$$I = \frac{2\omega^4}{3c^3}\alpha^2\mathbf{E}^2\cos^2\omega t.$$

To find the cross section, we should divide this quantity by the energy flux of the incident radiation. This energy flux is given by the Poynting vector $c\mathbf{E}\times\mathbf{H}/4\pi$, where **E** and **H** are the electric and magnetic fields of the electromagnetic wave. In our case, the energy flux has the magnitude $(c\mathbf{E}^2/4\pi)\cos^2\omega t$. When we define the scattering cross section as the ratio of the intensity of scattered light to the energy flux of the incident radiation, we obtain

$$\sigma = \frac{8\pi\omega^4}{3c^4}\alpha^2.$$

This agrees with the quantum result in Eq. (2.102). The advantage of the quantum-mechanical derivation is that it makes it possible to obtain the explicit expression for atomic polarizability. It is seen from the derivation that the scattering process considered is purely classical. A classical dipole moment radiates the same frequency that is induced by the electromagnetic wave. Such scattering is called Rayleigh scattering. It is interesting that both Rayleigh scattering ($\omega \ll \omega_{k0}$) and Thomson scattering ($\omega \gg \omega_{k0}$) are purely classical phenomena.

The maximum in the scattering of visible light by atoms with absorption frequencies in the ultraviolet range corresponds to the violet end of the spectrum, since the scattering cross section increases very strongly with frequency: as ω^4. The limit $\omega \ll \omega_{k0}$ holds true nevertheless. This explains the blue color of the sky. Sunset is of a red color for the same reason: the strong scattering of the violet part of the Sun's spectrum in the direct flux of the Sun's rays leaves a predominance of red in the remaining part of the sunlight.

The static polarizability can be exactly calculated for the ground state of the hydrogen atom. The result is that $\alpha = (\frac{9}{2})a_0^3$, where $a_0 = \hbar^2/(me^2)$ is the Bohr radius. Hence the cross section for the hydrogen ground state for low-energy photon scattering, with $\hbar\omega \ll \hbar^2/(ma_0^2)$, is given by

$$d\sigma = \frac{81}{4} r_e^2 \left(\frac{\hbar^3\omega}{me^4}\right)^4 (\mathbf{s}_1 \cdot \mathbf{s}_2)^2 \, d\Omega. \tag{2.103}$$

This expression describes accurately the elastic scattering cross section from zero frequency up to the frequency of the first resonance when $\hbar\omega = \hbar\omega_{21} = 3me^4/(8\hbar^2)$. The indices 1 and 2 refer, respectively, to the ground and first excited states of the hydrogen atom.

An additional problem with accounting for degeneracy of the state 0 with respect to magnetic quantum numbers appears in the case of nonzero angular momentum. If the low-energy photon is scattered by an atom in an excited state, then Raman scattering occurs as well as Rayleigh scattering, with the consequent transition of the atom to a lower lying state m.

Problem 2.29. Calculate the dependence of the intensity of induced Raman scattering on the propagation distance of the photon beam in the gas.

SOLUTION. We consider Eq. (2.95) for the Raman scattering cross section $d\sigma_c$ when an atomic electron makes a transition from the initial state 0 to the final state m. If we denote by N/V the density of atoms, then the quantity

$$g = \frac{N}{V}\sigma_c \tag{2.104}$$

represents the number of photons with frequency ω that is generated in a unit distance along the photon beam. The total Raman scattering cross section, σ_c, is obtained from Eq. (2.95) by performing the integration over the angles of the emitted photons of frequency ω_2.

We now have $n_{\omega_2} \neq 0$, since the photons go from an incident beam of frequency ω_1 to photons of scattered light with frequency ω_2. If we select coordinates with the z axis along the propagation direction of the incident beam, then by the requirement that each absorbed photon gives rise to a scattered photon, we have

$$n_{\omega_1}(z) + n_{\omega_2}(z) = \text{const} = n_{\omega_1}(0), \tag{2.105}$$

where $n_{\omega_1}(0)$ is the initial number of photons in the incident beam. In the usual scheme of quantization, each mode of oscillation is contained in the volume V, so we suppose that the typical characteristic length along the z axis is much greater than $V^{1/3}$.

We can now write balance equations that determine the change with z of the quantities $n_{\omega_1}(z)$ and $n_{\omega_2}(z)$. Equation (2.104) establishes the number of photons that appear in a unit length along the photon beam, under the condition that there is one photon of frequency ω_1 and none of frequency ω_2. However, if we have at the coordinate z the number n_{ω_1} photons with frequency ω_1 and n_{ω_2} photons of frequency

ω_2, then Eq. (1.46) says that the number of photons appearing in a unit length along the beam with frequency ω_2 is

$$\alpha(z) = g n_{\omega_1}(z)\left[1 + n_{\omega_2}(z)\right]. \tag{2.106}$$

Hence, the balance equations are of the simple form

$$\frac{d}{dz} n_{\omega_1}(z) = -\frac{d}{dz} n_{\omega_2}(z) = -\alpha(z). \tag{2.107}$$

The solution of the system (2.107) under the conditions (2.105) is elementary. We write it in the form

$$n_{\omega_2}(z) = \frac{\exp(Gz) - 1}{[\exp(Gz)/n_{\omega_1}(0)] + 1}, \tag{2.108}$$

where G is defined as

$$G = g\left[1 + n_{\omega_1}(0)\right] = \frac{N}{V}\sigma_c\left[1 + n_{\omega_1}(0)\right]. \tag{2.109}$$

The quantity G is called the increment coefficient. We see that at first the number of scattered photons increases linearly with z. This corresponds to the general theory developed in Problem 2.26. This linear increase occurs when $Gz \ll 1$. When $Gz \sim 1$, the linear increase becomes an exponential increase. Finally, when $Gz \gg 1$, saturation takes place, so that all photons from the incident beam (ω_1) are replaced by photons in the scattered state (ω_2), or $n_{\omega_2}(z) \approx n_{\omega_1}(0)$.

The intensity of induced Raman scattering for photons with frequency ω_2 is given by

$$I_2(z) = c\hbar\omega_2 n_{\omega_2}(z)/V,$$

where $n_{\omega_2}(z)$ is determined by Eq. (2.108). In the linear regime, Eq. (2.108) becomes

$$n_{\omega_2}(z) = \frac{Gz}{1 + n_{\omega_1}^{-1}(0)}, \tag{2.110}$$

which is in good agreement with Eq. (2.95). If we take the volume V to have the length z in the direction of the photon beam with frequency ω_1, where N the number of atoms in this volume, then the cross section of the volume V is V/z. We now calculate the energy flux through this cross section for the photons of frequency ω_1, and obtain

$$\begin{aligned} I_1(z)\frac{V}{z} &= c\hbar\omega_1 n_{\omega_1}(z)\frac{1}{V}\frac{V}{z} \\ &= -G\left[1 + \frac{1}{n_{\omega_1}(0)}\right]^{-1} c\hbar\omega_1 + \frac{c\hbar\omega_1 n_{\omega_1}(0)}{z}. \end{aligned} \tag{2.111}$$

Using Eq. (2.109) we rewrite the first term in Eq. (2.111) in the form

$$-\frac{N}{V}\sigma_c n_{\omega_1}(0)c\hbar\omega_1. \tag{2.112}$$

To calculate the cross section we divide Eq. (2.112) by the particle density N/V and by the photon flux for the incident photons. As should be expected, we obtain σ_c, the Raman scattering cross section given in Eq. (2.95).

In the nonlinear regime, the increment coefficient is more useful than the cross section σ_c.

2.6 TWO-PHOTON ABSORPTION

We consider here the case where the transition frequency ω_{m0} of an atomic electron from state 0 to state m is approximately twice the frequency ω of the incident photons. That is, we treat the case where $\omega_{m0} \approx 2\omega$.

The electric field vector of the electromagnetic wave is taken to be of the form

$$\mathbf{E}(t) = 2\mathbf{E}\cos kz \cos \omega t. \tag{2.113}$$

That is, the electromagnetic field is that of a standing wave. The notation here is that $k = \omega/c$ is the wave number, and $z = vt$, where v is the projection of the atomic velocity on the z axis. Equation (2.113) can be expressed as a superposition of two traveling waves with equal amplitudes, but with the different frequencies $\omega + kv$ and $\omega - kv$. Specifically, we can write

$$\mathbf{E}(t) = \mathbf{E}\cos(\omega + kv)t + \mathbf{E}\cos(\omega - kv)t. \tag{2.114}$$

Radiative transitions with two photons were studied in Section 1.6. Problems in this section are devoted to nonuniform broadening of spectral lines due to thermal motion of the atoms.

Problem 2.30. Find the frequency dependence of the two-photon absorption coefficient when the absorbing atoms move in a resonator under the influence of the standing light wave. The velocity distribution of the atomic velocities can be taken to be Maxwellian. The density of atoms is presumed to be small, so that atomic collisions can be neglected.

SOLUTION. We use Eq. (1.48) for the two-photon transition rate. Equation (2.114) gives the electric field for the standing wave of photons with frequencies $\omega + kv$ and $\omega - kv$. The following processes can occur: (1) absorption of two photons with frequency $\omega + kv$; (2) absorption of two photons with frequency $\omega - kv$; and (3) absorption of a photon of frequency $\omega + kv$ followed by absorption of a $\omega - kv$ photon, or vice versa.

The rate (1.48) contains the delta function

$$\delta(\omega_{m0} - \omega_1 - \omega_2) = \tfrac{1}{2}\delta\left[\tfrac{1}{2}\omega_{m0} - \tfrac{1}{2}(\omega_1 + \omega_2)\right].$$

In the present problem, we have $\omega_1 = \omega \pm kv$ and $\omega_2 = \omega \mp kv$. We can alter Eq. (2.9) to reflect the content of this delta function to obtain the photon frequency distribution function

$$a^{\tau}_{\omega} = \frac{1}{\pi\tau_{m0}} \frac{1}{(\omega_{m0} - \omega_1 - \omega_2)^2 + [1/(2\tau_{m0})]^2}, \tag{2.115}$$

where τ_{m0} is the reduced spontaneous lifetime for the transition $m \to 0$. This lifetime is determined by single-photon spontaneous transitions into other atomic states, rather than by the two-photon spontaneous decay to the state 0.

To take Doppler broadening into account, we must combine the Lorentz line shape (2.115) with the Doppler line shape. First, we consider the absorption of photons from different beams. The distribution function (2.115) then does not depend on atomic velocity. Consequently, averaging over the thermal distribution of velocities does not alter this function.

To calculate the corresponding two-photon transition rate according to Eq. (1.48), we suppose that both waves have the same polarization. We place the z axis along the common direction of the polarization vectors $\mathbf{s}_1$ and $\mathbf{s}_2$. The two-photon matrix element is defined by the relation

$$D^{(2)}_{m0} = \sum_k \frac{(D_z)_{mk}(D_z)_{k0}}{\omega_{k0} - \omega}. \tag{2.116}$$

The contribution to the two-photon transition rate is, from Eq. (1.48),

$$\frac{\pi E^4}{4\hbar^4} \left|D^{(2)}_{m0}\right|^2 a^{\tau}_{\omega}, \tag{2.117}$$

where a^{τ}_{ω} is determined by Eq. (2.115) with $\omega_1 + \omega_2 = 2\omega$.

Now we consider the absorption of photons from one beam. The Lorentz profile (2.115) that contains the velocity of the atom should be integrated with the Doppler shape. Actually, the Doppler width

$$\Delta\omega_D = \omega\left(\frac{2T}{Mc^2}\right)^{1/2} = \omega_{m0}\left(\frac{T}{2Mc^2}\right)^{1/2}$$

is large compared to the spontaneous width τ^{-1}_{m0}. We know from the solution of Problem 2.4 that the resulting contour is Doppler-like, so we have

$$a^{D}_{2\omega} = \frac{1}{\sqrt{\pi}\Delta\omega_D} \exp\left[-\left(\frac{2\omega - \omega_{m0}}{2\Delta\omega_D}\right)^2\right]. \tag{2.118}$$

We shall now calculate the corresponding rate for the two-photon transition. In Eq. (1.48) we have only one term instead of two in the two-photon matrix element as a result of the similarity of the two photons. Consequently, the rate should be multiplied by two because of the presence of two photon beams. When we employ the delta function $\delta(\omega_{m0}/2 - \omega \pm kv)$ in the Doppler shape (2.118), we obtain the contribution to the two-photon transition

$$\frac{\pi E^4}{8\hbar^4}\left|D_{m0}^{(2)}\right|^2 a_{2\omega}^D. \tag{2.119}$$

The total two-photon transition rate is the sum of the individual contributions in Eqs. (2.117) and (2.119). To obtain the absorption rate σ_a we divide this sum by the flux of incident photons $cE^2/(4\pi\hbar\omega)$. From Eq. (2.87), we must then multiply σ_a by the density of atoms N_0 in the initial state 0 in order to obtain the two-photon absorption coefficient

$$k_\omega = \frac{\pi^2 N_0 E^2 \omega_{m0}}{c\hbar^3}\left|D_{m0}^{(2)}\right|^2 \left|2a_\omega^\tau + a_{2\omega}^D\right|. \tag{2.120}$$

Equation (2.120) describes a narrow resonance superimposed on a broad Doppler shape. The ratio of the maximum peak height to the height of the background is

$$8\Delta\omega_D \tau_{m0}/\sqrt{\pi} \gg 1.$$

The dependence of the absorption coefficient on ω is thus of the form of a narrow high maximum amid a broad background. The width of the narrow resonance is determined by the reduced lifetime for the two-photon $0 \to m$ transition and is very small. Determination of the position of this resonance requires Doppler-free high-resolution spectroscopy.

We have seen that the two-photon absorption coefficient depends on the intensity of the photon beam, unlike the single-photon absorption coefficient. In consequence, its value is significant only at the high intensities available with lasers.

We note, finally, that if we considered a traveling wave instead of the standing wave, the two-photon absorption is determined only by the broad Doppler shape corresponding to the second term of Eq. (2.120).

Problem 2.31. Under the conditions of Problem 2.30, examine how an intermediate resonance level affects the two-photon absorption coefficient.

SOLUTION. Suppose that some level k is in resonance with the initial state 0, and thus also in resonance with the final state m. Both resonances are of single-photon type. Hence, in the sum over k in Eq. (2.116) for the two-photon matrix element, we should retain only that one state that corresponds to the resonance. In the vicinity of the resonance we can write the resonant denominator in the form

$$\omega_{k0} - \omega \pm kv - i/(2\tau_{k0}),$$

where the term $i/(2\tau_{k0})$ takes into account the spontaneous lifetime of the state k. We begin by considering the first term in Eq. (2.120). The quantity a_ω^τ does not contain the velocity v of the atom and is not altered by the Doppler averaging. However, the resonance factor

$$\left|D_{m0}^{(2)}\right|^2 = \frac{|(D_z)_{mk}(D_z)_{k0}|^2}{(\omega_{k0} - \omega \pm kv)^2 + 1/(2\tau_{k0})^2},$$

when Doppler averaged under the condition $\Delta\omega_D \tau_{k0} \gg 1$, takes the form

$$2\pi\tau_{k0}|(D_z)_{mk}(D_z)_{k0}|^2 a_\omega^D, \tag{2.121}$$

where the Doppler line shape a_ω^D is given by

$$a_\omega^D = \frac{1}{\sqrt{\pi}\Delta\omega_D} \exp\left[-\frac{(\omega - \omega_{k0})^2}{(\Delta\omega_D)^2}\right].$$

The quantity $\Delta\omega_D$ coincides with that defined in Problem 2.30. In resonance, the first term in Eq. (2.120) is larger than in the nonresonant case by a factor of the order of magnitude

$$\Delta\omega_D \tau_{k0} \left(\frac{\omega_{k0}}{\Delta\omega_D}\right)^2 \gg 1.$$

We now consider the second term in Eq. (2.120). Here the product of the line shape (2.115) and the resonance factor in the two-photon matrix element

$$\frac{1}{\pi\tau_{m0}} \frac{1}{(\omega_{m0} - 2\omega \mp 2kv)^2 + 1/(2\tau_{m0})^2} \frac{|(D_z)_{mk}(D_z)_{k0}|^2}{(\omega_{k0} - \omega \pm kv)^2 + 1/(2\tau_{k0})^2}$$

is subjected to Doppler averaging. If we take into account that $\Delta\omega_D \gg \tau^{-1}$, this averaging can be done by following the procedure employed in Problem 2.2, where the product of two Lorentz line shapes was averaged. In this way, we are led to the result

$$\frac{|(D_z)_{mk}(D_z)_{k0}|^2}{(\omega_{k0} - \omega_{m0}/2)^2 + 1/(2\tau_{mk})^2} \left(a_{2\omega}^D + \frac{\tau_{k0}}{2\tau_{m0}} a_\omega^D\right), \tag{2.122}$$

where we use the definition

$$\frac{1}{\tau_{mk}} = \frac{1}{2\tau_{m0}} + \frac{1}{\tau_{k0}},$$

and where $a_{2\omega}^D$ is given by Eq. (2.118).

Summing the terms given in Eqs. (2.121) and (2.122), we find the two-photon absorption coefficient in the resonant case to be

$$
\begin{aligned}
k_\omega = {} & \frac{\pi^2 N_0 E^2 \omega_{m0}}{4c\hbar^3} |(D_z)_{mk}(D_z)_{k0}|^2 \\
& \times \left\{ \frac{\tau_{k0}/\tau_{m0}}{(\omega_{k0} - \omega_{m0}/2)^2 + 1/(2\tau_{m0})^2} a_\omega^D \right. \\
& \left. + \frac{1}{(\omega_{k0} - \omega_{m0}/2)^2 + 1/(2\tau_{mk})^2} \left(a_{2\omega}^D + \frac{\tau_{k0}}{2\tau_{m0}} a_\omega^D \right) \right\}. \qquad (2.123)
\end{aligned}
$$

We can draw a number of conclusion from Eq. (2.123). The first term corresponds to absorption of photons from different beams. As in the nonresonant case, it is of the form of a narrow peak superimposed on a broad background. However, that part of the second term corresponding to photon absorption from only one beam has a height of the order of the height of the narrow peak when level k is exactly (to within an accuracy of the spontaneous width $1/\tau_{mk}$) halfway between levels 0 and m. Tuning of the intermediate level to resonance, on the one hand, increases the two-photon absorption coefficient, but on the other hand it also increases the height of the Doppler background with respect to the narrow peak.

The second term in Eq. (2.123) is connected with two-photon absorption when, after the first photon is absorbed, the atom goes to the "real" state k. For sufficiently exact tuning of the state k midway between the 0 and m states, the two-step absorption process and the two-photon absorption process described by the first term in Eq. (2.123) are of comparable importance in the two-photon absorption coefficient.

When the condition $|\omega_{k0} - \omega_{m0}/2| \ll \Delta\omega_D$ is satisfied, the distribution functions a_ω^D and $a_{2\omega}^D$ coincide. We observe that the inequalities $|\omega_{k0} - \omega_{m0}/2| > 1/\tau$ and $|\omega_{k0} - \omega_{m0}/2| < 1/\tau$ are both possible. However, if $|\omega_{k0} - \omega_{m0}/2| \gg \Delta\omega_D$, we then obtain two Doppler profiles: one with the maximum at the frequency $\omega_{m0}/2$, and the other with a maximum at the frequency ω_{k0}. In the latter case, the height of the peak is much larger than the background.

Finally, from the first term of Eq. (2.123) we can conclude that when there is large detuning from the resonant location of the state k at the midpoint between states 0 and m, the background is asymmetrical, since the peak of the narrow resonance at $\omega = \omega_{m0}/2$ and the maximum of the background contour at $\omega = \omega_{k0}$ do not coincide.

Problem 2.32. Calculate the two-photon matrix element in Eq. (2.116) for transitions between highly excited states under the conditions of Problem 2.30.

SOLUTION. We first consider a simplified case in which two-photon transitions occur between highly excited atomic states with zero orbital angular momentum quantum numbers. We take the electromagnetic field to be linearly polarized. We adopt the atomic system of units in which $e = \hbar = m = 1$.

We start by calculating the two-photon matrix element required for Eq. (2.116). That is, we need to evaluate the matrix element

$$
D_{m0}^{(2)} = \sum_k \frac{z_{mk} z_{k0}}{\omega_{k0} - \omega}, \qquad (D_z = z),
$$

where z_{mk}, z_{k0} are dipole matrix elements of $s \to p$ transitions. For the quantum numbers of the initial state 0, the intermediate state k, and the final state m, respectively, we have for state 0: $n, 0, 0$; for state k: $n', 1, 0$; and for state m: $n+1, 0, 0$. The first quantum numbers in each set: $n,\ n',\ n+1$, are the principal quantum numbers for these highly excited states. The second number in each set is the orbital angular momentum quantum number, and the third number in each set is the magnetic quantum number.

After the straightforward integration over angles, we express the dipole matrix element in terms of the radial part,

$$z_{n00}^{n'10} = \frac{1}{\sqrt{3}} R_{n0}^{n'1}.$$

The semiclassical expression for the radial dipole matrix element $R_{n0}^{n'1}$ under the conditions $n, n' \gg 1$ and $\Delta n = |n - n'| \ll n, n'$ (this constraint on the values of Δn is essential in the sum over intermediate states) in a Coulomb potential is found with the help of the Coulomb radial wave functions to be

$$R_{n0}^{n'1} \underset{\Delta n \neq 0}{=} \frac{3^{1/6}}{2^{1/3}} \Gamma\left(\frac{2}{3}\right) \frac{nn'}{(\Delta n)^{5/3}}.$$

If $\Delta n = 0$, we then obtain

$$R_{n0}^{n1} = -3n^2/2.$$

With the energy of the state with principal quantum number n denoted as ε_n, we introduce the resonance detuning

$$\xi = \varepsilon_{n+1} - \varepsilon_n - 2\omega \approx \frac{1}{n^3} - 2\omega$$

and presume that $\xi \ll n^{-3}$ so that a two-photon resonance is possible. On the other hand, we also impose the requirement that $\xi \gg n^{-4}$ so that we can neglect anharmonicity effects. This last condition can be understood as follows. In the summation over n' in the two-photon matrix element, we have states symmetrically disposed about the initial and final states of the transition. That is, we sum over states with the principal quantum numbers $n' = n + \Delta n + 1$ and $n' = n - \Delta n$, where $\Delta n = 1, 2, 3, \ldots$. The contributions from these symmetrically placed states tend to cancel each other. This cancellation is broken both by the resonance detuning ξ and by the anharmonicity of the highly excited atomic spectrum. We will require that the detuning ξ should be much less than the n^{-4} anharmonicity. If this condition is not satisfied, the calculations are greatly complicated by the necessity of introducing corrections of the next order in the semiclassically small parameter $1/n$. We recall the analogous situation in the semiclassical evaluation of the sum rule in Problem 1.4.

The summation over the intermediate states n' gives the two-photon matrix element in the form

$$D_{m0}^{(2)} = \xi n^5(n+1)^5\Gamma\left(\frac{2}{3}\right)$$
$$\times\left\{\frac{1}{3^{1/6}2^{4/3}} + \frac{\Gamma(2/3)}{3^{2/3}2^{8/3}}\sum_{\Delta n=1}^{\infty}\frac{1}{[\Delta n(\Delta n+1)]^{5/3}(\Delta n+1/2)^2}\right\}.$$

The first term within the brace in this expression comes from the contribution of the intermediate states $n' = n$ and $n' = n + 1$. If the two-photon resonance is exact ($\xi = 0$), this two-photon matrix element vanishes within the framework of this order of the semiclassical approximation. The sum in the second term within the brace in this expression converges quickly. Furthermore, its magnitude is only about 5% of that of the first term in the brace. Our final result is

$$D_{m0}^{(2)} = 1.87\left[n(n+1)\right]^5(\varepsilon_{n+1} - \varepsilon_n - 2\omega).$$

This two-photon matrix element grows as about n^{10}, a very strong dependence on the principal quantum number.

The results we have obtained can be applied also to complex atoms. In that case, n and $n + 1$ should be replaced by effective quantum numbers of the corresponding states, deduced from experimental values of the atomic energy levels using the hydrogenlike formula $\varepsilon_n = -\frac{1}{2}n^2$. Corrections to this expression are the subject of quantum defect theory.

We also point out that the results obtained are applicable only if the external electromagnetic field is sufficiently weak that it is valid to neglect the effects of Stark shifts and Stark splittings of highly excited states. These effects increase with growing values of n.

According to Eq. (2.120), the vanishing of the two-photon matrix element in the exactly resonant case strongly diminishes the height of the narrow resonance peak compared to the broad Doppler background.

We now move on to the consideration of two-photon transitions between states with nonzero orbital angular momentum quantum numbers l. If $l \sim n$, the semiclassical dipole moment is very small, and hence the two-photon absorption coefficient is very small. We may therefore confine our attention only to the case where $l \ll n$. We shall assume that the value of l is the same for initial and final states in the two-photon transition.

The matrix element $z_{n\,l\,m}^{n'\,l-1\,m}$ is connected to the radial dipole matrix element by the relation

$$z_{n\,l\,m}^{n'\,l-1\,m} = \left[\frac{l^2 - m^2}{(2l-1)(2l+1)}\right]^{1/2} R_{n\,l}^{n'\,l-1},$$

where m is the magnetic quantum number of the initial (and final) state. The radial dipole matrix element is independent of l for small values of l, as we have observed above.

By analogy, we can also write the matrix element $z_{nlm}^{n'l+1m}$. Then we must average the two-photon transition rate over the magnetic quantum number m of the initial state. The factor by which the two-photon transition rate for $l = 0$ must be multiplied is

$$f(l) = \frac{9(8l^2 + 8l - 5)}{15(2l-1)(2l+3)}.$$

For $l = 0$, we obtain the expected result $f(0) = 1$. The value of $f(l)$ is near unity even for $l \neq 0$.

Problem 2.33. Alkali atoms are excited by two oppositely directed laser beams from the $S_{1/2}$ ground state to an excited $D_{5/2}$ state via an intermediate $P_{3/2}$ state. The lifetime of the $P_{3/2}$ state is τ_1, and that of the $D_{5/2}$ state is τ_2. Find the dependence of the photon absorption coefficient on the frequencies ω_1 and ω_2 of the lasers for the first and second stages of the two-photon excitation.

SOLUTION. The frequencies ω_1 and ω_2 are different by several Doppler widths from the quantity $\omega_{m0}/2$, where 0 is the $S_{1/2}$ state and m is the $D_{5/2}$ state. Two-photon absorption from only one laser beam, with both photons of frequency ω_1 or both photons of frequency ω_2, is thus impossible. Only the absorption of a single photon from each of the beams can take place. We presume also that the energy denominators in the matrix element in Eq. (1.48) for two-photon transition rates are large compared to the spontaneous widths of the atomic levels. Hence we can neglect spontaneous broadening. Moreover, we suppose that resonance detunings are very large compared to the Doppler width of the intermediate state. The two-photon matrix element is constant in the presence of the stated conditions.

Since $\omega_1 \neq \omega_2$, the linear Doppler effect is only partly compensated. We will fix one of the frequencies (ω_2, e.g.), and consider the photon distribution function a_{ω_1}. From Eq. (2.115), it is of the form

$$a_{\omega_1}^{\tau} = \frac{1}{2\pi\tau_2}\left[\left(\omega_{m0} - \omega_1 - \omega_2 - \frac{\omega_1 - \omega_2}{c}v\right)^2 + \frac{1}{4\tau_2^2}\right]^{-1}. \tag{2.124}$$

When this expression is averaged over a Maxwellian distribution of atomic velocities v, it produces the so-called Voigt spectral line (see Problem 2.4),

$$a_{\omega_1} = \frac{1}{2\pi\tau_2}\left(\frac{M}{2\pi T}\right)^{1/2}\int_{-\infty}^{\infty}\exp\left(-\frac{Mv^2}{2T}\right)$$
$$\times\left[\left(\omega_{m0} - \omega_1 - \omega_2 - \frac{\omega_1 - \omega_2}{c}v\right)^2 + \frac{1}{(2\tau_2)^2}\right]^{-1} dv.$$

The Doppler width $(\omega_1 - \omega_2)\sqrt{T/(Mc^2)}$ is of the same order of magnitude as the spontaneous width $1/\tau_2$, so that it cannot be neglected in this integral. We thus cannot go to Lorentz or Doppler limiting distributions, unlike Problem 2.4. However, we do assume that the quantity $(2\omega_1 - \omega_{m0}) \sim (\omega_1 - \omega_2)$ is large compared to the Doppler width in order to be able to neglect absorption of photons from only one of the beams. This is valid under the obvious condition $T \ll Mc^2$.

It is clear that the greater the difference $|\omega_1 - \omega_2|$, the larger is the width of the spectral line. By the appropriate choice of the frequency ω_2, we can achieve a resonantly large two-photon matrix element. We are thus restricted to only one intermediate state: In the present case, this is the state $k = P_{3/2}$. The nearby state $P_{1/2}$ does not contribute because the transition $P_{1/2} \to D_{5/2}$ is forbidden by the dipole selection rules. We conclude that the decrease of the absorption rate due to the increase in line width that accompanies growth of the frequency difference $|\omega_1 - \omega_2|$ can be compensated by resonantly increasing the two-photon matrix element. This phenomenon is called resonant increase of two-photon absorption.

The two-photon cross section can be of the same order of magnitude as the single-photon cross section for the allowed $S \to P$ transition.

3

ATOMIC PHOTOPROCESSES INVOLVING FREE PARTICLES

Radiative transitions between discrete levels of atomic systems were treated in Chapter 2. We now proceed to the study of radiative transitions when either the initial or the final state of the atomic system possesses a continuous spectrum. Treatment of the intense-field environment will be deferred until Chapter 5, so we shall treat here only single-photon processes and only those problems in which the number of particles in initial and final states differ by no more than one.

3.1 DECOMPOSITION OF ATOMIC SYSTEMS

The process of photodecomposition can be described generically by

$$XY + \hbar\omega \longrightarrow X + Y,$$

where the initial state of the system XY is a bound state of the atomic particles X and Y, whereas in the final state these particles are free. Examples of processes of this kind are photodetachment of a negative ion,

$$A^- + \hbar\omega \longrightarrow A + e;$$

photoionization of an atom (or of a positive ion),

$$A + \hbar\omega \longrightarrow A^+ + e;$$

and photodissociation of a molecule,

$$XY + \hbar\omega \longrightarrow X + Y.$$

The special feature of the photodecomposition process is that the transition is from a state in the discrete spectrum to a state in the continuum. It is thus necessary to modify Eq. (1.9) for the absorption probability per unit time to take into account the fact that the frequency of the absorbed photon lies in a continuous range of possible values and is not confined to a unique frequency corresponding to the transition between a pair of discrete levels. Because continuum wave functions are not square-integrable in the conventional sense, it is necessary to define an alternative normalization scheme for continuum wave functions. It is convenient to adopt the normalization

$$\int \psi_{\mathbf{q}}(\mathbf{r})\psi^*_{\mathbf{q}'}(\mathbf{r})\,d\mathbf{r} = (2\pi)^3\delta(\mathbf{q}-\mathbf{q}'), \tag{3.1}$$

where $\mathbf{r}$ is the distance between particles X and Y with wave vectors $\mathbf{q}$ and $\mathbf{q}'$, respectively, and $\psi_{\mathbf{q}}(\mathbf{r})$, $\psi_{\mathbf{q}'}(\mathbf{r})$ are the wave functions that describe the motion of these particles.

We now consider the particles to be contained within a large spatial volume Ω, and we wish to find the connection between the wave function $\psi_{\mathbf{q}}(\mathbf{r})$ and the wave function $\psi_k(\mathbf{r})$, which is normalized to unity according to

$$\int_{\Omega} |\,\psi_k(\mathbf{r})\,|^2\,d\mathbf{r} = 1.$$

These wave functions are the same to within a multiplicative constant C: $\psi_k = C\psi_{\mathbf{q}}$. For the determination of C we use the orthonormalization conditions

$$\int \psi^*_k(\mathbf{r})\psi_m(\mathbf{r})\,d\mathbf{r} = \delta_{km},$$

$$\sum_m \int \psi^*_k(\mathbf{r})\psi_m(\mathbf{r})\,d\mathbf{r} = 1.$$

If we relate the index k to the wave vector $\mathbf{q}$, and the index m to the wave vector $\mathbf{q}'$, these conditions yield

$$\sum_m \int \psi^*_k(\mathbf{r})\psi_m(\mathbf{r})\,d\mathbf{r} = 1 = C^2 \int \frac{\Omega\,d\mathbf{q}'}{(2\pi)^3} \int \psi_{\mathbf{q}}(\mathbf{r})\psi^*_{\mathbf{q}'}(\mathbf{r})\,d\mathbf{r} = \mathrm{C}^2\Omega.$$

This requires that $C = \Omega^{-1/2}$.

The wave function $\psi_{\mathbf{q}}(\mathbf{r})$ is describable by a plane wave when the distance between the atomic particles is so great that the interaction between them can be neglected. Then, with the normalization given by (3.1), one has

$$\psi_{\mathbf{q}}(\mathbf{r}) \underset{r\to\infty}{=} \exp(i\mathbf{q}\cdot\mathbf{r}).$$

(For interactions that decline as slowly with distance as the Coulomb interaction, this expression is correct only to within a logarithmic term which must be added to the phase.)

The corresponding plane wave limit for the wave function that is normalized to unity is

$$\psi_k(\mathbf{r}) \underset{r\to\infty}{=} C \exp(i\mathbf{q}\cdot\mathbf{r}),$$

along with the normalization condition

$$\int_\Omega |\psi_k(\mathbf{r})|^2\, d\mathbf{r} = 1 = C^2 \int_\Omega d\mathbf{r} = \mathbf{C}^2\Omega,$$

so that we have $C = \Omega^{-1/2}$ as before.

Problem 3.1. Obtain the expression for the photodetachment cross section of an atomic particle.

SOLUTION. Rewrite expression (1.9) for the absorption probability per unit time by using the wave functions $\psi_\mathbf{q}(\mathbf{r})$ in the transition matrix element in place of $\psi_k(\mathbf{r})$. This corresponds to a change of the matrix element $|\mathbf{D}_{0k}|^2$ to $|\mathbf{D}_{0\mathbf{q}}|^2/\Omega$, where the final-state wave function is normalized by condition (3.1). The statistical weight g_k of the final state is now given by

$$g_\mathbf{q} = \frac{\Omega\, d\mathbf{q}}{(2\pi)^3} = \frac{\Omega q^2\, dq}{8\pi^3}\, d\Omega_\mathbf{q},$$

where $d\Omega_\mathbf{q}$ is the element of solid angle that characterizes the relative motion of the final-state particles. Now employ the condition of energy conservation

$$\hbar\omega = I + \hbar^2 q^2/(2\mu), \tag{3.2}$$

where I is the binding energy of the atomic particles X and Y in their initially bound state, and μ is the reduced mass of these two particles. This gives

$$q\, dq = \frac{\mu}{\hbar}\, d\omega.$$

When the new expression for the statistical weight of the final state is inserted into expression (1.9), the result obtained for the probability per unit time of radiative transitions is

$$dw_i = \frac{4\omega^3}{3\hbar c^3} \frac{|\mathbf{D}_{0\mathbf{q}}|^2}{\Omega} n_\omega \frac{\Omega q}{8\pi^3} \frac{\mu}{\hbar}\, d\omega\, d\Omega_\mathbf{q}.$$

Naturally, the arbitrary reference volume Ω disappears from this expression. Dividing the probability dw_i by the photon flux (2.84)

$$dj_\omega = n_\omega \left(\frac{\omega}{\pi c}\right)^2 d\omega,$$

an expression for the photodetachment cross section is obtained that is independent of the number of photons n_ω,

$$d\sigma_i = \frac{\mu q \omega}{6\pi\hbar^2 c} \mid \mathbf{D}_{0\mathbf{q}} \mid^2 d\Omega_{\mathbf{q}}. \tag{3.3}$$

Note that in this form the photodetachment cross section corresponds to a single internal state of the final particles X and Y.

Problem 3.2. Determine the photodetachment cross section for a negative ion, or the atomic photoionization cross section, assuming that the field of the neutral or positive atomic core in which the valence electron resides is spherically symmetrical. In addition, take the state of the atomic core to be unchanged after the radiative transition of the valence electron.

SOLUTION. Use formula (3.3) for the photodetachment cross section. Because of the symmetry of the field of the atomic core, the state of the valence electron can be described by the principal quantum number n, the angular momentum quantum number l, and the projection m of l in a given direction. Based on this description, the wave function of the initial bound state of the electron is

$$\psi_0(\mathbf{r}) = \frac{1}{r} u_{nl}(r) Y_{lm}(\theta, \varphi).$$

Here r, θ, φ are the spherical polar coordinates of the electron, $u_{nl}(r)$ is the radial wave function of the electron, normalized by the condition

$$\int_0^\infty u_{nl}^2(r)\, dr = 1,$$

and $Y_{lm}(\theta, \varphi)$ is a normalized spherical harmonic.

The final state of the detached electron with momentum $\hbar\mathbf{q}$ can be expressed in terms of spherical functions in the form

$$\psi_{\mathbf{q}}(\mathbf{r}) = \frac{1}{r} \sum_{l=0}^{\infty} (2l+1) i^l u_l(q, r) P_l(\cos\theta_{\mathbf{qr}}).$$

That is, the polar axis of spherical coordinates is determined by the direction of $\mathbf{q}$. Here $\theta_{\mathbf{qr}}$ is the angle between the vectors $\mathbf{q}$ and $\mathbf{r}$, and $u_l(q, r)$ is the radial wave function of a free electron with angular momentum l. Far from the atomic core, this wave function has the asymptotic form

$$u_l(q, r) \underset{r\to\infty}{=} \frac{1}{q} \sin\left(qr - \frac{\pi l}{2} + \delta_l\right),$$

where δ_l is the phase shift for scattering of an electron by the atomic core. This expression is valid for photodetachment of a negative ion. In the case of atomic photoionization, the phase shift δ_l becomes r dependent.

If the detached electron is sufficiently energetic, its interaction with the atomic core can be neglected, and we can take

$$\psi_{\mathbf{q}}(\mathbf{r}) = \exp(i\mathbf{q} \cdot \mathbf{r}).$$

In this case,

$$u_l(q, r) = \sqrt{\frac{\pi r}{2q}} J_{l+(1/2)}(qr),$$

where $J_{l+(1/2)}$ is the ordinary Bessel function, or

$$u_l(q, r) = r j_l(qr)$$

in terms of the spherical Bessel function j_l.

The general form of the photodetachment cross section can be expressed by means of integrals of radial wave functions. According to Eq. (3.3), the photodetachment cross section has the form

$$\sigma_i = \frac{q\omega}{6\pi a_0 c} \int | \mathbf{r}_{0\mathbf{q}} |^2 \, d\Omega_{\mathbf{q}}. \tag{3.4}$$

Here $a_0 = \hbar^2/(\mu e^2)$ is the Bohr radius, and we used $\mathbf{D} = e\mathbf{r}$.

Now we shall calculate the integral in (3.4).

We introduce the unit vector $\mathbf{s}$ such that projection of the electron momentum in a given direction in the initial state is zero for this particular direction. Then, since $m = 0$, the angular wave function of the initial state simplifies to

$$Y_{l0}(\theta, \varphi) = \left[(2l + 1)/(4\pi)\right]^{1/2} P_l(\cos\theta_{\mathbf{rs}}),$$

where $\theta_{\mathbf{rs}}$ is the angle between the vectors $\mathbf{r}$ and $\mathbf{s}$. With the above expressions for the wave functions, the integral can be written in the form

$$\int | \mathbf{r}_{0\mathbf{q}} |^2 \, d\Omega_{\mathbf{q}} = \int d\Omega_{\mathbf{q}} \left| \int\int dr \, d\Omega_{\mathbf{r}} \, u_{nl}(r) \sqrt{\frac{2l+1}{4\pi}} P_l(\cos\theta_{\mathbf{rs}}) \mathbf{r} \right.$$
$$\left. \times \sum_{l'=0}^{\infty} (2l' + 1) i^{l'} u_{l'}(q, r) P_{l'}(\cos\theta_{\mathbf{qr}}) \right|^2 .$$

Upon carrying out the square indicated in the integrand, we obtain

$$\int d\Omega_{\mathbf{q}} \int\int\int\int dr \, dr' \, d\Omega_r \, d\Omega_{r'} (\mathbf{r} \cdot \mathbf{r}') u_{nl}(r) u_{nl}(r') \frac{2l+1}{4\pi} P_l(\cos\theta_{\mathbf{rs}})$$
$$\times P_l(\cos\theta_{\mathbf{r's}}) \sum_{l',l''=0}^{\infty} i^{l'-l''} (2l' + 1)(2l'' + 1) P_{l'}(\cos\theta_{\mathbf{qr}})$$
$$\times P_{l''}(\cos\theta_{\mathbf{qr'}}) u_{l'}(q, r) u_{l''}(q, r').$$

The integration over the solid angle $\int d\Omega_{\mathbf{q}}$ can be accomplished by using the theorem for Legendre polynomials

$$P_{l''}(\cos\theta_{\mathbf{qr}'}) = P_{l''}(\cos\theta_{\mathbf{rr}'})P_{l''}(\cos\theta_{\mathbf{qr}}) + 2\sum_m \frac{(l''-m)!}{(l''+m)!} \times P_{l''}^m(\cos\theta_{\mathbf{rr}'})P_{l''}^m(\cos\theta_{\mathbf{qr}})\cos\left[m(\varphi_{\mathbf{r}} - \varphi_{\mathbf{r}'})\right].$$

The indices on the polar angle θ identify the vectors whose angle of intersection defines θ, and φ is the azimuthal angle.

The sum over l'' is carried out using the orthogonality of Legendre polynomials with different indices,

$$\int_{-1}^{1} d(\cos\theta)P_l(\cos\theta)P_{l'}(\cos\theta) = \frac{2}{2l+1}\delta_{ll'}.$$

We obtain thereby

$$\int |\,\mathbf{r}_{0\mathbf{q}}\,|^2\, d\Omega_{\mathbf{q}} = \sum_{l'=0}^{\infty}|R_{ll'}|^2 \int\int d\Omega_r\, d\Omega_{r'} \cos\theta_{\mathbf{rr}'}P_l(\cos\theta_{\mathbf{rs}})P_l(\cos\theta_{\mathbf{r}'\mathbf{s}}) \times P_{l'}(\cos\theta_{\mathbf{r}'\mathbf{r}})(2l+1)(2l'+1),$$

where the notation is used such that the radial matrix element is defined by

$$R_{ll'} \equiv \int u_{nl}(r)ru_{l'}(q,r)\,dr.$$

In order to make further use of the orthogonality of the Legendre polynomials, we introduce the recursion relation

$$(2l'+1)xP_{l'}(x) = (l'+1)P_{l'+1}(x) + l'P_{l'-1}(x).$$

This leads to the form

$$\int |\,\mathbf{r}_{0\mathbf{q}}\,|^2\, d\Omega_{\mathbf{q}} = \sum_{l'=0}^{\infty}|R_{ll'}|^2 \int\int d\Omega_r\, d\Omega_{r'}(2l+1)P_l(\cos\theta_{\mathbf{r}'\mathbf{s}}) \times [(l'+1)P_{l'+1}(\cos\theta_{\mathbf{r}'\mathbf{r}}) + l'P_{l'-1}(\cos\theta_{\mathbf{r}'\mathbf{r}})]P_l(\cos\theta_{\mathbf{rs}}).$$

By using the summation theorem for the Legendre polynomials, we get

$$\begin{aligned}\int |\,\mathbf{r}_{0\mathbf{q}}\,|^2\, d\Omega_{\mathbf{q}} &= \sum_{l'=0}^{\infty}|R_{ll'}|^2 \int\int d\Omega_r\, d\Omega_{r'}(2l+1) \\ &\quad\times \left[(l'+1)P_{l'+1}(\cos\theta_{\mathbf{r}'\mathbf{s}})P_{l'+1}(\cos\theta_{\mathbf{rs}}) \right. \\ &\quad \left. + l'P_{l'-1}(\cos\theta_{\mathbf{r}'\mathbf{s}})P_{l'-1}(\cos\theta_{\mathbf{rs}})\right]P_l(\cos\theta_{\mathbf{rs}})P_l(\cos\theta_{\mathbf{r}'\mathbf{s}}) \\ &= \frac{(4\pi)^2}{(2l+1)}\left[l|R_{l,l-1}|^2 + (l+1)|R_{l,l+1}|^2\right].\end{aligned}$$

We now employ orthonormality for the Legendre polynomials to obtain the cross section for photodetachment of a single-electron ion or atom in the form

$$\sigma_i = \frac{8\pi q\omega}{3a_0c(2l+1)}\left[lR_{l,l-1}^2 + (l+1)R_{l,l+1}^2\right]. \tag{3.5}$$

As is shown by the above manipulations, radiative transitions of electrons into the continuum require fulfillment of the selection rules that the electron orbital angular momentum must change by unity, and the parity of the state must change.

The proportionality $\sigma_i \sim q$ is the threshold dependence of Wigner. It is the usual result, explained by the statistical weight of the final state. When an increase of the photon energy causes a departure from threshold conditions, the dependence of σ_i on q becomes complicated by a dependence of $R_{ll'}$ on q. As will be seen below, the matrix elements $R_{ll'}$ for the Coulomb potential depend strongly on q.

Note that formula (3.5) relates to the case of an absorbed photon of definite polarization.

Later, in Section 3.5, in addition to the process of photodetachment just treated, we shall consider as well the inverse process—photorecombination. This is the process in which an electron makes a transition from a state in the continuum to a bound state.

Problem 3.3. Determine the cross section for photoionization of an atom or photodetachment of a negative ion on the basis of the shell-model atom (see Appendix F). Adopt the condition that it corresponds to a transition into the continuous spectrum of one of N valence electrons, and neglect spin-orbit interactions within the atom.

SOLUTION. This problem is the generalization of the preceding ones for the case when one has N valence electrons in place of a single valence electron. Analyze first the initial state of the atom. The wave function of this state is determined by the coordinates of N valence electrons in an atom at rest. We represent this function in the form

$$\Psi_{LM_LSM_S}(1,\ldots,N).$$

Here, L is the orbital angular momentum, M_L is its projection in a given direction, S is the spin of the atom, and M_S is its projection on the same direction as for M_L. We write this function as

$$\Psi_{LM_LSM_S}(1,\ldots,N) = \widehat{P}\sum_{\substack{L'\mu m\\ s\sigma m_s}} G_{LS}^{L's}\langle L'l\mu m \mid LM_L\rangle \times \left\langle \tfrac{1}{2}s\sigma m_s \mid SM_S\right\rangle \psi_{lm\sigma}(N)\Psi_{L'\mu s m_s}(1,2,\ldots,N-1),$$

where $\Psi_{L'\mu s m_s}$ is the wave function of $N-1$ electrons, which are characterized by quantum numbers L', μ, s, m_s; and $\psi_{lm\sigma}(N)$ is the wave function of the Nth valence electron of orbital angular momentum l, with projection m on the given direction, and spin $\frac{1}{2}$ with spin projection σ. The operator $\widehat{P}$ permutes the positions of the electrons, $G_{LS}^{L's}$ is the fractional parentage coefficient, and the angle brackets $\langle \cdots \rangle$ are the Clebsch–Gordan coefficients, which account for the sum of the spin and orbital angular momenta of a valence electron and the corresponding angular momenta of the remaining $N - 1$ electrons. Here we assume that all of these electrons have the same n and l quantum numbers, that is, that they all relate to one electron shell.

Now consider the wave function of the final state. It is characterized by the release of an electron, and therefore is described by a parameter that is the momentum of this electron, $\hbar q$. Because this electron departs to large distances from the remainder of the atomic system, we do not sum the momentum of this electron and that of the remnant atom. Thus the wave function of the final state has the form

$$\Psi_{\mathbf{q}} = \frac{1}{\sqrt{N}} \widehat{P} \Psi_{L'\mu s m_s}(1, 2, \ldots, N - 1) \psi_{\mathbf{q}}(\mathbf{r}) \chi_\sigma,$$

where $\psi_{\mathbf{q}}(\mathbf{r})$ is the wave function of a free electron that is normalized to unity by condition (3.1), and χ_σ is the spin wave function of an electron with spin projection σ on a given direction.

The dipole moment operator of an atom is

$$\mathbf{D} = \sum_i e\mathbf{r}_i,$$

where the index i numbers the electrons. Using the above expressions for the electrons, summing over final states (the projections of the orbital momenta and spins for the system of $N - 1$ electrons, and the spin of the free electron), and averaging over the possible initial states (the projections of the orbital and spin momenta of the atom), we obtain

$$\begin{aligned} |\mathbf{D}_{0\mathbf{q}}|^2 &= N \left(G_{LS}^{L's} \right)^2 \frac{e^2}{2L + 1} \sum_{m\mu ML} \langle L' l \mu m \mid L M_L \rangle^2 \\ &\quad \times \sum_{m_s \sigma} \left\langle s \frac{1}{2} m_s \sigma \mid S M_S \right\rangle^2 |\langle lm \mid \mathbf{r} \mid \mathbf{q} \rangle|^2 \\ &= \frac{e^2 N}{2L + 1} \left(G_{LS}^{L's} \right)^2 \sum_m |\langle lm \mid \mathbf{r} \mid \mathbf{q} \rangle|^2. \end{aligned} \tag{3.6}$$

Here $\langle lm \mid \mathbf{r} \mid \mathbf{q} \rangle$ is the matrix element of the dipole moment of the electron that makes the transition from the bound lm shell of the atom to the unbound state of momentum q.

Using the expression for the one-electron matrix element obtained in the preceding problem, we find on the basis of the formulas (3.5) and (3.6) the photodetachment cross section

$$\sigma_i = \frac{8\pi q \omega}{3 a_0 c} \frac{N}{2L + 1} \left(G_{LS}^{L's} \right)^2 \left[l R_{l,l-1}^2 + (l + 1) R_{l,l+1}^2 \right]. \tag{3.7}$$

Problem 3.4. Determine the photodetachment cross section for an s electron from a negative ion in a ^1S state. Assume the interaction between the valence electron and the atom to be important only in a region near the atom which is small compared to the size of a negative ion.

SOLUTION. In a negative ion, there will be two valence electrons in s states. The photodetachment cross section for each of these can be obtained on the basis of the results of the preceding problem by using the following parameters: $l = 0, N = 2, G_{00}^{0(1/2)} = 1$. From (3.7) we have

$$\sigma_i = \frac{16\pi q\omega}{3a_0 c} R_{01}^2. \tag{3.8}$$

This result can also be found directly from Eq. (3.5) by taking into account the fact that the photodetachment cross section for two valence electrons is twice that obtained for a single electron.

Now we will calculate the radial matrix element. We first give the expression for the radial wave function of a valence electron in a negative ion. The electron binding energy will be written as $\hbar^2\gamma^2/(2m)$. The Schrödinger equation for the radial function $u_0(r)$ with $l = 0$ has the form $u_0'' = \gamma^2 u_0$. In obtaining this equation, we neglect the interaction between the detachment electron and the atom core because of the short range of this interaction. Thus for a short-range interaction between an electron and the remaining atom, the wave function obtained as the solution of the Schrödinger equation is

$$u_0 = (2\gamma)^{1/2} \exp(-\gamma r).$$

Because of the absence of residual interaction between the ionized electron and the remnant atom, the wave function of a free electron with $l = 1$ is the first harmonic in the expansion of the plane wave $\exp(i\mathbf{q} \cdot \mathbf{r})$ in Legendre polynomials

$$u_1(q, r) = rj_1(qr) = \frac{1}{q}\left(\frac{\sin qr}{qr} - \cos qr\right).$$

Using this function, we obtain for the radial matrix element

$$R_{01} = \int_0^\infty u_0(r) r u_1(q, r)\, dr = \frac{2(2\gamma)^{1/2} q}{(\gamma^2 + q^2)^2}.$$

This then gives the result for the cross section (3.8) for the photodetachment of a negative ion

$$\sigma_i = \frac{8\pi}{3}\frac{e^2}{\hbar c}\frac{\gamma v^3}{\omega^3}, \tag{3.9}$$

where

$$\hbar\omega = \frac{\hbar^2}{2m}(\gamma^2 + q^2)$$

is the energy of the absorbed photon, and $v = \hbar q/m$ is the velocity of the detached electron. The validity of the expression obtained is based on the assumption that the size $1/\gamma$ of the negative ion is large compared with the region of interaction of an outer electron with a neutral atom. Therefore it is valid for small binding energies of an electron in a negative ion. The expression is not valid for large values of q because the principal contribution in the matrix element R_{01} implies electron distances from the atom $\sim 1/q$, and these distances must be large compared with atomic sizes.

If $q \gg \gamma$ (but, nevertheless, q is not large), formula (3.9) for the photodetachment cross section has the asymptotic form

$$\sigma_i \underset{q \gg \gamma}{=} \frac{64\pi}{3} \frac{e^2}{\hbar c} \frac{\gamma}{q^3}.$$

Near the threshold, where $q \ll \gamma$, this expression gives

$$\sigma_i \underset{q \ll \gamma}{=} \frac{64\pi}{3} \frac{e^2}{\hbar c} \frac{q^3}{\gamma^5}.$$

The dependence $\sigma_i \sim q^3$ is determined by the fact that the final electron state is a p state (not an s state), and at small values of q the wave function is proportional to q (while for an s state of an electron it does not depend on q).

Problem 3.5. Determine the photoionization cross section of the hydrogen atom in its ground state.

SOLUTION. According to formula (3.5), the photoionization cross section for the ground state of hydrogen has the form

$$\sigma_i = \frac{8\pi q\omega}{3a_0 c} R_{01}^2,$$

and our goal is the calculation of the matrix element R_{01}. For simplicity, we introduce the atomic system of units where $\hbar = m = e = 1$. The radial wave function of an electron with $l = 1$ and momentum $\hbar q$ in the continuum spectrum of a Coulomb field is

$$u_1(q, r) = \frac{2}{3}\left[\frac{q(1+q^2)}{1-\exp(-2\pi/q)}\right]^{1/2} r^2 e^{-iqr} F\left(\frac{i}{q} + 2, 4, 2iqr\right),$$

where $F(\alpha, \gamma, x)$ is the confluent hypergeometric function. Using the known expression for the radial wave function of the ground state of hydrogen, we obtain the matrix element R_{01} in the form

$$R_{01} = \int_0^\infty 2re^{-r} u_1(q, r)\, dr.$$

To evaluate this integral, we use the formula

$$\int_0^\infty e^{-\lambda z} z^{\gamma-1} F(\alpha, \gamma, kz)\, dz = \Gamma(\gamma)\lambda^{\alpha-\gamma}(\lambda - k)^{-\alpha}.$$

This gives the result

$$R_{01} = \frac{8\sqrt{2\pi}}{\sqrt{q(1+q^2)^5}} \frac{\exp[-(2/q)\arctan q]}{\sqrt{1-\exp(-2\pi/q)}}.$$

The photoionization cross section for the ground state of hydrogen is then

$$\sigma_i = \frac{2^9\pi^2}{3c} \frac{\exp[-(4/q)\arctan q]}{(1+q^2)^4[1-\exp(-2\pi/q)]}. \tag{3.10}$$

Now consider some special cases of Eq. (3.10). For small values of q this formula leads to the threshold value of the photoionization cross section

$$\sigma_i = \frac{2^9\pi^2}{3e^4}\alpha a_0^2, \tag{3.11}$$

where $\alpha = e^2/hc$ is the fine structure constant, and we have departed from atomic units to return to conventional units. It is seen that the threshold cross section does not approach zero as would follow from the Wigner threshold formula. The result can be understood in terms of the behavior of the wave function of an electron moving slowly in a Coulomb field. Under these circumstances, the matrix element R_{01} tends to infinity near the threshold, with $q^{-1/2}$ dependence. Indeed, near the threshold, we have

$$R_{01} \cong \frac{8}{e^2}\sqrt{\frac{2\pi}{q}},$$

where here $e = 2.718\ldots$ is the base of the natural logarithm. This dependence compensates for the small statistical weight of a slow electron. By contrast, in the preceding problem we had $R_{01} \sim q$, which led to a threshold cross section $\sigma_i \sim q^3$. Thus the Coulomb interaction yields a nontypical result due to the singularity of the Coulomb potential at $r \to 0$.

For large q ($q \gg 2\pi$), for the photon energy significantly in excess of the ionization potential I of ground-state hydrogen, that is,

$$\hbar\omega \cong \hbar^2 q^2/(2m) \gg 4\pi^2 I \approx 40I,$$

Eq. (3.10) gives

$$\sigma_i = \frac{256\pi}{3}\frac{\alpha a_0^2}{(qa_0)^7} = \frac{256\pi}{3}\alpha a_0^2\left(\frac{I}{\hbar\omega}\right)^{7/2}. \tag{3.12}$$

Note that $I = me^4/(2\hbar^2)$.

From (3.10) it follows that the photoionization cross section is maximal at threshold. As the photoelectron momentum increases, the cross section falls linearly at first, behaving as $\sigma_i/\sigma_{i0} \approx 1 - \frac{8}{3}q^2$ (where we have reverted again to atomic units for simplicity), with the threshold value of σ_i designated as σ_{i0}. As (3.12) shows, the photoionization cross section for hydrogen falls rapidly as q increases. One might expect that at $\hbar^2 q^2/(2m) = I$ the photoionization cross section would be close to its threshold value. In fact, it differs from the threshold value by a factor of $e^4/(2^5 \sinh \pi) \approx 0.15$, which is about an order of magnitude. There is then a very extended transition to the region $q \gg 1$, and the behavior $\sigma_i \sim q^{-7} \sim \omega^{-7/2}$ of Eq. (3.12) holds true. The transition to the asymptotic dependence (3.12) is so extensive that only at $[\hbar^2 q^2/(2m)]/I \approx (10\pi)^2$ does σ_i reach 90% of (3.12). The behavior of σ_i/σ_{i0} is illustrated in Figure 3.1, where the limiting analytical forms are shown.

The decrease of the photoionization matrix element and cross section according to a power law for $q \gg 1$ follows from the singularity of the Coulomb potential at $r \to 0$. In fact the matrix element R_{01} for $q \gg 1$ is the Fourier component for large values of the argument. Without the singularity in the interaction potential, the Fourier component would have an exponential dependence rather than a power-law dependence on the argument.

Formula (3.12) is valid up to values of the frequency $\hbar\omega \approx mc^2$, when relativistic effects become essential.

Problem 3.6. Find the angular distribution of the photoelectrons from the ionization of ground-state hydrogen.

SOLUTION. According to Eq. (3.12),

$$d\sigma_i \sim (\mathbf{n} \cdot \mathbf{s})^2 \, d\Omega_{\mathbf{q}},$$

where $\mathbf{n}$ is the unit vector in the direction of emission of the ionized electron, and $\mathbf{s}$ is the unit vector along the direction of polarization of the absorbed photon. One can determine the normalization factor by requiring that the integration over solid angle

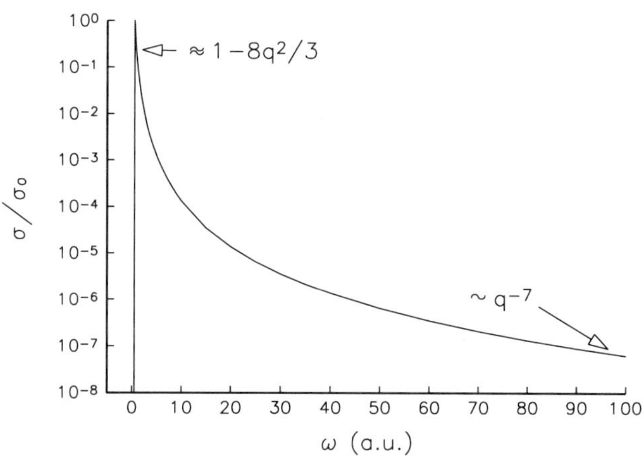

Figure 3.1. Cross section for the photoionization of hydrogen from the ground state, compared to the threshold cross section. Behavior near threshold and at high photoelectron momenta are shown on the figure.

of this expression should give the total cross section (3.10). This then yields

$$d\sigma_i = \frac{3}{4\pi}\sigma_i(\mathbf{n}\cdot\mathbf{s})^2\,d\Omega_{\mathbf{q}}. \tag{3.13}$$

The prediction of (3.13) is that most of the ionized electrons are emitted in a direction close to that of the polarization vector of the incident radiation. Perpendicular to this direction, including the direction of propagation of the incident radiation, the photodetachment cross section is zero.

If the incident radiation is unpolarized, Eq. (3.13) must be averaged over polarization directions $\mathbf{s}$ for a fixed direction $\mathbf{n}_0$ of the wave vector of the incident radiation. Then, from the relations $\cos\theta_{\mathbf{ns}} = \sin\theta_{\mathbf{nn}_0}\cos\varphi_{\mathbf{s}}$ and $\overline{\cos^2\varphi_{\mathbf{s}}} = \frac{1}{2}$, we have

$$d\sigma_i = \frac{3}{8\pi}\sigma_i(\mathbf{n}\times\mathbf{n}_0)^2\,d\Omega_{\mathbf{q}}. \tag{3.14}$$

In the case of photoionization from the innermost, or K shell, of an atom, one can use the above formulas; but it is necessary to double the cross sections since the K shell normally contains two electrons.

Sometimes the absorption coefficient k_ω is used instead of the cross section [see Eq. (2.87)]. The absorption coefficient is found by multiplying the cross section by the number density of atoms,

$$k_\omega = (N/V)\sigma_i.$$

Values of the absorption coefficient can be obtained in this fashion from the above cross-section formulas.

One can generalize the above formulas to hydrogenlike atoms or to the K shell of any atoms that are similar to the hydrogen atom. Then it is necessary to change the Bohr radius a_0 to a_0/Z, where Z is the nuclear charge. Then, according to (3.11) we have $\sigma_i \sim Z^{-2}$ near threshold, while from (3.12) it follows that $\sigma_i \sim Z^5$ for large photon energies.

Problem 3.7. Determine the frequency dependence of the photoionization cross section of an atomic state with angular momentum l for large photon energy.

SOLUTION. As indicated by the preceding problem, it is necessary to analyze the matrix element of the radial dipole moment to solve the present problem. For the case $\omega \to \infty$, the principal contribution to the matrix element will arise from small distances from the nucleus because $r \sim 1/q$, and q will be large for high photon frequency. This can be seen from Problem 3.5 for a transition $S \to P$ in the hydrogen atom. Let us analyze the radial wave function $u_l(q, r)$ of the continuous spectrum. It is a standard result in quantum mechanics that the radial wave function of a particle with angular momentum l at small distances r from the center has the dependence $u_l \sim r^l$. Therefore an additional (as compared with Problem 3.5) multiplier r^l will give an additional multiplier q^{-l} in the matrix element. Thus, for determining the dependence of the photoionization cross section on the frequency, it is necessary to multiply the result of Problem 3.5 [Eq. (3.12)] by the value q^{-2l}. As a result we have

$$\sigma_i \sim q^{-7-2l} \sim \left[I/(\hbar\omega)\right]^{7/2+l}. \tag{3.15}$$

It is seen that there is a strong decrease in the photoionization cross section with an increase in the angular momentum of a bound electron when that electron is ionized by an energetic photon. The asymptotic limit (3.15) corresponds to very high frequencies ($\hbar\omega \gg 40I$), as follows from Eq. (3.12).

Problem 3.8. For atomic photoionization or negative ion photodetachment near threshold, find the dependence of the cross section on the momentum of the detached electron.

SOLUTION. According to the general formula (3.5), the solution of this problem requires that one determine the dependence of the radial dipole moment matrix element $R_{l,l\pm1}$ on the wave vector q (or the momentum $\hbar q$) of a detached slow electron.

The matrix element of the radial dipole moment is given by the relation

$$R_{l,l\pm1} = \int u_{nl}(r)u_{l\pm1}(q,r)r\,dr,$$

and is determined by the behavior of the wave functions within a region of size $r \approx 1$ (in the atomic system of units). It is such distances that characterize the behavior of the radial wave function u_{nl} of the ground state. We obtain thereby the estimate

$$R_{l,l\pm1} \approx u_{l\pm1}(q, r \sim 1)$$

for the matrix element.

First, we consider the photoionization process for a neutral atom or positive ion. An ionized electron moves in the field of a positively charged ion. Because of the hypothesis that $q \ll 1$, for $r \sim 1$ the de Broglie wavelength has the form

$$\lambda = \left[q^2 + 2/r - l(l+1)/r^2\right]^{-1/2} \sim 1.$$

Therefore, at $r \approx 1$ we have

$$d\lambda/dr \sim 1,$$

that is, one can use a quasi-classical approach for the determination of qualitative behavior. The quasi-classical radial wave function of the final electron state has the form

$$u_{l\pm1}(q,r) = \frac{1}{q}\sqrt{\frac{q}{k(r)}}\sin\left(\int^r k\,dr - \varphi\right), \tag{3.16}$$

where the notation $k = 1/\lambda$ is used. The normalization multiplier is found from the asymptotic behavior of the wave function (see Problem 3.2)

$$u_{l\pm1}(q,r) = \frac{1}{q}\sin(qr - \varphi),$$

which corresponds to the normalization condition (3.1) for radial wave functions of states in the continuum.

For $r \sim 1$ we obtain the estimate $u_{l\pm 1} \sim q^{-1/2} \gg 1$. Therefore, on the basis of this and of Eq. (3.5) we have $R_{l,l\pm 1} \sim q^{-1/2}$ and $\sigma_i = \text{const}$. Thus the threshold photoionization cross section is a constant for atoms and positive ions. This result is corroborated by the work on the photoionization of hydrogen near the threshold, as obtained in Eq. (3.11). However, the photodetachment of a negative ion gives a qualitatively different result because of the short-range character of the interaction between the detached electron and the remaining neutral atom. Outside of this limited range, the wave function of the detached electron is described by a plane wave, and its decomposition into spherical harmonics (see Problem 3.2) gives the expression

$$u_l(q, r) = r j_l(qr)$$

for the radial wave function, where j_l is the spherical Bessel function. For $q \ll 1$ and $r \sim 1$ we obtain the estimate

$$u_l(q, r) \sim q^l.$$

Therefore,

$$R_{l,l+1} \sim q^{l+1}, \ R_{l,l-1} \sim q^{l-1}.$$

Equation (3.5) thus leads to the threshold dependence for the photodetachment cross section of a negative ion

$$\sigma_i = \begin{cases} q^3 & l = 0; \\ q^{2l-1}, & l \geq 1. \end{cases}$$

In particular, for $l = 0$, this agrees with a special case of Problem 3.4.

Problem 3.9. Determine the general character of the electron angular distribution resulting from photoionization.

SOLUTION. According to Eq. (1.8) the photoionization cross section is determined by the matrix element

$$\int \psi_{\mathbf{q}}^*(\mathbf{r})(\mathbf{r} \cdot \mathbf{s})\psi_{nlm}(\mathbf{r})\, d\mathbf{r}. \tag{3.17}$$

Here, the vector **s** characterizes the polarization of the incident radiation, and **q** is the wave vector of the detached electron. Further, we select the direction **q** as the quantum axis for the quantum numbers *nlm* of the initial state of the atom, and the differential cross section must be averaged over all projections m of the angular momentum.

The angle between the vectors **s** and **q** is labeled ϑ. Averaging over m is similar to averaging over directions of the quantum axis **r**. We label with ϑ' the angle between **r** and **q**. Further, we use the summation theorem for trigonometric functions that here has the form

$$\mathbf{r} \cdot \mathbf{s} = r(\cos\vartheta \cos\vartheta' + \sin\vartheta \sin\vartheta' \cos\varphi'). \tag{3.18}$$

Using (3.18) in (3.17), and averaging over angles (with $\overline{\cos\varphi'} = 0$), the photoionization cross section is obtained in the form

$$\overline{d\sigma_i} = (a + b\cos^2\vartheta)\, d\Omega_{\mathbf{q}}. \tag{3.19}$$

The fact that the photoionization cross section does not contain odd powers of $\cos\vartheta$ means that it does not depend on the sign of the momentum of a detached electron, or on the sign of $\mathbf{q}$. It can be obtained from general considerations. Indeed, a change of the sign of $\mathbf{q}$ is analogous to a change of the sign of $\mathbf{r}$ in $\psi_{\mathbf{q}}(\mathbf{r})$. But the substitution $\mathbf{r} \to -\mathbf{r}$ leads to the multiplication of ψ_{nlm} by a factor $(-1)^l$. Since the photoionization cross section is a quadratic function of the dipole matrix element, the cross section does not change as the result of such a transformation.

If the initial state nl is an S state ($l = 0$), the angular dependence of the photoionization cross section is simplified. In this case, we shall expand the wave function of the final state in Legendre polynomials as a function of the angle ϑ' between the vectors $\mathbf{q}$ and $\mathbf{r}$ to obtain

$$\psi_{\mathbf{q}}(\mathbf{r}) = \sum_{l=0}^{\infty} u_l(q, r) P_l(\cos\vartheta'). \tag{3.20}$$

The wave function of the initial state does not depend on angles. Since the function (3.20) does not depend on the angle φ', the second term of Eq. (3.18) is zero due to the relation

$$\int_0^{2\pi} \cos\varphi' d\varphi' = 0.$$

Thus the dipole matrix element is proportional to $\cos\vartheta$, and the cross section has the form

$$d\sigma_i = b\cos^2\vartheta\, d\Omega_{\mathbf{q}}. \tag{3.21}$$

This dependence for the photoionization cross section of the hydrogen atom was used at the end of Problem 3.5.

Problem 3.10. Obtain the expression for the photodissociation cross section of a diatomic molecule, viewing this process as the photoexcitation of a diatomic molecule to a repulsive state.

SOLUTION. The mechanism of the process we consider follows from Figure 3.2, where the electronic potential energy curves 1 and 2 for a diatomic molecule are shown as a function of the distance R between nuclei. Upon absorbing a photon of frequency ω, a diatomic molecule makes a transition from the ground state 1 to a repulsive or antibonding state 2. The atoms would then continue to separate, leading to dissociation of the diatomic molecule. We designate by D the dissociation energy of the molecule. According to the law of energy conservation, the value $\hbar\omega - D$ must be equal to the sum of the kinetic energies of the atoms that are formed, $\hbar^2 q^2/(2\mu)$ (μ

is the reduced mass of the atoms) and the value $\Delta U(\infty)$, which is the asymptotic (for $R \to \infty$) difference of the energies for the bound and repulsive states of the diatomic molecule. Note that the value $\hbar\mathbf{q}$ is the momentum of the atoms in the center-of-mass system. Thus, we have

$$\hbar\omega = D + \Delta U(\infty) + \hbar^2 q^2/(2\mu). \tag{3.22}$$

For determination of the photodissociation cross section we use Eq. (3.3). The initial and final state wave functions in the transition are given as the products of electron and nuclear wave functions. Then we have

$$\mathbf{D}_{0\mathbf{q}} = \langle 0|\mathbf{D}_{12}|\mathbf{q}\rangle,$$

where $\mathbf{D}_{12}$ is the matrix element of the dipole moment operator taken between electron wave functions, and the indices 0 and $\mathbf{q}$ correspond to nuclear wave functions for the lower and upper electron states of Figure 3.2.

When we take into account the small amplitude of the nuclear vibrations as compared with typical electron motions, we conclude that the matrix element $\mathbf{D}_{12}$ does not depend on the distance between the nuclei. Thus we obtain

$$\mathbf{D}_{0\mathbf{q}} = \mathbf{D}_{12} \cdot S_{0\mathbf{q}},$$

where $S_{0\mathbf{q}}$ is the integral of the overlap of the nuclear wave functions Φ_0, $\Phi_{\mathbf{q}}$, which are, respectively, the nuclear functions for the initial and final states of the diatomic molecule. They are not orthogonal because they relate to different Hamiltonians. These Hamiltonians have as interaction potentials the terms $U_1(\mathbf{R})$ and $U_2(\mathbf{R})$ shown in Figure 3.2. The value of $S_{0\mathbf{q}}$ is nonzero in the range of $\mathbf{q}$ that corresponds to the classical region of motion for the lower of the two potentials. See Figure 3.2. Outside of this range the overlap of the nuclear wave functions declines sharply.

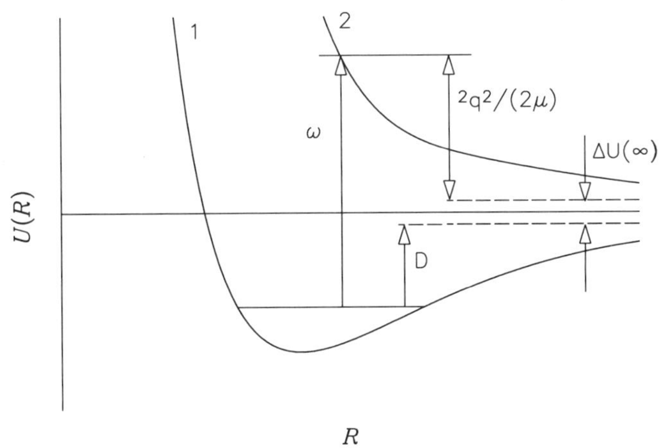

Figure 3.2. Electron potential energy for a diatomic molecule as a function of the distance R between the atoms. The curve labeled 1 corresponds to the ground state and that labeled 2 represents an excited, repulsive state.

The bound-state wave function Φ_0 is normalized to unity, whereas the unbound wave function $\Phi_{\mathbf{q}}$ is normalized by the condition (3.1), that is,

$$\int \Phi_{\mathbf{q}'}^*(\mathbf{R})\Phi_{\mathbf{q}}(\mathbf{R})\,d\mathbf{R} = (2\pi)^3\delta(\mathbf{q}-\mathbf{q}'),$$

and therefore it satisfies the relation

$$\int \Phi_{\mathbf{q}}^*(\mathbf{R}')\Phi_{\mathbf{q}}(\mathbf{R})\,d\mathbf{q} = (2\pi)^3\delta(\mathbf{R}-\mathbf{R}').$$

We thus find the condition

$$\int |S_{0\mathbf{q}}|^2\,d\mathbf{q} = \int \Phi_0(\mathbf{R})\Phi_0^*(\mathbf{R}')\Phi_{\mathbf{q}}^*(\mathbf{R})\Phi_{\mathbf{q}}(\mathbf{R}')\,d\mathbf{q}\,d\mathbf{R}\,d\mathbf{R}'$$

$$= (2\pi)^3 = \int |S_{0\mathbf{q}}|^2 q^2\,dq\,d\Omega_{\mathbf{q}} \tag{3.23}$$

for the normalization of the overlap integral $S_{0\mathbf{q}}$.

On the basis of the above relations, we rewrite the expression for the photodissociation cross section as

$$\sigma_D = \frac{\mu q}{4\pi m}\frac{e^2}{\hbar c}f_{12}\int |S_{0\mathbf{q}}|^2\,d\Omega_{\mathbf{q}}. \tag{3.24}$$

The quantity

$$f_{12} = \frac{2}{3}\frac{m\omega_{21}}{\hbar e^2}|\mathbf{D}_{12}|^2 \tag{3.25}$$

is the oscillator strength for the electron transition [see Eq. (1.18)], m is the electron mass, and ω_{21} is the transition frequency for the equilibrium distance $|\mathbf{R}|$ between the nuclei. This distance can be taken to have a definite value because the amplitude of nuclear vibrations is so small as compared to typical electron motions.

We shall now obtain an integral relation for the photodissociation cross section of a diatomic molecule. It is akin to a sum rule. We take into account that the oscillator strength does not depend on the nuclear coordinate $\mathbf{R}$. We multiply expression (3.23) by $d\omega$, introduce the integral over ω, and use the relation $d(\hbar\omega) = \hbar q dq/\mu$, which follows from (3.22), to obtain

$$\int \sigma_D\,d\omega = \frac{2\pi^2 e^2}{mc}f_{12}. \tag{3.26}$$

This relation is a particular case of a general sum rule found in Section 1.2. Note that the integrand as a function of ω is concentrated in the range of frequencies near the electron transition ω_{21} because the photodissociation cross section falls so steeply outside of this range. The sum rule (3.26) makes it possible to estimate the photodissociation cross section under certain conditions.

3.2 PHOTOEXCITATION OF RYDBERG STATES OF ATOMS

Highly excited states of atoms have the behavior of the outermost electron determined primarily by the Coulomb field of the atomic core. For this reason, such highly excited states exhibit the same structure as the hydrogen atom. In particular, the energy of a highly excited atom has the simple form $\varepsilon = -1/(2n^2)$ to within an accuracy of the order of $1/n^3$. Throughout this section we use the atomic system of units with $\hbar = m = e = 1$. Atoms in highly excited states are usually called Rydberg atoms (see also Problem 1.8). In this section, we consider the photoionization of a Rydberg atom.

Problem 3.11. Find the connection between the excitation cross section for transition of an atom to a Rydberg state and the photoionization cross section of this atom near threshold.

SOLUTION. According to the result of Problem 1.8, the oscillator strength for transition to a highly excited state with principal quantum number n is estimated to be proportional to n^{-3}. From Eq. (2.85), the photoexcitation cross section to such a state has the form

$$\sigma_n \sim \frac{1}{\omega^2}\omega_{n0}^3|\mathbf{D}_{n0}|^2 a(\omega - \omega_{n0}) \sim f_{n0}a(\omega - \omega_{n0}),$$

that is,

$$\sigma_n = \frac{C}{n^3}a(\omega - \omega_{n0}).$$

Here $a(\omega - \omega_{n0})$ is the distribution function for absorbing photons of frequency ω, and the quantity

$$\omega_{n0} = -\frac{1}{2n^2} - \varepsilon_0$$

is the frequency (or energy) difference in initial and final states.

The width of the distribution function is determined by the mechanism of spectral line broadening. We use this concept in the limiting case that the width of the spectral lines significantly exceeds the distance between neighboring levels, in which case the photoexcitation cross section coincides with the photoionization cross section. Indeed, in this case the discrete spectrum of excited electrons appears to an incoming photon to be the same as a continuous spectrum, and the behavior of weakly bound electrons and low-energy free electrons is similar in the field of the atomic core. Thus we have

$$\sigma_i = \sum_n \sigma_n = C\sum_n \frac{1}{n^3}a(\omega - \omega_{n0}). \tag{3.27}$$

Note that in this photoionization cross section those parts of the smoothed Rydberg states that are connected with a particular orbital angular momentum correspond directly to those parts of the continuum for the same orbital angular momentum. To determine the normalization constant C, we make use of the hypothesis that n is large to replace the summation in Eq. (3.27) by an integration. Then we have

$$\sigma_i = C \int a\left(\omega + \frac{1}{2n^2} + \varepsilon_0\right) d\left(\frac{1}{2n^2}\right).$$

From condition (2.2) for normalization of the wave function, we see that the value of this integral is unity, and so $\sigma_i = C$. Because the photoionization cross section is approximately constant near threshold, the relation $\sigma_i = C$ is approximately valid for any relation between the width of the spectral line and the distance between neighboring excited states of the Rydberg atom. Thus we have for the photoionization cross section of a highly excited level

$$\sigma_n = \frac{\sigma_i}{n^3} a(\omega - \omega_{n0}). \tag{3.28}$$

Problem 3.12. Determine the dependence of the photoexcitation cross section on photon frequency near the photoionization threshold. Assume Lorentz broadening of the spectral lines.

SOLUTION. Now allow the width of the spectral lines Γ to be either larger or smaller than the distance n^{-3} between neighboring levels. The Lorentz frequency distribution of photons is

$$a(\omega - \omega_{n0}) = \frac{1}{2\pi} \frac{\Gamma}{(\omega - \omega_{n0})^2 + \Gamma^2/4}.$$

According to Eq. (3.28), the photoabsorption cross section expressed as a sum of excitations of high-lying states is

$$\sigma_a(\omega) = \sum_n \sigma_n = \sigma_i \frac{\Gamma}{2\pi} \sum_n \frac{1}{n^3} \frac{1}{(\omega - \omega_{n0})^2 + \Gamma^2/4}. \tag{3.29}$$

The main contribution to the sum in (3.29) can be written in terms of values n^* (which are not necessarily integers) for which $\omega_{n^*0} = \omega$. We expand ω_{n0} for n in the vicinity of n^* in the Taylor series

$$\omega_{n0} - \omega = (n - n^*) \left.\frac{d\omega_{n0}}{dn}\right|_{n=n^*} = \frac{n - n^*}{(n^*)^3}. \tag{3.30}$$

Using (3.30) in (3.29), we have

$$\sigma_a(\omega) = \sigma_i \frac{\Gamma}{2\pi (n^*)^3} \sum_{n=1}^{\infty} \left[\frac{(n - n^*)^2}{(n^*)^6} + \frac{\Gamma^2}{4}\right]^{-1}. \tag{3.31}$$

Because of the strong convergence of the series (3.31), we can extend the lower limit of the range of summation to $-\infty$. The Mittag–Leffler theorem provides the result

$$\sum_{n=-\infty}^{\infty} \frac{1}{(n - x)^2 + y^2} = \frac{\pi \sinh(2\pi y)}{y[\cosh(2\pi y) - \cos(2\pi x)]}.$$

Using this, we can rewrite Eq. (3.31) as

$$\sigma_a(\omega) = \sigma_i \frac{\sinh(\pi n^{*3}\Gamma)}{\cosh(\pi n^{*3}\Gamma) - \cos(2\pi n^*)}. \tag{3.32}$$

The quantity n^* in this equation can be expressed in terms of the frequency of the absorbed photon by

$$n^* = \left[2|\varepsilon_0 + \omega|\right]^{-1/2},$$

with $\omega < |\varepsilon_0|$.

We shall now consider special cases of Eq. (3.32). If $\Gamma \gg (n^*)^{-3}$ and the energy levels overlap, then $\sigma_a(\omega) = \sigma_i$. In this case photoabsorption corresponds to photoionization of an atom.

In the opposite special case $\Gamma \ll (n^*)^{-3}$, we have

$$\sigma_a(\omega) = \sigma_i \frac{\pi(n^*)^3\Gamma}{1 - \cos(2\pi n^*) + (\pi n^{*3}\Gamma)^2/2}. \tag{3.33}$$

As n^* changes, the photoabsorption cross section oscillates from the minimum value

$$\sigma_a^{\min} = \sigma_i \frac{\pi(n^*)^3\Gamma}{2} \ll \sigma_i$$

at half-integer values of n^* to the maximum value of the photoabsorption cross section

$$\sigma_a^{\max} = \sigma_i \frac{2}{\pi(n^*)^3\Gamma} \gg \sigma_i$$

for integer n^*. Figure 3.3 shows the dependence of $\sigma_a(\omega)$ on frequency.

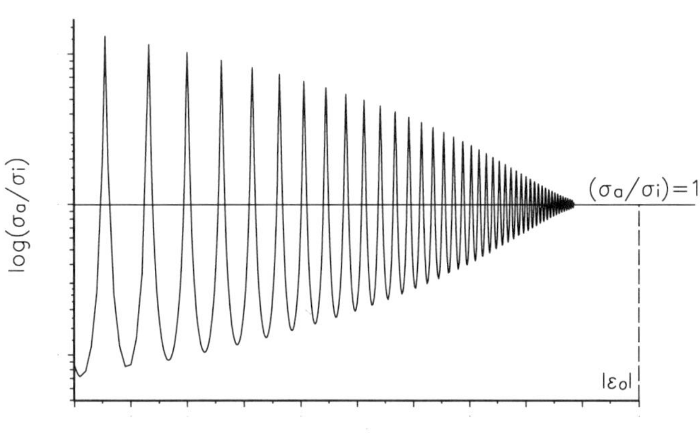

Figure 3.3. Cross section for excitation of a Rydberg state as a function of photon frequency, as calculated from Eq. (3.32). The maximum frequency indicated by the vertical dashed line corresponds to the binding energy of the initial atomic state.

The values $\sigma_a^{\max}$ correspond to resonance excitation of discrete levels with a principal quantum number n^*. In the vicinity of whole numbers n, where $n = [n^*]$ is the largest integer contained in n^*, we can write $n^* = n + (n^* - n)$, and then

$$1 - \cos(2\pi n^*) = 2\pi^2(n - n^*)^2,$$

which gives the photoabsorption cross section

$$\sigma_a(\omega) = \frac{\sigma_i}{n^3}\frac{1}{2\pi}\frac{\Gamma}{(\omega - \omega_{n0})^2 + \Gamma^2/4}. \qquad (3.34)$$

This equation coincides with (3.28).

After integration over the frequencies ω, we obtain the sum rule

$$\int \sigma_a(\omega)\, d\omega = \frac{\sigma_i}{n^3}. \qquad (3.35)$$

Problem 3.13. Determine the oscillator strength for transitions between the Rydberg states n, l and the states $n', l \pm 1$ if $\Delta n = n' - n \ll n, n'$.

SOLUTION. As a first step it is necessary to determine matrix elements for a given transition. Because both initial and final states are quasi-classical, the correspondence principle of quantum mechanics can be invoked, which gives these matrix elements as the Fourier components of the corresponding time-dependent classical coordinates. For Coulomb bound states with principal quantum number n and orbital angular momentum quantum number l, it is convenient to parameterize the electron coordinates $x(t)$ and $y(t)$ for motion in the xy plane in terms of a variable ξ such that

$$x = n^2(\cos\xi - \varepsilon); \qquad y = n^2\sqrt{1 - \varepsilon^2}\sin\xi; \qquad t = n^3(\xi - \varepsilon\sin\xi).$$

This describes an ellipse with eccentricity ε given by

$$\varepsilon = \sqrt{1 - (l/n)^2}.$$

We calculate first the matrix element of the y coordinate,

$$y_{nn'} = \frac{1}{T}\int_0^T y(t)\exp\left(\frac{i\Delta nt}{n^3}\right) dt,$$

where $T = 2\pi n^3$ is the period of the motion. This integral is equal to

$$y_{nn'} = \frac{in^2\sqrt{1 - \varepsilon^2}}{\Delta n\varepsilon} J_{\Delta n}(\varepsilon\Delta n).$$

In the same fashion, the matrix element of the x coordinate is found to be

$$x_{nn'} = \frac{n^2}{\Delta n} J'_{\Delta n}(\varepsilon\Delta n).$$

In the above expressions, J and J' denote the Bessel function and its derivative, respectively. The evaluation of the integrals is accomplished through an integration by parts and the use of the integral representation of the Bessel function

$$J_n(z) = \frac{1}{2\pi}\int_0^{2\pi} e^{i(n\xi - z\sin n\xi)}\,d\xi.$$

The operator $x+iy$ corresponds to a lowering operator for decreasing the magnetic quantum number by unity. In the classical case, the orbital angular momentum then also decreases by unity. In the same way, $x - iy$ is a raising operator for increasing the orbital angular momentum by unity. Thus we have

$$(x \pm iy)_{n,l;n',l\mp 1} = \frac{n^2}{\Delta n}\left[J'_{\Delta n}(\varepsilon\Delta n) \mp \frac{\sqrt{1-\varepsilon^2}}{\varepsilon}J_{\Delta n}(\varepsilon\Delta n)\right].$$

According to Eq. (1.18), the oscillator strength for the transition $n, l \to n, l \pm 1$ when $n, n' \gg 1$ and $\Delta n = (n' - n) \ll n, n'$, is given by

$$f(n, l \to n', l \pm 1) = \frac{2}{3}\omega_{n'n}\left|\frac{1}{\sqrt{2}}(x \mp iy)_{n,l;n',l\pm 1}\right|^2,$$

that is,

$$f(n, l \to n', l \pm 1) = \frac{n}{3\Delta n}\left[J'_{\Delta n}(\varepsilon\Delta n) \pm \frac{\sqrt{1-\varepsilon^2}}{\varepsilon}J_{\Delta n}(\varepsilon\Delta n)\right]^2. \tag{3.36}$$

This expression can be summed over the final orbital angular momentum quantum numbers $l' = l \pm 1$ to give

$$f(n, l \to n', l') = \frac{2n}{3\Delta n}\left[J'^2_{\Delta n}(\varepsilon\Delta n) + \frac{1-\varepsilon^2}{\varepsilon^2}J^2_{\Delta n}(\varepsilon\Delta n)\right]. \tag{3.37}$$

It is seen from Eq. (3.37) that, for $n \gg 1$, one has

$$f(n, l \to n', l') \sim n.$$

The oscillator strength decreases strongly with an increase of Δn, as is characteristic of classical matrix elements.

Problem 3.14. Determine the average oscillator strength for transition between Rydberg states with principal quantum numbers n and n'. Assume that the initial state has substates that are equally populated.

SOLUTION. In this case, we average the oscillator strength (3.37) over all the available substates of the initial state with principal quantum number n. The expression for the oscillator strength then has the form

$$f(n - n') = \frac{1}{n^2}\sum_{l=0}^{n-1}(2l+1)f(n, l \to n', l' = l \pm 1).$$

We have used the fact that there is a $(2l + 1)$-fold degeneracy in terms of orbital angular momentum quantum numbers l, and the total number of substates for a principal quantum number n is n^2. Based on the large magnitude of n, we change from summation to integration, choose the variable ε to replace l, and use the relation $n^2\varepsilon\, d\varepsilon = -l\, dl$, which follows from the definition of ε

$$\varepsilon^2 = 1 - l^2/n^2,$$

to obtain

$$f(n - n') = \frac{4n}{3\Delta n}\int_0^1 \varepsilon\left[J'^2_{\Delta n}(\varepsilon\Delta n) + \frac{1-\varepsilon^2}{\varepsilon^2}J^2_{\Delta n}(\varepsilon\Delta n)\right] d\varepsilon.$$

Now use the property of the Bessel functions

$$z\left[J'^2_p(z) + \left(\frac{p^2}{z^2} - 1\right)J^2_p(z)\right] = \frac{d}{dz}\left[zJ_p(z)J'_p(z)\right]$$

to find that

$$f(n - n') = \frac{4n}{3(\Delta n)^2}J_{\Delta n}(\Delta n)J'_{\Delta n}(\Delta n). \tag{3.38}$$

Expression (3.38) can be simplified if $\Delta n = n' - n$ is sufficiently large. Then we obtain the asymptotic forms

$$J_{\Delta n}(\Delta n) \approx \frac{1}{2\pi}\int_{-\infty}^{\infty} e^{i\Delta n\xi^2/6}\, d\xi = \frac{1}{2\pi\sqrt{3}}\left(\frac{6}{\Delta n}\right)^{1/3}\Gamma\left(\frac{1}{3}\right),$$

$$J'_{\Delta n}(\Delta n) \approx -\frac{i}{2\pi}\int_{-\infty}^{\infty} \xi e^{i\Delta n\xi^2/6}\, d\xi = \frac{1}{2\pi\sqrt{3}}\left(\frac{6}{\Delta n}\right)^{2/3}\Gamma\left(\frac{2}{3}\right).$$

Thus we have, from (3.38),

$$f(n \to n') = \frac{2n}{3\pi^2(\Delta n)^3}\Gamma\left(\frac{1}{3}\right)\Gamma\left(\frac{2}{3}\right).$$

Using the property of the gamma functions

$$\Gamma(x)\Gamma(1-x) = \frac{\pi}{\sin \pi x},$$

we obtain the expression for the averaged oscillator strength in the form

$$f(n \to n') \underset{\Delta n \gg 1}{=} \frac{4n}{3^{3/2}\pi(\Delta n)^3}. \tag{3.39}$$

Problem 3.15. Determine the photoionization cross section for the Rydberg states of an atom near the ionization threshold.

SOLUTION. We use the concept of Problem 3.11 that in the limit of a broad spectral line the photoabsorption cross section coincides with the photoionization cross section near the ionization threshold. Therefore it is necessary to determine the photoabsorption cross section for the transition between two Rydberg states. We can then usc Eq. (3.39) assuming $\Delta n \gg 1$. According to Eq. (2.85) the photoabsorption cross section is

$$\sigma_a = 2\pi^2 f(n \to n')\frac{a(\omega - \omega_{n'n})}{c}.$$

When we use formula (3.39) for the oscillator strength in this expression, we obtain

$$\sigma_a = \frac{8\pi}{3^{3/2}c\omega^3 n^8}a(\omega - \omega_{n'n}), \tag{3.40}$$

where we have used $\Delta n = \omega n^3$.

For evaluation of the photoionization cross section σ_i, we can make use of Eq. (3.35). When we integrate Eq. (3.40) over ω, and account for the normalization condition (2.2) on the photon frequency distribution function $a(\omega - \omega_{n'n})$, we obtain Kramers's formula for the photoionization process

$$\sigma_i = \frac{8\pi}{3^{3/2}}\frac{1}{cn^5\omega^3}. \tag{3.41}$$

This expression can be understood in terms of the fact that the state n' is in a part of the spectrum where the width of the distribution function $a(\omega - \omega_{n'n})$ is greater than the distance between neighboring levels.

If the ionization potential of the final state is small as compared to that of the initial state, then $\omega = 1/(2n)^2$, and Eq. (3.41) gives

$$\sigma_i = \frac{64\pi}{3^{3/2}}\frac{e^2}{\hbar c}na_0^2. \tag{3.42}$$

In this last equation, we have returned from atomic units to the usual units, and a_0 is the Bohr radius. Thus the photoionization cross section for a highly excited atom near the ionization threshold, which is the maximum atomic photoionization cross section, changes proportionally to n as the principal quantum number increases, whereas the area of a highly excited atom is proportional to $n^4a_0^2$.

3.3 AUTOIONIZING STATES OF ATOMS

To this point, we have considered photoionization as a single-electron process, while real atoms typically contain other electrons in addition to the valence electron. These remaining electrons have ionization threshold energies larger than that of the valence electron. Nevertheless, the atomic core has excited states, and excitation energies for some of these states may exceed the ionization potential of the atom. Such states are called autoionizing states. Autoionizing states are characterized by a certain lifetime, and therefore the autoionizing levels have a width Γ corresponding to ionization of the atom with the release of a valence electron. This width is determined by the interaction of the valence electron with the remainder of the atom and is small compared to typical energies of the atom. The existence of autoionizing states leads to resonances in the photoionization cross section at energies that correspond to excitation of these states. We consider this phenomenon as well as giving examples of the influence of autoionizing states on the character of photoionization processes.

Problem 3.16. Determine the photoionization cross section in the energy neighborhood of the excitation of autoionizing states.

SOLUTION. When autoionizing states exist, photoionization can take place in two channels. In one, the atomic core is excited and this excitation transfers itself to the valence electron. In the second case, direct photoionization of the valence electron occurs. Since both processes involve the same initial and final states, they are coherent and the transition amplitudes add. The second process occurs in a single step and can be determined within first-order perturbation theory, while the first photoionization channel includes an interaction with the external electromagnetic field as well as with the internal electric field of the system.

It is essential that both channels possess the same final and initial states, and therefore one can sum the amplitudes (but not the probabilities) of these two channels. The initial state of the valence electron will be given the label 0, and the final state will be identified by $\hbar q$, the momentum of the photoionized electron. Equation (3.3) gives the cross section for this photoionization process as

$$d\sigma_i = \frac{mq\omega}{6\pi\hbar^2 c}|\mathbf{D}^*_{0\mathbf{q}}|^2\, d\Omega_{\mathbf{q}}. \tag{3.43}$$

Here, m is the electron mass, $d\Omega_{\mathbf{q}}$ is the differential solid angle describing the motion of the free electron, and $\mathbf{D}^*_{0\mathbf{q}}$ is the effective dipole moment of the transition. It has the form $\mathbf{D}^*_{0\mathbf{q}} = \mathbf{D}_{0\mathbf{q}} + \mathbf{D}^{(2)}_{0\mathbf{q}}$, where $\mathbf{D}_{0\mathbf{q}}$ is the dipole moment operator for the usual photoionization transition, while $\mathbf{D}^{(2)}_{0\mathbf{q}}$ is the dipole moment operator related to the two-step transition. Label by A and B the initial and final states of the atomic core, and the frequency of the transition between them as ω_{AB}. In the case of the two-step process, the external electromagnetic field first excites state B. Label by $\mathbf{D}_{AB}$ the dipole moment matrix element for this transition. The following step is the transition of the atomic core from state B to state A with transfer of the excitation energy to a valence electron. This electron then makes a transition from state 0 to the state with momentum $\hbar q$. The interaction matrix element between the valence electron and the ion is labeled $U^{BA}_{0\mathbf{q}}$. Then, from Eq. (1.44), within the framework of second-order perturbation theory we have

$$\mathbf{D}^{(2)}_{0\mathbf{q}} = \frac{\mathbf{D}_{AB} U^{BA}_{0\mathbf{q}}}{\hbar(\omega_{BA} - \omega)}. \tag{3.44}$$

Expression (3.44) is correct in the general vicinity of the resonance, where perturbation theory is valid, but it is not correct on resonance. To obtain the proper result at resonance, it is necessary to consider the following corrections within the framework of perturbation theory. First consider the energy denominator of Eq. (3.44). Because of interaction with the continuum states, level B experiences a second-order shift by the amount

$$\delta\varepsilon_B = \frac{1}{\hbar}\sum_{\mathbf{q}} \frac{|U_{\mathbf{q}0}^{AB}|^2}{\omega_{BA} - \omega_{\mathbf{q}0} + i\delta}. \tag{3.45}$$

When $\delta \to 0^+$ in the denominator, this corresponds to a path around the pole associated with damping in level B. The real part of the correction to ε_B is small, and we shall neglect it. The imaginary part of this energy has the form

$$\begin{aligned}\operatorname{Im}\delta\varepsilon_B &= -\frac{\pi}{\hbar}\sum_{\mathbf{q}} |U_{\mathbf{q}0}^{AB}|^2\delta(\omega_{BA} - \omega_{\mathbf{q}0})\\ &= -\frac{\pi}{\hbar}|U_{\mathbf{q}0}^{AB}|^2\rho(\mathbf{q}).\end{aligned} \tag{3.46}$$

Here, $\rho(\mathbf{q})$ is the density of final states for a valence electron.

The value (3.46) characterizes the width of the autoionizing level. The quantity $\operatorname{Im}\delta\varepsilon_B/\hbar$ is the probability per unit time of transition from state B as a result of interaction with the continuum spectrum due to the potential U, when this value is obtained within the context of second-order perturbation theory. We obtain

$$w_B = \frac{2\pi}{\hbar}|U_{\mathbf{q}0}^{AB}|^2\rho(\mathbf{q}).$$

This expression agrees with "Fermi's golden rule."

It is evident that there is no imaginary correction to level A. Also, the probability w_B significantly exceeds the corresponding probability for radiative decay of this state because of the absence from this expression of the small parameter $e^2/(\hbar c)$. Now we analyze the numerator of Eq. (3.44), where it is necessary to add a second-order correction. This correction takes into account the two-step transition through state B along with the direct one. According to Eq. (1.44), we have

$$\mathbf{D}_{AB} \to \mathbf{D}_{AB} + \frac{1}{\hbar}\sum_{\mathbf{q}} \frac{\mathbf{D}_{0\mathbf{q}}U_{\mathbf{q}0}^{AB}}{\omega_{\mathbf{q}0} - \omega - i\delta} \quad (\delta \to 0^+).$$

When we neglect real corrections and retain only the imaginary part, we obtain

$$\mathbf{D}_{AB} \to \mathbf{D}_{AB} + i\pi\mathbf{D}_{0\mathbf{q}}U_{\mathbf{q}0}^{AB}\rho(\mathbf{q})/\hbar. \tag{3.47}$$

The procedure for obtaining this result is similar to that used to find Eq. (3.46).

When Eqs. (3.46) and (3.47) are used in Eq. (3.44), and Eq. (3.44) is employed in (3.43), we obtain for the photoionization cross section in the vicinity of an autoionizing state the result

$$d\sigma_i = \frac{mq\omega}{6\pi\hbar^2 c}\left|\mathbf{D}_{0\mathbf{q}} + \frac{[\mathbf{D}_{AB} + i(\pi/\hbar)\mathbf{D}_{0\mathbf{q}}U_{\mathbf{q}0}^{AB}\rho]U_{0\mathbf{q}}^{BA}}{\hbar(\omega_{BA} - \omega) - i(\pi/\hbar)|U_{\mathbf{q}0}^{AB}|^2\rho}\right|^2 d\Omega_{\mathbf{q}}. \tag{3.48}$$

Simplifying this expression, and labeling by $d\sigma_i^{(0)}$ the photoionization cross section far away from resonance, Eq. (3.48) leads to

$$d\sigma_i = d\sigma_i^{(0)} \frac{|(\omega_{BA} - \omega) + \beta|^2}{(\omega_{BA} - \omega)^2 + \frac{1}{4}w_B^2}, \tag{3.49}$$

where $w_B = \Gamma/\hbar$ is the above probability per unit time for the decay of the autoionizing state, and the parameter β is

$$\beta = \frac{1}{\hbar}\frac{\mathbf{D}_{AB}}{\mathbf{D}_{0\mathbf{q}}}U_{0\mathbf{q}}^{BA}. \tag{3.50}$$

A number of conclusions can be drawn from Eq. (3.49). If $\omega_{BA} - \omega = -\beta$, then the photoionization cross section vanishes. The maximum cross section occurs at the frequency $\omega = \omega_{BA} - w_B^2/(4\beta)$, which gives a larger result than at resonance ($\omega = \omega_{AB}$). The ratio $d\sigma_i/d\sigma_i^{(0)}$ is shown in Figure 3.4. The maximum cross section exceeds the cross section far from resonance by the factor $(2\beta/w_B)^2$. Evidently, it disappears if $\mathbf{D}_{AB} = 0$. Figure 3.4 shows two different values of the parameter $\alpha = (2\beta/w_B)^2$. The larger the value of α, the more strongly the resonance is exhibited. The solid curve in the figure corresponds to $\alpha = 2$, and the dashed curve is for $\alpha = \frac{1}{2}$.

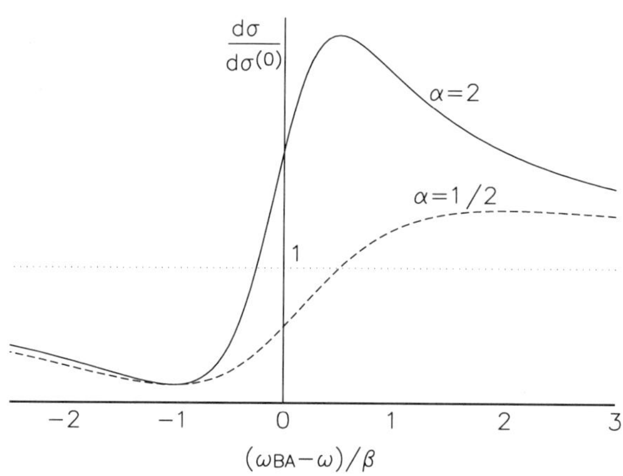

Figure 3.4. Ratio of the photoionization cross section in the vicinity of the excitation of an autoionizing state to the corresponding cross section far from the resonance.

3.4 BREMSSTRAHLUNG FROM SCATTERING OF AN ELECTRON BY ATOMS AND IONS

The photoabsorption processes studied so far correspond to the transition of an atomic electron from a state in a discrete spectrum of energies to a final state that may be either in a discrete or a continuum spectrum. Now we shall study radiative processes that result from the interaction of free atomic particles. The radiation arising from the deflection of a free, charged particle by an external field is called *bremsstrahlung*. In particular, we consider the radiation resulting from the interaction of an electron with atoms and ions. This is the principal bremsstrahlung process in atomic physics.

Problem 3.17. Determine the cross section for bremsstrahlung from the scattering of an electron by a center of force.

SOLUTION. Consider an electron with the initial wave vector $\mathbf{q}$ that, after being scattered from a center of force with the concomitant emission of a photon, has a final wave vector $\mathbf{q}'$. Conservation of energy yields the condition

$$\hbar\omega = \frac{\hbar^2}{2m}\left(\mathbf{q}^2 - \mathbf{q}'^2\right), \tag{3.51}$$

where m is the electron mass. We can use the general expression (1.13) for the radiation frequency in the case under consideration, which, because of the absence of photons in the initial state, has the form

$$dw_b = \frac{4\omega^3}{3\hbar c^3}|\mathbf{D}_{12}|^2 \frac{\Omega\, d\mathbf{q}'}{(2\pi)^3}.$$

Indices 1 and 2 refer to the initial and final states of the system, and Ω is a volume in which the system is contained. (We note that this quantity must not appear in the final result.) Using wave functions normalized by condition (3.1), the matrix element is

$$\mathbf{D}_{12} = \frac{1}{\Omega}\mathbf{D}_{\mathbf{q}\mathbf{q}'} = \frac{1}{\Omega}e\mathbf{r}_{\mathbf{q}\mathbf{q}'},$$

where $\mathbf{r}$ is the electron coordinate. The flux of incident electrons is

$$\mathbf{j} = \frac{\hbar}{2mi\Omega}\left(\psi_{\mathbf{q}}^*\nabla\psi_{\mathbf{q}} - \psi_{\mathbf{q}}\nabla\psi_{\mathbf{q}}^*\right) = \frac{\hbar\mathbf{q}}{m\Omega},$$

and, far from the scattering center, the electron wave function is

$$\psi_{\mathbf{q}} = \exp(i\mathbf{q}\cdot\mathbf{r}).$$

In addition, energy conservation gives

$$d\omega = \frac{\hbar}{m}q'\,dq'.$$

The transition probability per unit time, when divided by the incident flux, gives the cross section for bremsstrahlung

$$\frac{d\sigma_b}{d\omega} = \frac{e^2 m^2 \omega^3 q'}{6\pi^3 \hbar^3 c^3 q} \int |\mathbf{r}_{\mathbf{q}\mathbf{q}'}|^2 \, d\Omega_{\mathbf{q}'}, \tag{3.52}$$

where $d\Omega_{\mathbf{q}'}$ is the element of solid angle into which an electron is scattered after collision.

Problem 3.18. Obtain the cross section for bremsstrahlung for an electron scattering from a spherical atomic system.

SOLUTION. Our goal is to simplify Eq. (3.52) by using the symmetry of the problem, and to express the matrix elements of the dipole moment operator in terms of radial wave functions. We represent the electron wave function in the form of the spherical harmonic decomposition

$$\psi_{\mathbf{q}}(\mathbf{r}) = \frac{1}{r} \sum_{l=0}^{\infty} i^l (2l+1) P_l(\cos\theta_{\mathbf{qr}}) u_l(q, r).$$

A similar representation is used for the wave function $\psi_{\mathbf{q}'}$. These relations are then used in Eq. (3.52). For the quantity

$$I = \int |\mathbf{r}_{\mathbf{q}\mathbf{q}'}|^2 \, d\Omega_{\mathbf{q}'},$$

we find the expression

$$I = \sum_{l,l',n,n'} R_{ll'} R_{nn'} (2l+1)(2l'+1)(2n+1)(2n'+1)$$
$$\times \int \cos\theta_{\mathbf{rr}'} P_l(\cos\theta_{\mathbf{qr}}) P_{l'}(\cos\theta_{\mathbf{q}'\mathbf{r}}) P_n(\cos\theta_{\mathbf{qr}'}) P_{n'}(\cos\theta_{\mathbf{q}'\mathbf{r}'}) \, d\Omega_{\mathbf{r}} \, d\Omega_{\mathbf{r}'} \, d\Omega_{\mathbf{q}'}.$$

Here the subscript vectors associated with the angles θ define each such angle as the angle between the vectors cited, $d\Omega$ is the element of solid angle for the direction shown by the subscript vector, and the radial matrix element is

$$R_{ll'} = \int_0^{\infty} u_l(q, r) u_{l'}(q', r) r \, dr.$$

The final result will be expressed in terms of these matrix elements.

The integration over the solid angle $\Omega_{\mathbf{q}'}$ is carried out with the help of the summation theorem for the Legendre polynomials

$$P_{n'}(\cos\theta_{\mathbf{q}'\mathbf{r}'}) = P_{n'}(\cos\theta_{\mathbf{q}'\mathbf{r}}) P_{n'}(\cos\theta_{\mathbf{rr}'})$$
$$+ 2 \sum_m \frac{(n'-m)!}{(n'+m)!} P_{n'}^m(\cos\theta_{\mathbf{q}'\mathbf{r}}) P_{n'}^m(\cos\theta_{\mathbf{rr}'}) \cos(m\varphi_{\mathbf{q}'})$$

and their orthogonality property. We obtain in this fashion

$$I = 4\pi \sum_{l,l',n} (2l+1)(2l'+1)(2n+1) R_{ll'} R_{nl'}$$
$$\times \int \cos\theta_{\mathbf{rr'}} P_l(\cos\theta_{\mathbf{qr}}) P_{l'}(\cos\theta_{\mathbf{rr'}}) P_n(\cos\theta_{\mathbf{qr'}})\, d\Omega_{\mathbf{r}}\, d\Omega_{\mathbf{r'}}.$$

We integrate over the solid angle, use the recurrence relation for the Legendre polynomials

$$(2l'+1)xP_{l'}(x) = (l'+1)P_{l'+1}(x) + l'P_{l'-1}(x),$$

and the summation rule for the Legendre polynomials. The result is

$$\begin{aligned} I &= 16\pi^2 \int d\Omega_{\mathbf{r}} \sum_{l,l',n} (2l+1) P_l(\cos\theta_{\mathbf{qr}}) P_n(\cos\theta_{\mathbf{qr}}) \\ &\quad \times R_{ll'} R_{nl'} \left[(l'+1)\delta_{n,l'+1} + l'\delta_{n,l'-1} \right] \\ &= 64\pi^3 \sum_{l,l',n} \delta_{l,n} R_{ll'} R_{nl'} \left[(l'+1)\delta_{n,l'+1} + l'\delta_{n,l'-1} \right] \\ &= 64\pi^3 \sum_{l=0}^{\infty} \left[lR_{l,l-1}^2 + (l+1)R_{l,l+1}^2 \right] \\ &= 64\pi^3 \sum_{l=0}^{\infty} (l+1)(R_{l+1,l}^2 + R_{l,l+1}^2). \end{aligned}$$

When this expression is used in Eq. (3.52), we obtain the cross section for bremsstrahlung by an electron as a result of its collision with a spherical structureless atomic system as

$$\frac{d\sigma_b}{d\omega} = \frac{32 m\omega^3 q'}{3\hbar c^3 a_0 q} \sum_{l=0}^{\infty} (l+1)(R_{l+1,l}^2 + R_{l,l+1}^2). \tag{3.53}$$

Problem 3.19. Determine the cross section for emission of bremsstrahlung photons of long wavelength as a result of the scattering of an electron by an atom.

SOLUTION. Assume the energy of the emitted photon to be small compared to that of the electron, that is, assume $q - q' \ll q$. Use the asymptotic expression for the radial wave function of a scattered electron far from the scattering center,

$$u_l(q, r) = \frac{1}{q} \sin\left(qr - \frac{\pi l}{2} + \delta_l \right),$$

where δ_l is the scattering phase shift. In the case considered, the primary contribution to the bremsstrahlung corresponds to a large impact parameter (or large distance) of the electron with respect to the scattering center. Then the radial matrix elements are described by the expression

$$R_{l,l+1} = \int_0^\infty \frac{1}{qq'} \sin\left[qr - \frac{\pi l}{2} + \delta_l(q)\right] \times \sin\left[q'r - \frac{\pi(l+1)}{2} + \delta_{l+1}(q')\right] r\,dr.$$

When we exclude from this expression the strongly oscillating term, we have

$$R_{l,l+1} \approx \frac{1}{2q^2} \int_0^\infty \cos\left[\delta_l(q) - \delta_{l+1}(q) + (q-q')r + \frac{\pi}{2}\right] r\,dr,$$

where we have used the fact that q and q' are almost the same. From the form of the integral, it is evident that the main contribution to it comes from $r \sim 1/(q-q')$. This is large enough to validate our procedure of using asymptotic expressions for the radial wave functions. Thus we obtain

$$R_{l,l+1} \approx \frac{\sin\left[\delta_l(q) - \delta_{l+1}(q)\right]}{2q^2(q-q')^2}.$$

The same expression applies as well to $R_{l+1,l}$. Using these expressions in Eq. (3.53), we obtain the cross section for bremsstrahlung emission

$$\frac{d\sigma_b}{d\omega} = \frac{32m\omega^3}{3\hbar c^3 a_0} \cdot \frac{(q-q')^{-4}}{2q^4} \sum_{l=0}^{\infty} (l+1)\sin^2(\delta_l - \delta_{l+1}).$$

When the electron energy $\varepsilon = \hbar^2 q^2/(2m)$ is introduced, $q - q'$ is expressed in terms of the photon frequency $\omega = \hbar q(q-q')/m$, and the initial postulate $\hbar\omega \ll \varepsilon$ is used, then we obtain

$$\frac{d\sigma_b}{d\omega} = \frac{16\hbar e^2}{3m^2c^3} \frac{1}{\omega} \sum_{l=0}^{\infty} (l+1)\sin^2(\delta_l - \delta_{l+1}). \tag{3.54}$$

Note that if this cross section is integrated over ω, the result diverges. This is because perturbation theory is violated for slow electrons. Let us estimate the minimum emitted frequency $\omega_{\min}$ down to which expression (3.54) is valid. We have assumed that the emission of a single photon is more probable than the emission of two photons. However, at small photon frequencies, $\omega \to 0$, this is not true. Equation (3.54) gives an estimate for the probability of the emission of one photon with a large wavelength as

$$w \sim \left(\frac{\hbar}{mc}\right)^{-2} \sigma_b \sim \frac{e^2}{\hbar c} \ln \frac{\varepsilon}{\hbar\omega}.$$

Therefore the condition

$$\frac{e^2}{\hbar c} \ln \frac{\varepsilon}{\hbar\omega_{\min}} \lesssim 1 \qquad \text{or} \qquad \hbar\omega_{\min} \gtrsim \varepsilon \exp\left[-\frac{1}{e^2/(\hbar c)}\right]$$

gives the limit of the applicability of Eq. (3.54), which is based on perturbation theory.

Problem 3.20. Determine the cross section for the bremsstrahlung produced by the scattering of an electron by an ion.

SOLUTION. We use Eq. (3.52) and replace the matrix element of the electron position vector by the matrix element of the electron momentum,

$$\mathbf{p}_{\mathbf{qq}'} = im\omega\mathbf{r}_{\mathbf{qq}'}.$$

For the determination of the matrix element $\mathbf{p}_{\mathbf{qq}'}$, we employ the wave function in the momentum representation. Based on the premise of large electron velocity, we use perturbation theory for the wave function. This gives

$$\psi_{\mathbf{q}}(\mathbf{q}_1) = \delta(\mathbf{q} - \mathbf{q}_1) + \frac{2m}{\hbar^2(q^2 - q_1^2)} V(\mathbf{q} - \mathbf{q}_1),$$

where $V(\mathbf{q} - \mathbf{q}_1)$ is the Fourier transform of the interaction potential. We can use the same procedure to write the wave function for the electron in the final state. Within the restriction of first-order perturbation theory, we have

$$\begin{aligned}\mathbf{p}_{\mathbf{qq}'} &= \int \frac{2m\hbar}{\hbar^2(q'^2 - q_1^2)} V(\mathbf{q}' - \mathbf{q}_1)\mathbf{q}\delta(\mathbf{q} - \mathbf{q}_1)\, d\mathbf{q}_1 \\ &\quad + \int \delta(\mathbf{q}' - \mathbf{q}_1)\mathbf{q}' \frac{2m\hbar}{\hbar^2(q^2 - q_1^2)} V(\mathbf{q} - \mathbf{q}_1)\, d\mathbf{q}_1 \\ &= \frac{1}{\omega}(\mathbf{q} - \mathbf{q}')V(\mathbf{q} - \mathbf{q}'),\end{aligned}$$

where

$$\hbar\omega = \frac{\hbar^2}{2m}(q^2 - q'^2).$$

From this, we obtain the expression for the cross section of bremsstrahlung in the Born approximation

$$\frac{d\sigma_b}{d\omega} = \frac{e^2 m^2 \omega^3 q'}{6\pi^3 \hbar^3 c^3 q} \frac{1}{m^2\omega^2} \frac{1}{\omega^2} \int |\mathbf{q} - \mathbf{q}'|^2 |V(\mathbf{q} - \mathbf{q}')|^2 \, d\Omega_{\mathbf{q}'}.$$

For the problem under consideration, $V = -Ze^2/r$, and its Fourier transform is

$$V(\mathbf{q} - \mathbf{q}') = \frac{-4\pi e^2 Z}{|\mathbf{q} - \mathbf{q}'|^2}.$$

On the basis of this expression, we must determine the integral

$$\int |\mathbf{q} - \mathbf{q}'|^2 |V(\mathbf{q} - \mathbf{q}')|^2 d\Omega_{\mathbf{q}'} = (4\pi Z e^2)^2 \int \frac{d\Omega_{\mathbf{q}'}}{|\mathbf{q} - \mathbf{q}'|^2}$$
$$= Z^2 \int_{-1}^{1} \frac{32\pi^3 e^4 d(\cos\theta)}{q^2 + q'^2 - 2qq'\cos\theta}$$
$$= \frac{32\pi^3 e^4 Z^2}{qq'} \ln\left|\frac{q+q'}{q-q'}\right|.$$

Here θ is the angle between the vectors $\mathbf{q}$ and $\mathbf{q}'$. The result for the cross section for bremsstrahlung from an electron scattered by an ion is

$$\frac{d\sigma_b}{d\omega} = \frac{16 Z^2}{3q^2\omega}\left(\frac{e^2}{\hbar c}\right)^3 \ln\left|\frac{q+q'}{q-q'}\right|. \tag{3.55}$$

The validity of the Born approximation for the scattering of a particle in a Coulomb field is better the higher is the velocity of this particle. The criterion for the validity of the Born approximation has the form

$$\frac{e^2}{\hbar v} \ll 1,$$

where v is the electron velocity.

Note that, as in the preceding problem, the cross section for the emission of soft photons integrated over frequencies is divergent. In the special case $q' = 0$, where the total electron momentum is transferred to the photon, then the cross section for the process is zero. Outside the Born approximation, neither of these limiting cases gives these end results.

Expression (3.55) corresponds to bremsstrahlung produced by the scattering of a fast electron by the nucleus of an ion with a charge multiplicity of Z. We may also take into account the bremsstrahlung produced by this fast electron in interaction with the electrons bound to the ionic nucleus. This can be done by the same technique just employed.

Now the interaction potential includes the terms

$$\sum_i \frac{e^2}{|\mathbf{r} - \mathbf{r}_i|},$$

where the $\mathbf{r}_i$ are the coordinates of the ionic electrons, and $\mathbf{r}$ is the coordinate of the fast electron that is being scattered. The Fourier component of this potential is

$$\frac{4\pi e^2}{|\mathbf{q} - \mathbf{q}'|^2} \sum_i \exp\left[i(\mathbf{q} - \mathbf{q}') \cdot \mathbf{r}_i\right].$$

Quantum-mechanical averaging of this Fourier component over the wave function $\Psi(\mathbf{r}_1, \mathbf{r}_2, \ldots)$ of the ionic electrons produces the so-called form factor

$$F(\mathbf{q}-\mathbf{q}') = \int d\mathbf{r}_1\, d\mathbf{r}_2 \ldots \sum_i \exp\left[i(\mathbf{q}-\mathbf{q}')\cdot\mathbf{r}_i\right] |\Psi(\mathbf{r}_1, \mathbf{r}_2, \ldots)|^2 .$$

It should be noted that exchange effects can be neglected in view of the high velocity of the incident electron as compared to the velocities of the ionic electrons.

Combining the contribution of the ionic electrons with that from the ionic nucleus calculated above, we obtain the final result for the bremsstrahlung cross section

$$\frac{d\sigma_b}{d\omega} = \frac{8q'}{3\pi\omega q}\left(\frac{e^2}{\hbar c}\right)^3 \int \frac{d\Omega_{q'}}{|\mathbf{q}-\mathbf{q}'|^2} \left|Z - F(\mathbf{q}-\mathbf{q}')\right|^2 .$$

It is seen that the form factor diminishes the effective ionic charge. That is, it describes the screening of the ionic nucleus by its electrons.

All the above considerations apply as well to a neutral atom. In this case, at small momentum transfer when

$$|\mathbf{q}-\mathbf{q}'|a_0 \ll 1,$$

where a_0 is the atomic dimension, the form factor is equal to Z. This inequality is fulfilled for a sufficiently small deflection angle of the fast electron,

$$\theta \ll \theta_a = (qa_0)^{-1}.$$

It occurs when the impact parameter of the collision is large. We can conclude that small deflection angles $\theta \ll \theta_a$ do not make an essential contribution to the bremsstrahlung cross section for neutral atoms. This can be explained simply. A fast electron with a large impact parameter does not feel the electron structure of the neutral atom, and hence the incident fast electron experiences virtually no braking acceleration from the atom, that is, there is no bremsstrahlung.

Problem 3.21. Determine the cross section for bremsstrahlung from the scattering of a slow electron by an atom.

SOLUTION. Our task is the determination of the sum

$$\sum_{l=0}^{\infty}(l+1)\left(R^2_{l+1,l} + R^2_{l,l+1}\right)$$

in Eq. (3.53). The value of $\mathbf{q}-\mathbf{q}'$ is not small now compared with $\mathbf{q}$, so we can not use the results of Problem 3.19. For the determination of the above sum, we must take into account the fact that the main contribution to the matrix element $R_{l,l+1}$ comes from distances $r \sim 1/q \gg 1$. Then one can use asymptotic formulas for the electron wave functions. These wave functions are close to the wave functions of a free electron, for which we label the corresponding matrix elements as $R^{(0)}_{l,l+1}$. Because electrons free of external influences do not radiate, we have

$$\sum_{l=0}^{\infty}(l+1)\left(R^{(0)2}_{l,l+1}+R^{(0)2}_{l+1,l}\right)=0. \tag{3.56}$$

Note that the maximum difference between the matrix elements $R^{(0)}_{l,l+1}$ and $R_{l,l+1}$ relates to small values of l. At small electron energies, only $l = 0$ will give a contribution to the cross section, so we make this restriction. Further, because of Eq. (3.56), only differences between $R_{l,l+1}$ and $R^{(0)}_{l,l+1}$ will contribute to the required summation. The sum is then

$$\sum_{l=0}^{\infty}(l+1)\left(R^2_{l,l+1}+R^2_{l+1,l}\right)\approx R^2_{01}-R^{(0)2}_{01}+R^2_{10}-R^{(0)2}_{10}.$$

Now we evaluate the matrix element $R^{(0)}_{10}$. The relevant wave functions for the electron are

$$u_0(q,r)=\sqrt{\frac{\pi r}{2q}}J_{1/2}(qr)=rj_0(qr)=\frac{1}{q}\sin qr;$$

$$u_1(q'r)=\sqrt{\frac{\pi r}{2q'}}J_{3/2}(q'r)=rj_1(q'r)=\frac{1}{q'}\left(\frac{\sin q'r}{q'r}-\cos q'r\right).$$

On the basis of these wave functions we have

$$\begin{aligned}R^{(0)}_{01}&=\frac{1}{qq'}\int_0^{\infty}r\sin qr\left(\frac{\sin q'r}{q'r}-\cos q'r\right)dr\\&=\frac{1}{2qq'}\int_0^{\infty}\left\{\frac{1}{q'}\left[\cos(q-q')r-\cos(q+q')r\right]\right.\\&\qquad\left.-r\left[\sin(q+q')r-\sin(q'-q)r\right]\right\}dr.\end{aligned}$$

Each of the four terms in this integral gives zero after integration because of the strong oscillations of the trigonometric functions. Thus, one obtains $R^{(0)}_{01} = 0$. For evaluation of the matrix elements R_{01} and R_{10}, we use the asymptotic expression of the wave function u_0, which is

$$u_0(q,r)=\frac{1}{q}\sin(qr+\delta_0),$$

where δ_0 is the s-wave phase shift for the scattering of an electron by the atom. For a small electron energy, this phase shift is $\delta_0 = -Lq$, where L is the scattering length of an electron scattering from the atom. L does not depend on q. For the state with

$l = 1$ we use the expression for a free electron. Thus we have

$$\begin{aligned} R_{01} &= \frac{1}{2qq'}\int_0^\infty \{(1/q')\left[\cos\left((q-q')r+\delta_0\right)-\cos\left((q+q')r+\delta_0\right)\right] \\ &\quad - r\left[\sin\left((q+q')r+\delta_0\right)-\sin\left((q'-q)r-\delta_0\right)\right]\}\,dr \\ &= \frac{1}{2qq'}\left\{-\frac{1}{q'}\left[\frac{\sin\delta_0}{q-q'}-\frac{\sin\delta_0}{q+q'}\right]+\left[\frac{\sin\delta_0}{(q+q')^2}+\frac{\sin\delta_0}{(q'-q)^2}\right]\right\} \\ &= \frac{2q'}{q}\frac{\sin\delta_0}{(q^2-q'^2)^2} = -\frac{2Lq'}{(q^2-q'^2)^2}. \end{aligned}$$

The result for R_{10} can be obtained from R_{01} by the substitutions $q \to q'$ and $q' \to q$. In this way, we have

$$R_{10} = -\frac{2Lq}{(q^2-q'^2)^2},$$

and thus obtain

$$\begin{aligned} \sum_{l=0}^{\infty}(l+1)(R_{l,l+1}^2 + R_{l+1,l}^2) &= \frac{4L^2(q^2+q'^2)}{(q^2-q'^2)^4} \\ &= \sigma_e \frac{q^2+q'^2}{\pi(q^2-q'^2)^4}, \end{aligned}$$

where $\sigma_e = 4\pi L^2$ is the cross section for the elastic scattering of a slow electron by an atom.

Using these expressions in Eq. (3.53), we obtain for the cross section for the emission of bremsstrahlung arising from the scattering of a slow electron by an atom, the result

$$\frac{d\sigma_b}{d\omega} = \frac{32m\omega^3 q'}{3\hbar c^3 a_0 q}\frac{q^2+q'^2}{\pi(q^2-q'^2)^4}\sigma_e.$$

In this equation, we now replace q and q', the electron wave vectors in initial and final states, by the energy of the emitted photon

$$\omega = \frac{\hbar^2}{2m}(q^2-q'^2)$$

and the energy of the incident electron $\varepsilon = \hbar^2 q^2/(2m)$. Then we have

$$\frac{d\sigma_b}{d\omega} = \frac{4}{3\pi}\frac{e^2}{mc^3}\frac{(2-\hbar\omega/\varepsilon)\sqrt{1-\hbar\omega/\varepsilon}}{\hbar\omega/\varepsilon}\sigma_e. \tag{3.57}$$

The cross section for the emission of bremsstrahlung decreases monotonically as the frequency ω increases. For soft photons ($\hbar\omega \ll \varepsilon$), Eq. (3.57) gives

$$\frac{d\sigma_b}{d\omega} = \frac{8}{3\pi}\frac{e^2\varepsilon}{mc^3}\frac{\sigma_e}{\hbar\omega}. \tag{3.58}$$

This result follows also from Eq. (3.54) of Problem 3.19 if we use the same parameters. Note that, as in the preceding problem, the cross section for bremsstrahlung emission goes to zero if $q' = 0$, when the entire initial electron energy is transformed into the energy of the emergent photon.

Problem 3.22. Calculate the cross section for the emission of low-frequency bremsstrahlung resulting from the scattering of fast electrons from an atom, that arises as a result of the dynamic polarizability of the atom.

SOLUTION. In previous problems in this section, the bremsstrahlung studied was a result of the deceleration of an electron as a consequence of scattering from the atom. For complex atoms, another mechanism exists that can produce bremsstrahlung. The electron incident on the atom can produce oscillations of the electrons bound to the atom. These induced electron oscillations can lead to the radiation of photons, which can then be regarded as bremsstrahlung. The goal of this problem is to calculate quantum mechanically the cross section for this process.

This problem will be solved on the basis of the Born approximation applied to the interaction between the incident fast electron and the atomic electrons. We considered such an interaction in Problem 3.20. The form factor calculated in the solution of Problem 3.20 was diagonal with respect to the wave function of the ground state of the atomic electrons. Here we take into account that the atom can make a transition to an excited state and back; that is, we consider the nondiagonal part of the form factor, also called the inelastic form factor. Such transitions are taken into account in terms of the polarizability of the atom by going to second order in perturbation theory.

We express the induced dipole moment of an atom as $\mathbf{D} = \beta(\omega)\mathbf{E}$, where $\beta(\omega)$ is the atomic polarizability of the atom at frequency ω (the so-called dynamic polarizability), and $\mathbf{E}$ is the electric field strength produced by the fast electron. According to the Coulomb law, we have $\mathbf{E} = e\mathbf{r}/r^3$, where $\mathbf{r}$ is the distance between the incident fast electron and the atom. We suppose that the magnitude of this distance r is large compared to the atomic dimension a_0, so that the coordinates of the atomic electrons are not essential in the solution of the problem.

As in the solution of Problem 3.17 [see Eq. (3.52)], the bremsstrahlung cross section is of the form

$$\frac{d\sigma_t}{d\omega} = \frac{m^2\omega^3 q'}{6\pi^3\hbar^3c^3q}\int |\mathbf{D}_{\mathbf{qq}'}|^2 d\Omega_{\mathbf{q}'}.$$

When we substitute in this equation the expression for the dipole moment $\mathbf{D}$, we find

$$\frac{d\sigma_t}{d\omega} = \frac{e^2m^2\omega^3 q'\beta^2(\omega)}{6\pi^3\hbar^3c^3q}\int \left|\left(\frac{\mathbf{r}}{r^3}\right)_{\mathbf{qq}'}\right|^2 d\Omega_{\mathbf{q}'}.$$

Calculation of the Fourier component yields

$$\left|\left(\frac{\mathbf{r}}{r^3}\right)_{\mathbf{qq}'}\right| = \frac{4\pi}{|\mathbf{q}-\mathbf{q}'|}.$$

Hence the bremsstrahlung cross section can be written as

$$\frac{d\sigma_t}{d\omega} = \frac{8e^2m^2\omega^3 q'\beta^2(\omega)}{3\pi\hbar^3c^3q}\int\frac{d\Omega_{\mathbf{q}'}}{|\mathbf{q}-\mathbf{q}'|^2}.$$

Unlike Problem 3.20, integration over angles of the wave vector $\mathbf{q}'$ is limited because of the restriction that the incident fast electrons experience only small deflections from the original direction. Indeed, were it presumed that $|\mathbf{q}-\mathbf{q}'|a_0 \geq 1$, where a_0 is a typical dimension of the atom, then such transferred momenta would imply distances between the fast electron and the atom of the order of $r < a_0$. However, we have assumed that $r \gg a_0$, so the foregoing presumption about the transferred momenta cannot be correct. Thus the upper limit of integration over the angle of deflection θ is of the order of $\theta_0 = (qa_0)^{-1} \ll 1$. This inequality follows from the assumption that the velocity of the fast electron is large as compared to typical velocities of atomic electrons. When the integration over the range of angles $0 \leq \theta \leq \theta_0$ is performed, the result is

$$\int\frac{d\Omega_{\mathbf{q}'}}{|\mathbf{q}-\mathbf{q}'|^2} = 2\pi\int_0^{\theta_0}\frac{d(\cos\theta)}{q^2+q'^2-2qq'\cos\theta}$$

$$= \frac{\pi}{qq'}\ln\frac{(q-q')^2+qq'\theta_0^2}{(q-q')^2}.$$

In the numerator of the logarithm in this expression the second term is large compared to the first, since $qq'\theta_0^2 \sim a_0^{-2}$, and

$$(q-q')^2 \approx \frac{m^2\omega^2}{\hbar^2q^2} \ll a_0^{-2},$$

where ω is the frequency of the radiated photon. The last relation follows from conservation of energy during the emission of the low-frequency photon, which is

$$\omega = \frac{\hbar q^2}{2m} - \frac{\hbar q'^2}{2m} \approx \frac{\hbar^2}{m}q(q-q').$$

The inequality

$$\omega \ll \hbar q/ma_0 = v/a_0,$$

where v is the velocity of the fast electron, represents the condition that the radiated photon should be soft, that is, of low energy. This condition is a requirement in this problem since the argument of the logarithm would be of order unity at higher frequencies, and the cross section would be very small. It should be noted that the photon energy is negligibly small as compared to the energy of the fast electron, so that the scattering of this electron is almost elastic. Hence, under the conditions stated for this problem, we have (with $q \approx q'$)

$$\int \frac{d\Omega_{\mathbf{q}'}}{|\mathbf{q} - \mathbf{q}'|^2} = \frac{2\pi}{q^2} \ln \frac{v}{\omega a_0}.$$

This calculation is, of course, correct only to within logarithmic accuracy, since the dimension of an atom a_0 is expressed here only as an order of magnitude.

Finally, the bremsstrahlung cross section is, within logarithmic accuracy,

$$\frac{d\sigma_t}{d\omega} = \frac{16e^2\omega^3\beta^2(\omega)}{3\hbar c^3 v^2} \ln \frac{v}{\omega a_0}.$$

This type of bremsstrahlung is called polarization bremsstrahlung. It vanishes in the low-frequency limit $\omega \to 0$. The polarization bremsstrahlung cross section becomes large at photon frequencies ω corresponding to frequencies of bound–bound transitions between ground and excited states of the atom, since at such frequencies the dynamic polarizability $\beta(\omega)$ is resonantly large.

It should be noted that, in addition to the polarization bremsstrahlung evaluated here, there also exists the usual bremsstrahlung due to the simple scattering of a fast electron by an atom. This was calculated in Problem 3.20. The ratio of the polarization bremsstrahlung to the usual bremsstrahlung [see Eq. (3.55)] is approximately

$$\left[\frac{\beta(\omega)}{e^2/(m\omega^2)}\right]^2.$$

The denominator of this estimate expresses the polarizability of a free electron. We conclude that the dynamic polarizability produces resonant dependence of the cross section on the photon frequency ω. The influence of dynamic polarizability vanishes at small frequencies since polarization bremsstrahlung does not exhibit the infrared catastrophe, unlike the usual bremsstrahlung.

It is important to note that conventional bremsstrahlung and polarization bremsstrahlung do not interfere with each other. This can be explained by the fact that the first is nonzero in the range of scattering angles of a fast electron $\theta > \theta_0$, while the other is nonzero in the complementary range of angles $\theta < \theta_0$.

Problem 3.23. Calculate the cross section for bremsstrahlung resulting from the scattering of fast electrons by a negative ion.

SOLUTION. On first inspection it might seem that the cross section for bremsstrahlung arising from the scattering of a fast electron should be the same if the scatterer is either a positive or a negative ion, since according to the solution of Problem 3.20 the cross section does not depend on the sign of the ionic charge. However, negative ions are distinguished from positive ions because they possess a large polarizability arising from the small binding energy of the excess electron. Therefore the solution of the preceding problem is appropriate to this case, but only with some modifications incorporated. If the energy of the radiated photon is large as compared to the photodetachment energy I for the negative ion, then the scattering of the fast electron on the excess electron is analogous to the scattering of one free electron by another. It will be shown in Problem 3.25 that, in the dipole approximation, no bremsstrahlung arises from such a process. There will then exist only the small cross section for bremsstrahlung of a fast electron on a neutral atom.

The solution of the preceding problem can be used, after modification to the form

$$\frac{d\sigma_t}{d\omega} = \frac{16e^2\omega^3}{3\hbar c^3 v^2}\left[\beta(\omega) + \frac{e^2}{m\omega^2}\right]^2 \ln\frac{v}{\omega a_0}. \tag{3.59}$$

Here $\beta(\omega)$ is the polarizability of the negative ion, and the second term in the square brackets is associated with the Coulomb scattering arising from the excess negative charge of the ion (albeit without taking its structure into account). This second term is found from Eq. (3.55) with $Z = -1$, but with a different argument for the logarithmic function, since the impact parameter of the fast electron cannot be small.

It should be noted that we add the amplitudes of two processes and not the cross sections. This follows from the fact that the perturbation is the sum of two potentials in the Born approximation: the potential of the interaction between the fast electron and the induced dipole moment of the negative ion, and the potential of the interaction of the fast electron with the negative charge of the ion. That is, the transition amplitude is the sum of two processes in first-order perturbation theory. If $\hbar\omega \gg I$, then

$$\beta(\omega) = -e^2/m\omega^2,$$

and the cross section vanishes, as was pointed out above.

On the other hand, if $\omega \to 0$, then the contribution from $\beta(\omega)$ disappears, and only the second term in the square brackets remains in Eq. (3.59).

The bremsstrahlung cross section as a function of ω has a maximum when the dynamic polarizability is largest. This occurs in the frequency range corresponding to the photodetachment threshold, since that is where both the value of $\beta(\omega)$ and the bremsstrahlung cross section have broad maxima.

Problem 3.24. Obtain the expression for the power emitted in the classical limit for the case of bremsstrahlung caused by an electron incident on a spherical atomic system.

SOLUTION. The value

$$\int \hbar\omega\, d\sigma_b$$

characterizes the power radiated during the process. For small ω, Eq. (3.58) gives the behavior $d\sigma_b \sim d\omega/\omega$. This means that the emitted power has a bounded value. From Eq. (3.52) we have

$$\hbar\omega\, d\sigma_b = \int \frac{e^2 m^2 \omega^4 q'}{6\pi^3\hbar^2 c^3 q}|\mathbf{r}_{\mathbf{q}\mathbf{q}'}|^2\, d\omega\, d\Omega_{\mathbf{q}'},$$

and on the basis of the relations $\ddot{\mathbf{r}}_{\mathbf{qq}'} = -\omega^2 \mathbf{r}_{\mathbf{qq}'}$ and $d\omega = \hbar q' dq'/m$, we obtain

$$\int \hbar\omega \, d\sigma_b = \frac{e^2 m}{6\pi^3 c^3 \hbar q} \int |\ddot{\mathbf{r}}_{\mathbf{qq}'}|^2 d\mathbf{q}'. \tag{3.60}$$

We use here the usual notation that $d\mathbf{q} = dq_x \, dq_y \, dq_z = q^2 \, dq \, d\Omega_{\mathbf{q}}$. The integral stated above is then

$$\begin{aligned} \int |\ddot{\mathbf{r}}_{\mathbf{qq}'}|^2 \, d\mathbf{q}' &= \int_{q'<q} \psi_{\mathbf{q}}^*(\mathbf{r})\ddot{\mathbf{r}}(\mathbf{r})\psi_{\mathbf{q}'}(\mathbf{r}) \, d\mathbf{r} \psi_{\mathbf{q}}(\mathbf{r}')\ddot{\mathbf{r}}(\mathbf{r}')\psi_{\mathbf{q}'}^*(\mathbf{r}') \, d\mathbf{r}' \, d\mathbf{q}' \\ &= \frac{1}{2} \int_{all\ q,q'} \psi_{\mathbf{q}}^*(\mathbf{r})\ddot{\mathbf{r}}(\mathbf{r})\psi_{\mathbf{q}'}(\mathbf{r})\psi_{\mathbf{q}}(\mathbf{r}')\ddot{\mathbf{r}}(\mathbf{r}')\psi_{\mathbf{q}'}^*(\mathbf{r}') \, d\mathbf{r}' \, d\mathbf{q}' \, d\mathbf{r} \\ &= \frac{(2\pi)^3}{2} \int \delta(\mathbf{r} - \mathbf{r}')\psi_{\mathbf{q}}^*(\mathbf{r})\ddot{\mathbf{r}}(\mathbf{r})\psi_{\mathbf{q}}(\mathbf{r}')\ddot{\mathbf{r}}(\mathbf{r}') \, d\mathbf{r}' \, d\mathbf{r} \\ &= 4\pi^3 \int \psi_{\mathbf{q}}^*(\mathbf{r}') \left|\ddot{\mathbf{r}}(\mathbf{r}')\right|^2 \psi_{\mathbf{q}}(\mathbf{r}') \, d\mathbf{r}' \\ &= 4\pi^3 \int \left|\ddot{\mathbf{r}}(\mathbf{r}')\right|^2 W(\mathbf{r}') d\mathbf{r}', \end{aligned} \tag{3.61}$$

and the quantity

$$W(\mathbf{r}') = |\psi_{\mathbf{q}}(\mathbf{r}')|^2$$

is the relative probability amplitude for the particle to be at the point $\mathbf{r}'$.

Assume the interaction potential of the electron with the atomic system to be spherically symmetrical, and consider the classical motion of the electron flux relative to the scatterer. We suppose that the electrons occupy a volume element $2\pi\rho \, d\rho \, ds$ far from the scattering center, where ρ is the impact parameter and ds is the length of a volume element. At some time the electrons will occupy a volume

$$2\pi\rho' \, d\rho' \, ds' = W(\mathbf{r}') \, d\mathbf{r}'.$$

If v' is the velocity of the electron at point $\mathbf{r}'$, conservation of electron flux gives

$$v \, 2\pi\rho \, d\rho = v' \, 2\pi\rho' \, d\rho',$$

where v is the electron velocity far from the scattering center. Also, $ds' = v' \, dt$, where dt is the time required to move the distance ds'. From this we obtain

$$2\pi\rho' \, d\rho' \, ds' = 2\pi\rho \, d\rho \, v \, dt = W(\mathbf{r}') \, d\mathbf{r}',$$

and from Eqs. (3.60) and (3.61) we have

$$\int \hbar\omega \, d\sigma_b = \frac{2e^2 m v}{3c^3 q \hbar} \int |\ \ddot{\mathbf{r}}\ (t)|^2 2\pi\rho \, d\rho \, dt. \tag{3.62}$$

We shall explain below that $q = mv/\hbar$.

We now Fourier analyze $\mathbf{r}(t)$, which gives

$$\mathbf{r}(t) = \int_{-\infty}^{\infty} \mathbf{r}_\omega e^{-i\omega t}\, d\omega; \qquad \mathbf{r}_\omega = \frac{1}{2\pi}\int_{-\infty}^{\infty} \mathbf{r}(t) e^{i\omega t}\, dt; \qquad \ddot{\mathbf{r}}_\omega = -\omega^2 \mathbf{r}_\omega.$$

This yields the result

$$\begin{aligned}\int_{-\infty}^{\infty} |\ddot{\mathbf{r}}(t)|^2\, dt &= \iiint_{-\infty}^{\infty} e^{i(\omega-\omega')t} \omega^2 \mathbf{r}_\omega^* \omega'^2 \mathbf{r}_{\omega'}\, d\omega\, d\omega'\, dt \\ &= 2\pi \iint_{-\infty}^{\infty} \delta(\omega-\omega')\omega^2 \mathbf{r}_\omega^* \mathbf{r}_{\omega'} \omega'^2\, d\omega\, d\omega' \\ &= 2\pi \int_{-\infty}^{\infty} |\mathbf{r}_\omega|^2 \omega^4\, d\omega = 4\pi \int_0^{\infty} |\mathbf{r}_\omega|^2 \omega^4\, d\omega.\end{aligned}$$

When we use this expression in Eq. (3.62), we obtain

$$\int \hbar\omega\, d\sigma_b = \frac{2}{3}\frac{e^2}{c^3} 4\pi \iint_0^{\infty} |\mathbf{r}_\omega|^2 \omega^4\, d\omega\, 2\pi\rho\, d\rho. \tag{3.63}$$

When we compare the expressions contained in the integral, we find that

$$\frac{d\sigma_b}{d\omega} = \frac{8\pi e^2 \omega^3}{3\hbar c^3} \int_0^{\infty} |\mathbf{r}_\omega|^2 2\pi\rho\, d\rho. \tag{3.64}$$

The presence of the Planck constant in this equation does not imply a quantum character for the result. In fact, $\hbar\omega\, d\sigma_b/d\omega$ is a classical quantity that does not contain quantum parameters. It is to be noted that Eq. (3.64) can be derived without using the apparatus of quantum mechanics. In a classical approach, one uses the classical expression for the intensity of radiation, and determines the total electron energy lost to radiation as a result of a single collision of the electron with the scattering center. The criterion for the applicability of the classical calculation is the condition that the photon energy is small as compared to the energy of the incident electron, that is, that $\hbar\omega \ll \varepsilon$. However, the photon energy $\hbar\omega$ can be of the order of magnitude of the kinetic energy $mv^2/2$ of the impact electron. At first sight, it is therefore unclear whether the electron described in terms of its trajectory $\mathbf{r}(t)$ in Eq. (3.62) and its Fourier components $\mathbf{r}_\omega$ in Eqs. (3.63) and (3.64) is the electron with the initial velocity v or the final velocity v'. Since these velocities are so different from each other, then also the trajectories of the classical motion will be very different. The resolution of this apparent ambiguity is that it is only the region of the hyperbolic trajectory near the perihelion (where the distance between the electron and the atomic system has its minimum) that is essential to the problem considered. It should be kept in mind that hyperbolic trajectories relate only to Coulomb scattering. In the general case, the trajectories are more complex. However, the trajectory near perihelion does not depend on the velocity of the electron and is determined only by the centrifugal potential. Clearly, these considerations are correct only if the atomic potential does not contain a strong singularity at small distances r, so that $\mathbf{r}(t)$ in Eq. (3.62) does not depend on the electron velocity in the essential range of the integration over the time t.

Problem 3.25. Prove that there is no dipole radiation in the collision of two electrons.

SOLUTION. The operator for the dipole moment for two electrons has the form

$$e(\mathbf{r}_1 + \mathbf{r}_2) = 2e\mathbf{R},$$

where $\mathbf{R}$ is the coordinate of the center of mass of the electrons. We can express the coordinates for the two-electron problem as center-of-mass coordinates and relative coordinates $\mathbf{r} = \mathbf{r}_1 - \mathbf{r}_2$. We express the wave function as a product of center-of-mass wave functions and relative wave functions. Bremsstrahlung leads to a change of the electron energy as viewed in the center-of-mass system, that is, the electron energy as expressed in the relative coordinates. The bremsstrahlung does not lead to a change in $\mathbf{R}$. That is, the matrix element of the dipole operator will vanish because the matrix element is between different (and hence orthogonal) states of the relative wave function, with no change in the dipole operator. Thus there is no dipole component to bremsstrahlung from the interaction of two electrons.

Problem 3.26. Estimate the dependence of classical bremsstrahlung on the parameters of the problem of an electron scattering from an ion. Assume that the energy of the incident electron is very much smaller than typical atomic energies, and the energy of the photons emitted significantly exceeds the value $\hbar m v^3/e^2$.

SOLUTION. The conditions stated for this problem imply that the interaction potential between the electron and the ion in the region where the radiation is created is very much larger than the electron kinetic energy far from the ion. Thus it is necessary to analyze the region of strong interaction of an electron with the ion.

The classical cross section (3.64) for the bremsstrahlung cross section can be represented by the analytical estimate

$$\frac{d\sigma_b}{d\omega} \sim \frac{e^2\omega^3}{\hbar c^3}\rho^2\frac{r^2}{\omega^2}. \tag{3.65}$$

Here, ρ is the collision impact parameter, which is determined by the cross section for bremsstrahlung for a given frequency ω, and r is a typical distance between an electron and an ion that gives rise to this radiation. Evidently we should choose as r a value $r_{\min}$ that corresponds to the closest approach of the electron to the ion on a given trajectory. The distance can be obtained by setting the interaction energy equal to the centrifugal energy, or

$$\frac{e^2}{r_{\min}} \sim \frac{mv^2\rho^2}{2r_{\min}^2},$$

where v is the electron velocity far from the ion. We find from this that $r_{\min} \sim \rho^2 \varepsilon / e^2$, where $\varepsilon = mv^2/2$ is the initial electron energy.

Next we estimate a value for the impact parameter that gives the principal contribution in the radiation of a photon of frequency ω. This frequency has the order of magnitude $\omega \sim v_{\max}/r_{\min}$, where $v_{\max}$ is the electron velocity at the distance of closest approach. We establish this value from the principle of momentum conservation

$$mv\rho = mv_{\max} r_{\min}, \quad \text{i.e.,} \quad v_{\max} = v\rho / r_{\min}.$$

Thus we have

$$\omega \sim \frac{v\rho}{r_{\min}^2} \sim \frac{e^4 v}{\varepsilon^2 \rho^3},$$

which gives

$$\rho \sim \left(\frac{e^4 v}{\varepsilon^2 \omega} \right)^{1/3} \quad \text{and} \quad r_{\min} \sim \frac{\rho^2 \varepsilon}{e^2} \sim \left(\frac{e^2 v^2}{\varepsilon \omega^2} \right)^{1/3}.$$

Using these relations in Eq. (3.65), we obtain

$$\frac{d\sigma_b}{d\omega} \sim \frac{e^2 \omega^3}{\hbar c^3} \left(\frac{e^4 v}{\varepsilon^2 \omega} \right)^{2/3} \frac{1}{\omega^2} \left(\frac{e^2 v^2}{\varepsilon \omega^2} \right)^{2/3} \sim \frac{e^6 v^2}{\hbar \omega^3 \varepsilon^2}$$
$$\sim \frac{e^6}{\hbar \omega c^3 m^2 v^2} \sim \left(\frac{e^2}{\hbar c} \right)^3 \frac{\hbar^2}{m^2 v^2 \omega}. \tag{3.66}$$

This estimate for $d\sigma_b/d\omega$ is valid with the conditions that $r_{\min} \ll \rho$ and that the centrifugal energy is greatly in excess of the kinetic energy. These were the conditions employed to arrive at the estimates given above. Thus we have $\rho \ll e^2/\varepsilon$, and therefore

$$\omega \gg mv^3/e^2.$$

This is a criterion for a lower bound on the frequency. The upper limit for the frequency comes from the condition for the validity of classical mechanics. For this purpose, we require that the main contribution to the bremsstrahlung should arise from large values of the collisional angular momentum, or $l \gg 1$. That is, we demand that

$$l \sim \frac{m\rho v}{\hbar} \sim \frac{mv}{\hbar} \left(\frac{e^4 v}{\varepsilon^2 \omega} \right)^{1/3} \sim \left(\frac{me^4}{\hbar^3 \omega} \right)^{1/3} \gg 1.$$

It follows from this that the frequency of the bremsstrahlung emitted must be much less than typical atomic frequencies, or $\omega \ll me^4/\hbar^3$. Thus the criteria for the validity of Eq. (3.66) are

$$me^4/\hbar^3 \gg \omega \gg mv^3/e^2.$$

From this it follows that $e^2/(\hbar v) \gg 1$, that is, the electron velocity must be small.

Problem 3.27. Under the conditions of Problem 3.26, find the specific numerical coefficients that belong in Eq. (3.66).

SOLUTION. The solution of this problem requires the evaluation of the Fourier transform $\mathbf{r}_\omega$ for the Kepler scattering problem. As in the preceding problem, the electron motion can be considered to be classical. The motion of two particles subject to an attractive Coulomb potential is equivalent to the motion of a single particle of reduced mass describing a hyperbolic orbit under the influence of a fixed center of force. Problem 3.26 provided us with all the dimensional information necessary for the cross section, so we now simplify matters by using the special system of units with $m = v = e = 1$. Consider the particle with the reduced mass to be moving in the xy plane. The coordinates for the motion can be written in the parametric form

$$x = \sqrt{1+\rho^2} - \cosh\xi; \qquad y = \rho\sinh\xi; \qquad t = \sqrt{1+\rho^2}\sinh\xi - \xi,$$

where ξ is a variable parameter. In addition, it is convenient to evaluate the Fourier transforms of $\dot{x}$ and $\dot{y}$ rather than x and y. Because of the relation $\dot{\mathbf{r}}_\omega = -i\omega\mathbf{r}_\omega$, the cross section for bremsstrahlung given in Eq. (3.64) can be written

$$\frac{d\sigma_b}{d\omega} = \frac{8\pi\omega}{3\hbar c^3}\int_0^\infty \left(|\dot{x}_\omega|^2 + |\dot{y}_\omega|^2\right) 2\pi\rho\, d\rho. \tag{3.67}$$

The Fourier transform of $\dot{x}$ is

$$\begin{aligned}\dot{x}_\omega &= \frac{1}{2\pi}\int_{-\infty}^{\infty} \dot{x}e^{i\omega t}\, dt = \frac{1}{2\pi}\int_{-\infty}^{\infty}\frac{dx}{d\xi}e^{i\omega t(\xi)}\, d\xi \\ &= -\frac{1}{2\pi}\int_{-\infty}^{\infty}\sinh\xi \exp\left[i\omega\left(\sqrt{1+\rho^2}\sinh\xi - \xi\right)\right] d\xi.\end{aligned}$$

The physical problem we are considering is one in which the bremsstrahlung has $\omega \gg 1$. The impact parameters are such that $\rho \ll 1$. Therefore only values of the parameter $\xi \ll 1$ will contribute significantly to the integral. This enables us to simplify the integral to

$$\begin{aligned}\dot{x}_\omega &= -\frac{1}{2\pi}\int_{-\infty}^{\infty}\xi\exp\left[i\omega\left(\frac{\xi^3}{6} + \frac{\rho^2\xi}{2}\right)\right] d\xi \\ &= -\frac{i}{\pi}\int_{-\infty}^{\infty}\xi\sin\left(\frac{\omega\rho^2\xi}{2} + \frac{\omega\xi^3}{6}\right) d\xi.\end{aligned}$$

In the same way, the expression for $\dot{y}_\omega$ is

$$\dot{y}_\omega = \frac{\rho}{\pi}\int_{-\infty}^{\infty}\cos\left(\frac{\omega\rho^2\xi}{2} + \frac{\omega\xi^3}{6}\right) d\xi.$$

This integral is the Airy integral. It can be expressed on terms of modified Bessel functions of the second kind (also referred to as Macdonald functions) $K_{1/3}$ as

$$\dot{y}_\omega = \frac{\rho^2}{\pi\sqrt{3}} K_{1/3}\left(\frac{\omega\rho^3}{3}\right).$$

The quantity $\dot{x}_\omega$ can be evaluated in an analogous way to yield

$$\dot{x}_\omega = \frac{\rho^2}{\pi\sqrt{3}} K_{2/3}\left(\frac{\omega\rho^3}{3}\right).$$

Using these expressions in Eq. (3.67), we find

$$\begin{aligned}\frac{d\sigma_b}{d\omega} &= \frac{16\omega}{9\hbar c^3}\int_0^\infty \left[K_{1/3}^2\left(\frac{\omega\rho^3}{3}\right) + K_{2/3}^2\left(\frac{\omega\rho^3}{3}\right)\right]\rho^5\,d\rho \\ &= \frac{16}{3\hbar c^3\omega}\int_0^\infty \left[K_{1/3}^2(z) + K_{2/3}^2(z)\right] z\,dz.\end{aligned}$$

The dimensionless integrals are equal to, respectively, $\pi/3^{3/2}$ and $2\pi/3^{3/2}$. Restoring the dimensional factors m, e, v in the cross section for bremsstrahlung, we obtain

$$\frac{d\sigma_b}{d\omega} = \left(\frac{e^2}{\hbar c}\right)^3 \frac{16\pi\hbar^2}{3\sqrt{3}m^2 v^2}\frac{Z^2}{\omega}. \tag{3.68}$$

Equation (3.68) is called Kramers's formula. It describes the principal part of the classical bremsstrahlung spectrum resulting from the scattering of a charged particle in a Coulomb field.

In Eq. (3.68), the value Z is introduced, which represents the number of elementary charges contained on the scattering center. In other words, the interaction parameter is now Ze^2 instead of e^2, and the cross-section expression depends on Z^2.

The range of frequencies where Eq. (3.68) is valid is as determined in the previous problem, and the classical case corresponds to low velocities, $v \ll e^2/\hbar$.

For highly excited states the concepts of photoabsorption and photoionization are closely linked, as analyzed earlier, and the cross sections are related as shown in Eq. (3.28). In an analogous fashion, the bremsstrahlung process considered here is related to photorecombination when the electron passes through a highly excited level. This connection will be analyzed further.

3.5 PHOTORECOMBINATION OF ATOMIC SYSTEMS

The process of photorecombination of atomic systems is inverse to that of photodetachment and proceeds according to the scheme

$$X + Y \rightarrow XY + \hbar\omega.$$

According to the principle of detailed balance, the ratio of the cross sections for these processes has the form

$$\frac{\sigma_r}{\sigma_i} = \frac{g_r}{g_i}\frac{j_{\mathrm{ph}}}{j_p}. \tag{3.69}$$

Here, g_r, g_i are the statistical weights of the final states for recombination and detachment processes, respectively, and j_{ph}, j_p are the fluxes of photons and constituent atomic particles for the corresponding channels of the processes, normalized to one photon per given volume.

Problem 3.28. Obtain the connection between the cross sections for photorecombination and photodetachment of atomic particles.

SOLUTION. The statistical weights of the final states considered are

$$g_r = g_{XY} 2\frac{4\pi k^2\, dk}{(2\pi)^3}; \qquad g_i = g_X g_Y \frac{4\pi q^2\, dq}{(2\pi)^3},$$

where g_X, g_Y, g_{XY} are statistical weights of internal states of the particles, $\mathbf{k}$ is the photon wave vector, and $\mathbf{q}$ is the wave vector of the relative motion of the particles. The photon flux for the case where there is one photon in a volume Ω is $j_{\mathrm{ph}} = c/\Omega$, and the particle flux in this case is $j_p = v/\Omega$. Here, $v = \hbar q/\mu$ is the relative velocity and μ is the reduced mass of the particles. From the dispersion relation for photons $k = \omega/c$, and the law of energy conservation (3.2), Eq. (3.69) leads to

$$\sigma_r = 2\frac{k^2}{q^2}\frac{g_{XY}}{g_X g_Y}\sigma_i. \tag{3.70}$$

From this it follows that $\sigma_r \ll \sigma_i$ because $k \ll q$. For example, if the frequency and wave vector of a photon in a photoionization process is of the order of a typical atomic value, we can make the estimate

$$\frac{k}{q} \sim \frac{\omega}{c\sqrt{2m\varepsilon/\hbar^2}} \sim \frac{e^2}{\hbar c} \sim 10^{-2},$$

and therefore the photorecombination cross section is approximately four orders of magnitude smaller than the photodetachment cross section.

Equation (3.70) describes total cross sections. An analogous expression for the differential cross sections is

$$\frac{d\sigma_r}{d\Omega_{\mathbf{k}}} = \frac{k^2}{q^2}\frac{g_{XY}}{g_X g_Y}\frac{d\sigma_i}{d\Omega_{\mathbf{q}}}. \tag{3.71}$$

Equation (3.71) applies to the radiation or absorption of photons of a given polarization, and thus it does not have a factor of 2, in contrast to Eq. (3.70).

Problem 3.29. Using the principle of detailed balance, calculate the cross section for the recombination of an electron with a hydrogenlike ion, ending in the ground state of the atom.

SOLUTION. As a prelude to the examination of the recombination process, we consider the cross section σ_i for the ionization of the ground state of hydrogen with the production of an electron moving with the speed v. This is given by Eq. (3.10). As written, this equation refers to hydrogen. It can be generalized to an atomic nucleus of charge Ze by replacing v with v/Z and replacing the Bohr radius $a_0 = \hbar^2/me^2$ with $\hbar^2/Zme^2$. We find

$$\sigma_i = (2^9\pi^2/3cZ^2)f(v), \tag{3.72}$$

with the notation

$$f(v) = \frac{Z^8 \exp[-(4Z/v)\arctan(v/Z)]}{(Z^2 + v^2)^4[1 - \exp(-2\pi Z/v)]}.$$

It is the cross section for the inverse process that is sought here, that is, photorecombination of an electron with speed v with transition to the ground state of an ion with charge $Z - 1$. We can obtain this by using Eq. (3.70), which follows from the principle of detailed balance. The statistical weight of the electron and the ion are $g_e = g_i = 1$, and the statistical weight of the ground ionic state is $g_a = n^2 = 1$. We obtain thereby

$$\sigma_r = (2k^2/v^2)\sigma_i,$$

where $k = \omega/c$ is the wave number of the photon radiated in the recombination process. Substituting Eq. (3.72) in this expression, we find

$$\sigma_r = \frac{2^{10}\pi^2\omega^2}{3c^3Z^3v^2} f(v). \tag{3.73}$$

Energy conservation gives

$$\omega = (Z^2 + v^2)/2,$$

so Eq. (3.73) can be rewritten as

$$\sigma_r = \frac{2^8\pi^2 Z^6}{3c^3}\varphi(v), \tag{3.74}$$

where

$$\varphi(v) = \frac{\exp[-(4Z/v)\arctan(v/Z)]}{v^2(Z^2 + v^2)^2[1 - \exp(-2\pi Z/v)]}.$$

In the WKB (Wentzel-Kramers-Brillouin) limit of slow electrons, when $v \ll Z$, Eq. (3.74) reduces to

$$\sigma_r = 2^8\pi^2 Z^2/3e^4c^3v^2,$$

where here e is the base of natural logarithms. It is seen that the cross section for recombination into the ground state increases with a decrease in the electron velocity.

In the opposite limit of fast electrons, when $v \gg Z$, one can also simplify the general expression (3.74). The result is

$$\sigma_r = 2^7 \pi Z^5 / 3c^3 v^5.$$

It is seen that here also the photorecombination cross section increases with a decrease in the electron velocity, but much more sharply than in the slow-electron case.

Problem 3.30. On the basis of the cross section for bremsstrahlung, determine the cross section for the recombination of a slow electron and an ion into an atom in a highly excited state.

SOLUTION. The states involved in the process under consideration have large statistical weights. We shall therefore use a classical description for the electron motion. Within the framework of classical electron motion, we consider this transition to be a result of a bremsstrahlung process. The atomic system of units with $m = e = \hbar = 1$ will be used.

According to Eq. (3.68), the cross section for the process being considered is

$$d\sigma_r = \frac{16\pi}{3\sqrt{3}} \frac{1}{v^2 c^3} \frac{d\omega}{\omega}. \tag{3.75}$$

Energy conservation gives

$$\omega = 1/(2n^2) + v^2/2,$$

and when we differentiate this, we get

$$d\omega = -dn/n^3.$$

The cross section for photorecombination of a slow electron with an ion, leading to a state with principal quantum number n, follows from these expressions with the use of $dn = 1$. (This takes into account that highly excited states with closely similar values of n are approximately equally spaced in energy.) Thus we have

$$\sigma_r = d\sigma_b(dn = 1) = \frac{16\pi}{3\sqrt{3}} \frac{1}{v^2 n^3 \omega c^3}$$
$$= \frac{16\pi}{3\sqrt{3}c^3} \frac{1}{n\omega(2n^2\omega - 1)}. \tag{3.76}$$

The boundary of photorecombination described in this way is determined by the condition $\omega \geq 1/(2n)^2$, which corresponds at $n \gg 1$ to the condition $\omega \ll 1$ for the "softness" of photons. Note that the classical description of electron motion requires that the condition $v \ll 1$ [or $e^2/(\hbar v) \gg 1$ in usual units] be fulfilled.

Problem 3.31. Compare intensities of bremsstrahlung and photorecombination radiation of electrons in multicharged ions. Assume that the Maxwell distribution function holds true for the electron velocities.

SOLUTION. The cross section σ_r for photorecombination of an electron, with transition to the ionic state with charge multiplicity $Z-1$ and principal quantum number $n \gg 1$, can be obtained from Eq. (3.76) by suitable introduction of Z^2. The cross section is

$$\sigma_r = \frac{16\pi Z^2}{3^{3/2}c^3}\frac{1}{n\omega[(2n^2\omega/Z^2)-1]}, \tag{3.77}$$

obtained by multiplication by Z^2 and replacement of the hydrogenic energy $1/2n^2$ by $Z^2/2n^2$ for the ion of charge multiplicity Z. The multiplier Z^2 arises from a combination of factors. There is a replacement of the cube of the coupling strength $(e^2/\hbar c)^3$ by $(Ze^2/\hbar c)^3$, a replacement of the square of the Bohr radius $a_0^2 = (\hbar^2/me^2)^2$ by $(\hbar^2/Zme^2)^2$, and a substitution of n/Z for n in the denominator. Altogether, there is a multiplication by Z^2. In this review of the origins of the extra factor we have departed from the atomic system of units for expository clarity, but Eq. (3.77) is again expressed in atomic units.

The rate of the recombination transition is given by $N_e v\sigma_r$, where v is the electron velocity and N_e is the electron concentration. At each transition, the photon energy ω is radiated, so the radiation intensity for a single ion is $N_e\langle v\sigma_r\omega\rangle$, where the brackets correspond to an averaging of the electron velocity in terms of the Maxwell distribution. The radiation intensity $S_r^{(n)}$ per unit volume is obtained by multiplying by the number N_i of ions in a unit volume, so that

$$S_r^{(n)} = N_i N_e\langle v\sigma_r\omega\rangle. \tag{3.78}$$

The cross section expression (3.77) can be rewritten with the help of the energy conservation expression

$$\omega = v^2/2 + Z^2/2n^2,$$

so the factor in brackets in the denominator of (3.77) becomes

$$(2n^2\omega/Z^2) - 1 = n^2v^2/Z^2.$$

The recombination cross section then becomes

$$\sigma_r = \frac{16\pi Z^4}{3^{3/2}c^3n^3\omega v^2}.$$

This, together with Eq. (3.78) gives the radiation intensity in the form

$$S_r^{(n)} = \frac{16\pi Z^4 N_e N_i}{3^{3/2}c^3n^3}\left\langle \frac{1}{v}\right\rangle.$$

It is seen from this formula that radiation to the ground state ($n=1$) predominates. However, it should be kept in mind that the WKB formula (3.76) of Kramers can produce errors for the ground state of 10–15%.

We now evaluate the radiation intensity resulting from recombination to all of the bound states. For this, the mean value of the inverse velocity

$$\langle v^{-1} \rangle = (2/\pi T)^{1/2}$$

is needed (T is the temperature), as well as the sum

$$\sum_{n=1}^{\infty} \frac{1}{n^3} = \zeta(3) \approx 1.202,$$

where $\zeta(x)$ is the Riemann function. The summed radiation intensity is

$$S_r = \sum_{n=1}^{\infty} S_r^{(n)} = \frac{16}{3} \left(\frac{2\pi}{3} \right)^{1/2} \frac{Z^4 N_e N_i}{c^3} \frac{\zeta(3)}{T^{1/2}}. \tag{3.79}$$

The intensity of bremsstrahlung from a unit volume of a medium produced by scattering of electrons on ions, designated $S_i(T)$, can be shown to be

$$S_i(T) = \frac{32}{3c^3} \left(\frac{2\pi}{3} \right)^{1/2} T^{1/2} Z^2 N_e N_i. \tag{3.80}$$

The ratio of this result to S_r is

$$\frac{S_i(T)}{S_r} = \frac{2T}{\zeta(3) Z^2}.$$

This expression indicates that if $T \gg Z^2$, then bremsstrahlung predominates; but if $T \ll Z^2$, then recombination radiation predominates. However, it must be recalled here that the condition for applicability of the WKB method, upon which both Eqs. (3.79) and (3.80) depend, is that the electron velocity should be small. This requires that $T \ll Z^2$ be fulfilled. Hence, photorecombination radiation predominates in the region of applicability of the expressions obtained.

Problem 3.32. Obtain the expression (3.41) for the photoionization cross section of a Rydberg atom on the basis of the photorecombination cross section (3.76) and the principle of detailed balance. Consider the accuracy of both formulas if the principal quantum numbers are not large.

SOLUTION. In order to use the principle of detailed balance, it is necessary to calculate the statistical weights for the channels involved. Because there are no electron or ion spins in the problem, one can take the statistical weights of the electron and the ion to be unity: $g_e = g_i = 1$. The statistical weight of an excited state is $g_a = n^2$. Using these expressions in Eq. (3.76) along with the principle of detailed balance (3.70), we obtain

$$\sigma_i = \frac{q^2}{2k^2} \frac{g_i g_e}{g_a} \sigma_p = \frac{8\pi}{3\sqrt{3}} \frac{1}{n^5 \omega^3 c}, \tag{3.81}$$

for the photoionization cross section of a Rydberg atom upon absorption of a photon of frequency ω. This is Eq. (3.41) for the photoionization cross section of a Rydberg atom with ejection of a slow electron near the threshold of the process. Equations (3.76) and (3.81) are called Kramers's formulas.

Kramers's formulas are adequate even at small values of n. Let us compare Eq. (3.81) with the accurate value of the threshold cross section for photoionization in the ground state ($\omega = \frac{1}{2}$) given by Eq. (3.11),

$$\sigma_{i,ac} = 2^9\pi^2/(3e^4c). \tag{3.82}$$

If we employ the values $n = 1$ and $\omega = \frac{1}{2}$ in Eq. (3.81), we obtain the ratio of the cross sections

$$\sigma_i/\sigma_{i,ac} = e^4/(8\pi\sqrt{3}) = 1.25.$$

As for the frequency dependence of the photoionization cross section near threshold, Eq. (3.81) gives $\sigma_i \sim \omega^{-3}$, while the accurate result according to Eq. (3.10) is $\sigma_{i,ac} \sim \omega^{-8/3}$. Thus the threshold behavior of the quasi-classical cross section is quite similar to the accurate cross section for the ground state. The accuracy of the quasi-classical cross section increases with an increase in n.

With an increase of photon frequency ω, the accuracy of quasi-classical formulas declines, while the applicability of the Born approximation, in which one neglects the interaction of the detached electron with the remaining ion, improves. In particular, for $n = 1$, the comparison of quasi-classical and accurate photoionization cross sections for large photon frequencies ($\omega \gg 1$) gives

$$\sigma_i/\sigma_{i,ac} = \sqrt{\omega}/(2\sqrt{6}) = 0.2\sqrt{\omega} \gg 1.$$

Problem 3.33. The recombination coefficient is defined as the recombination rate constant averaged over electron velocities. Determine the temperature dependence of the recombination coefficient α_r if the process leads to a highly excited state of the atom formed. Presume that the velocity distribution function for the electrons is the Maxwell distribution.

SOLUTION. The recombination coefficient can be found from Eq. (3.76) for the photorecombination of an electron with an ion, forming an atom in a Rydberg state. We find that

$$\alpha_r = \overline{v\sigma_r} = \frac{16\pi}{3\sqrt{3}}\frac{1}{n^3c^3}\overline{(v\omega)^{-1}}. \tag{3.83}$$

Atomic units are used here, and the bar over a quantity means that an average is to be taken over electron velocities.

With the expressions for the photon frequency

$$\omega = v^2/2 + 1/(2n^2),$$

and the electron energy, $\varepsilon = v^2/2$, we obtain for the recombination coefficient

$$\alpha_r = \frac{16\sqrt{2\pi}}{3\sqrt{3}} \frac{1}{c^3 n^3 T^{3/2}} \int_0^\infty \frac{\exp(-\varepsilon/T)}{\varepsilon + 1/(2n^2)}\, d\varepsilon.$$

With a simple change of variable, the integral converts to the exponential-integral function, and we obtain the form

$$\alpha_r = -\frac{16\sqrt{2\pi}}{3\sqrt{3}} \frac{1}{c^3 n^3 T^{3/2}} \exp\left(\frac{1}{2n^2 T}\right) \mathrm{Ei}\left(-\frac{1}{2n^2 T}\right). \tag{3.84}$$

Let us consider special cases of Eq. (3.84). If the binding energy for the Rydberg state is large compared with the energy of the thermal electron, $n^2 T \ll 1$ (the threshold case), we can use the asymptotic expression for the exponential-integral function,

$$\mathrm{Ei}(-z) \underset{z\to\infty}{\longrightarrow} -\frac{1}{z}\exp(-z)$$

and obtain thereby

$$\alpha_r = \frac{32\sqrt{2\pi}}{3\sqrt{3}} \frac{1}{c^3 n T^{1/2}}. \tag{3.85}$$

For the opposite case of $n^2 T \gg 1$, we use the small-argument limit of the Ei function

$$\mathrm{Ei}(-z) \underset{z\to 0}{\longrightarrow} \ln(\gamma z),$$

where $\gamma = \exp(C)$ and $C = 0.5772\ldots$ is Euler's constant. In this limit we have

$$\alpha_r = \frac{16\sqrt{2\pi}}{3\sqrt{3}} \frac{1}{c^3 n^3 T^{3/2}} \ln\left(\frac{2n^2 T}{\gamma}\right). \tag{3.86}$$

It is seen that the recombination coefficient decreases with an increase in the electron temperature. Note that when $n^2 T \gg 1$, the Kramers formulas are not valid, and the Born approximation becomes applicable.

We shall now appraise the validity of the quasi-classical expression (3.85) for the recombination coefficient if we use this equation for the hydrogen atom in its ground state. First we use Eq. (3.11) for the photoionization cross section near threshold to obtain the accurate expression (for $n = 1$)

$$\sigma_i = \frac{2^9 \pi^2}{3e^4 c}.$$

Assuming the electron temperature to be small compared with the ionization potential, Eq. (3.70) gives the photorecombination cross section

$$\sigma_r = 2\frac{k^2}{q^2}\sigma_i = \frac{1}{2v^2 c^2}\sigma_i,$$

where v is the velocity of a slow electron. This gives, for the case under consideration, the recombination coefficient

$$\alpha_{r,ac} = \overline{v\sigma_r} = \frac{1}{2c^2}\sigma_i\overline{v^{-1}}.$$

Since $\overline{v^{-1}} = \sqrt{2/(\pi T)}$ for a Maxwell distribution function describing the velocities of the electrons, we obtain the final result

$$\alpha_{r,ac} = \frac{2^8\pi\sqrt{2\pi}}{3e^4}\frac{1}{c^3T^{1/2}}. \tag{3.87}$$

The quasi-classical value of the recombination coefficient α_r corresponds to Eq. (3.85) with $n = 1$. The ratio of this value to the accurate value (3.87) is $\alpha_r/\alpha_{r,ac} = e^4/(8\pi\sqrt{3}) = 1.25$. Thus the quasi-classical approximation works rather well even at values of parameters where its validity conditions are violated. This conclusion relates both to the recombination coefficient obtained by averaging over electron velocities and also if this averaging is not done.

Problem 3.34. Calculate the recombination coefficient for an electron on an ion with charge multiplicity Z. Assume a hydrogen-like model for the ion.

SOLUTION. The photorecombination coefficient to a final state with the principal quantum number n for the case $Z = 1$ is given by Eq. (3.84). This expression can be extended in simple fashion to values $Z > 1$ by reference to the photorecombination cross section given in Problem 3.29. Introduction of a general Z value in Eq. (3.84) leads to

$$\alpha_r = -\frac{64\pi^{1/2}Z}{3^{3/2}c^3}b_n^{3/2}\exp(b_n)\,\mathrm{Ei}(-b_n), \tag{3.88}$$

where the quantity b_n is

$$b_n = Z^2/(2n^2T).$$

If the condition $Z^2 \gg n^2T$ is fulfilled (i.e., the electrons are slow), then the expression for α_r simplifies to

$$\alpha_r = \frac{32(2\pi)^{1/2}}{3^{3/2}}\frac{Z^2}{c^3nT^{1/2}}.$$

This result is the generalization of Eq. (3.85) to the case $Z \geq 1$. It is seen that α_r decreases slowly (as n^{-1}) as n increases.

Expression (3.88) for the photorecombination coefficient can be summed up to values of the principal quantum number $n_0 \sim Z/T^{1/2}$, after which α_r begins to decrease more sharply with further increases in n. According to Eq. (3.86), α_r then varies as n^{-3}. With the result that

$$\sum_{n=1}^{n_0} \frac{1}{n} \approx \ln(n_0),$$

we obtain (within logarithmic accuracy) the final expression for the photorecombination coefficient summed over all levels of a hydrogenlike ion with charge multiplicity Z, that

$$\alpha_r^s = \sum_{n=1}^{n_0} \alpha_r = \frac{32(2\pi)}{3^{3/2}c^3} \frac{Z^2}{T^{1/2}} \ln\left(\frac{Z}{T^{1/2}}\right).$$

If the opposite inequality $Z^2 \ll T$ is fulfilled for the ground state, then it is certainly satisfied for excited states with $n > 1$. In this case, the summed photorecombination coefficient is determined for all practical purposes entirely by the term with $n = 1$. Taking the appropriate limit in Eq. (3.88) gives

$$\alpha_r^s = \frac{16(2\pi)}{3^{3/2}c^3} \frac{Z^4}{T^{3/2}} \ln\left(\frac{\gamma Z^2}{2T}\right),$$

where γ is given by $\gamma = \exp(C)$, and C is Euler's constant, $C = 0.5772\ldots$.

Problem 3.35. Determine the recombination coefficient for two colliding atoms assuming that the recombination proceeds from a repulsive molecular state to highly excited vibrational levels of the final molecule formed.

SOLUTION. This process is inverse to that treated in Problem 3.10 and is described by the scheme

$$X + Y \to (XY)^* + \hbar\omega.$$

The molecular energies involved in the transition are illustrated in Figure 3.2. As in Problem 3.10, we assume the motions of the nuclei to be classical. We now introduce the quantity $w(R)$, defined as the probability per unit time for the spontaneous radiative recombination of the atoms as a function of the distance R between nuclei. It is given by Eq. (1.13) with $\overline{n_\omega} = 1$. Then the total probability for radiative recombination as a result of the collision of the atoms is

$$W = \int_{-\infty}^{\infty} w(R)\, dt. \tag{3.89}$$

This value is assumed to be less than unity. This assumption is usually well fulfilled because typical times for spontaneous radiative transitions are less than atomic collision times.

The probability W depends on the collision impact parameter ρ, so the cross section for radiative transition can be written

$$\sigma_r = \int_0^{\infty} 2\pi\rho\, d\rho \int_{-\infty}^{\infty} w(R)\, dt. \tag{3.90}$$

We evaluate the integral by using the classical connection between dt and dR,

$$dt = \frac{dR}{v\sqrt{1 - (\rho^2/R^2) - U_2(R)/\varepsilon}}. \tag{3.91}$$

Here v is the relative velocity of the nuclei of reduced mass μ, $\varepsilon = \mu v^2/2$ is their kinetic energy in the center-of-mass system, and $U_2(R)$ is the interaction potential of the atoms in the final channel of the process.

Using (3.91) in (3.90) and changing the order of integration, one can evaluate the integral over the impact parameter. This yields

$$\begin{aligned} \sigma_r &= \int \frac{dR}{v} w(R) 2\pi \int_0^\infty \frac{\rho\, d\rho}{\sqrt{1 - (\rho^2/R^2) - U_2(R)/\varepsilon}} \\ &= \frac{4\pi}{v} \int_{R_0}^\infty w(R) R^2 \sqrt{1 - U_2(R)/\varepsilon}\, dR, \end{aligned} \tag{3.92}$$

where R_0 is the distance of closest approach for collision with zero impact parameter, that is, $U_2(R_0) = \varepsilon$.

The recombination coefficient averaged over a Maxwell distribution for the velocity of the atoms follows from Eq. (3.92) as

$$\alpha_r = \frac{2}{\sqrt{\pi} T^{3/2}} \int_0^\infty \exp\left(-\frac{\varepsilon}{T}\right) \sqrt{\varepsilon} v \sigma_r \, d\varepsilon. \tag{3.93}$$

The integral over ε can be evaluated to give

$$\int_{U_2(R)}^\infty \exp\left(-\frac{\varepsilon}{T}\right) \sqrt{\varepsilon - U_2(R)}\, d\varepsilon = \frac{\sqrt{\pi}}{2} T^{3/2} \exp\left[-\frac{U_2(R)}{T}\right].$$

When this result is used in Eq. (3.93), we obtain the photorecombination coefficient

$$\alpha_r = 4\pi \int_0^\infty \exp\left[-\frac{U_2(R)}{T}\right] w(R) R^2\, dR. \tag{3.94}$$

The structure of this equation is such that the photorecombination coefficient is represented in the form of a probability per unit time as a function of the distance between the nuclei, weighted by the corresponding Boltzmann factor $\exp[-U_2(R)/T]$. Evidently, this result could have been obtained from very basic considerations.

Problem 3.36. Using the conditions of Problem 3.35, determine $\alpha_r(\omega)$, the spectral recombination coefficient of atoms, defined such that $\alpha_r(\omega)\, d\omega$ is the recombination coefficient corresponding to the emission of photons in the frequency interval from ω to $\omega + d\omega$.

SOLUTION. According to the definition of the spectral photorecombination coefficient $\alpha_r(\omega)$, the above value of the recombination coefficient is related to it by

$$\alpha_r = \int \alpha_r(\omega)\, d\omega.$$

From energy conservation and the classical character of the motion of the atoms, it follows that a photon of frequency ω can arise only at a particular distance between the nuclei that satisfies the condition

$$U_2(R_\omega) - U_1(R_\omega) = \hbar\omega. \tag{3.95}$$

Then, from (3.94) we have

$$\alpha_r(\omega) = 4\pi \exp\left[\frac{-U_2(R_\omega)}{T}\right] w(R_\omega) R_\omega^2 \frac{dR_\omega}{d\omega}.$$

We introduce the derivatives of the interatomic potential with respect to the distance between the atoms, $F_{1,2}(R) = dU_{1,2}(R)$, and arrive at the form

$$\alpha_r(\omega) = 4\pi w(R_\omega) R_\omega^2 \hbar |F_1(R_\omega) - F_2(R_\omega)|^{-1} \exp\left[-U_2(R_\omega)/T\right]. \tag{3.96}$$

In a similar way, the photorecombination cross section $\sigma_r(\omega)$ for frequency ω can be evaluated from the relation

$$\sigma_r = \int \sigma_r(\omega)\, d\omega.$$

The spectral properties of the cross section are given by

$$\sigma_r(\omega) = \sqrt{\frac{4\pi w(R_\omega) R_\omega^2 \hbar}{v|F_1(R_\omega) - F_2(R_\omega)|}} \sqrt{1 - \frac{U_2(R_\omega)}{\varepsilon}}. \tag{3.97}$$

4

COHERENT PHENOMENA IN RADIATIVE TRANSITIONS

Oscillations with the same frequency, and in a fixed-phase relationship are called *coherent*. Using lasers, we can produce mixtures of coherent oscillations with different phases. The development in time of such coherent states results in interference phenomena, and the investigation of these phenomena gives information about the properties of coherent states. Some of these interference phenomena, usually called *coherent spectroscopic phenomena*, will be considered in this chapter.

4.1 POLARIZATION EFFECTS IN RADIATIVE TRANSITIONS IN A UNIFORM MAGNETIC FIELD

In this section we shall consider phenomena associated with the appearance of polarization in the radiation scattered by an atomic electron due to the presence of a constant, uniform magnetic field. These phenomena arise from the interaction of atomic electrons with the radiation.

Problem 4.1. A beam of unpolarized light propagating in a direction normal to a magnetic field is resonantly absorbed by atoms. Transition from a ground s state to an excited p state takes place in this process. We wish to investigate the spontaneous radiation in the direction of the magnetic field and calculate the degree of polarization of the radiation, expressed by

$$P = \frac{I_x - I_y}{I_x + I_y}.$$

In this expression, I_x and I_y are the intensities for emission of photons with polarizations in the x and y directions, respectively, with the z axis taken to be in the direction of the magnetic field vector **H**. Magnetic sublevels with projections ± 1 of orbital angular momentum are excited in the transition. The energy spacing between these sublevels is $\hbar\omega_H$. The spontaneous lifetime of the excited p state of the atom is τ and is presumed to be large compared to the duration of the radiation pulse. Calculate also the mean value $\overline{P}$, averaged over the observation time.

SOLUTION. The scheme of the experiment and the corresponding coordinate system are shown in Figure 4.1. We take the y axis to be in the direction of propagation of the radiation. It can have polarization components along the z and x axes. The part of the radiation polarized along the z axis excites a magnetic sublevel of the p state with $m = 0$. The spontaneous radiation from the atom is observed along the z axis, and can have polarization in the xy plane. The matrix element of the transition $p \to s$ between sublevels with $m = 0$ resulting from the x or y operators is equal to zero. Therefore, in the process considered, only the part of the radiation propagating along the y axis and polarized along the x axis will contribute.

This part of the radiation excites sublevels of the p state with magnetic quantum numbers ± 1. Both matrix elements from the operator x are the same, as is well known. The wave function of the p state at $t = 0$ is of the form $\Psi(0) = \psi_1 + \psi_{-1}$, where $\psi_{\pm 1}$ are coordinate wave functions of the states considered. If the magnetic field is absent, then at $t > 0$, the wave function $\Psi(t)$ contains a phase factor common to the $m = \pm 1$ sublevels, since their energy is the same. The intensity of emission of a photon with a given polarization **s** is

$$I_s \sim |\langle \Psi_0 | \mathbf{s} \cdot \mathbf{D} | (\psi_1 + \psi_{-1}) \rangle|^2 . \tag{4.1}$$

Here **D** is the dipole moment operator of the atom, and Ψ_0 is the wave function of the s state. It is seen that only the intensity component I_x is nonzero, since the matrix elements from the coordinate y between the states Ψ_0, ψ_1 and between Ψ_0, ψ_{-1} have

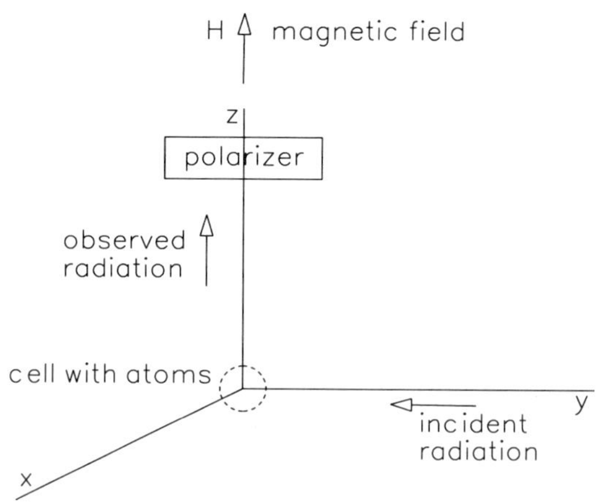

Figure 4.1. Geometry of the Hanle experiment and the axes employed.

the same moduli and opposite signs. Thus if the magnetic field is absent, then the radiation is emitted along the z axis, polarized only along the x axis. In this case, the polarization coefficient is obviously equal to 1.

Now we consider the case where the magnetic field is turned on. The Zeeman effect causes the sublevels with magnetic quantum numbers $+1$ and -1 to split symmetrically to opposite sides of the initial level. If we denote the energy interval between these sublevels as $\hbar\omega_H$, then instead of Eq. (4.1), we have at $t = 0$,

$$I_s \sim \left|\left\langle \Psi_0 \left| \mathbf{s} \cdot \mathbf{D} \right| \left[\psi_1 \exp(-i\omega_H t/2) + \psi_{-1} \exp(i\omega_H t/2)\right]\right\rangle\right|^2. \tag{4.2}$$

In consequence, we have

$$I_x \sim \left|\left\langle \Psi_0 \left| D_x \right| \left[\psi_1 \exp(-i\omega_H t/2) + \psi_{-1} \exp(i\omega_H t/2)\right]\right\rangle\right|^2 \sim \cos^2(\omega_H t/2),$$

since the matrix elements of the operator x are the same for both of the sublevels of the state. It follows from Eq. (4.2) that, for the intensity of radiation with polarization along the y axis, the result is

$$I_y \sim \sin^2(\omega_H t/2),$$

since matrix elements of the operator y are the same in modulus but opposite in sign for the states ψ_1 and ψ_{-1}. Hence, the polarization of the emitted radiation at time t is

$$P(t) = \frac{I_x - I_y}{I_x + I_y} = \cos(\omega_H t). \tag{4.3}$$

It is seen from Eq. (4.3) that the polarization of the emitted light oscillates depending on the time of emission of the photon. This is due to the interference of the two magnetic sublevels. It can be viewed as arising from an oscillation of an atomic electron in its classical motion in the field. The polarization returns to the initial value when the orbital motion completes an integer number of cycles.

Now we average the polarization coefficient (4.3) over the observation time, taking into account the exponential decrease of the population of the p state with time, to obtain

$$\overline{P} = \int_0^\infty P(t) \exp\left(-\frac{t}{\tau}\right) \frac{dt}{\tau} = \frac{1}{1 + \omega_H^2 \tau^2}. \tag{4.4}$$

Thus the polarization coefficient of the radiation decreases when the magnetic field is turned on. That is, depolarization of the radiation occurs. This phenomenon of depolarization of the spontaneous radiation of an atom in a magnetic field is called the Hanle effect. This effect was discovered in 1924, and it is one of the first of the interference phenomena discovered in atomic systems.

We have shown that the decrease of the polarization of the light emitted in the magnetic field is explicable from the classical point of view of the different precessions of the orbital momentum in the magnetic field for the sublevels of the p state of an atom.

Problem 4.2. Calculate the cross section for photon absorption under the conditions of Problem 4.1 when the emitted photons are polarized along the x axis. Assume that spontaneous decay from the p state to the s state is the only channel for radiative transitions.

SOLUTION. Start with Eq. (2.89) for the photon absorption cross section in resonance fluorescence. The following modifications must be made to this expression to make it applicable to the present problem:

1. Introduce the factor $\cos^2(\omega_H t/2)$ in accordance with Eq. (4.2) for calculation of the cross section associated with the emission of photons along the x axis.
2. Replace the factor $2J_k + 1$ by 1 in the numerator of Eq. (2.89), since the initial and final channels are determined by a fixed photon polarization.
3. Average over time t with the probability function $(1/\tau)\exp(-t/\tau)$ in analogy with the procedure in the preceding problem.

We obtain in this fashion

$$\begin{aligned}\sigma_x &= \frac{\pi c^2}{2\omega^2} \frac{\tau^{-2}}{(\omega_{k0} - \omega)^2 + (1/2\tau)^2} \int_0^\infty dt \cos^2(\omega_H t/2) \exp(-t/\tau) \\ &= \frac{\pi^2 c^2}{2\omega^2 \tau} \left(1 + \frac{1}{1 + \omega_H^2 \tau^2}\right) a_\omega. \end{aligned} \tag{4.5}$$

Here a_ω is the Lorentz distribution function for the absorption of photons of frequency ω, with the width determined by the spontaneous lifetime of the p state.

In an analogous fashion, we find that the corresponding cross section for photons emitted in the direction of the z axis with polarization along the y axis, is given by

$$\sigma_y = \frac{\pi^2 c^2}{2\omega^2 \tau} \left(1 - \frac{1}{1 + \omega_H^2 \tau^2}\right) a_\omega. \tag{4.6}$$

It follows from Eqs. (4.5) and (4.6) that, in the cross section for emission along the z axis of unpolarized photons,

$$\sigma_x + \sigma_y = \frac{\pi^2 c^2}{\omega^2 \tau} a_\omega,$$

the interference effect disappears. This cross section is in agreement with the general expression (2.85) for the photon absorption cross section.

Now we shall take into account the motion of the atoms that is responsible for the Doppler effect. Since the Doppler width is usually large as compared to the spontaneous width, then, in Eqs. (4.5) and (4.6), we should replace the Lorentz distribution function for the absorbed photons by the Doppler distribution of frequencies. When these expressions are averaged over atomic velocities, it is found that the interference structure is retained.

Problem 4.3. Obtain Eqs. (4.5) and (4.6) in the framework of a model of a damped classical dipole in a constant magnetic field.

SOLUTION. We posit as above that the incident radiation propagates along the y axis and is polarized along the x axis. In the classical model, the oscillations of an atomic electron along the x axis take place during the excitation process. The interaction of this dipole with the emitted radiation is given by the expression $-\mathbf{d} \cdot \mathbf{E}$, where $\mathbf{d}$ is the dipole moment and $\mathbf{E}$ is the electric field vector. Hence, the emitted radiation is also polarized along the x axis. As above, we take the direction of observation of the emitted radiation to be coincident with the x axis.

Now add to the above problem a constant magnetic field, with the magnetic field vector $\mathbf{H}$ along the z axis. After the dipole moment is excited, it begins to rotate in the xy plane, in accordance with the principles of classical electromagnetic theory. This is the plane perpendicular to the magnetic field vector. It is known that the precession frequency (the so-called Larmor frequency) is given by $\omega_0 = eH/(2mc)$. Let the time $t = 0$ be the moment of excitation of the dipole. As above, the time of duration of the incident radiation is presumed to be small as compared to the relaxation time of the dipole, that is, as compared to its lifetime for emission of spontaneous radiation. During the time t, the dipole $\mathbf{d}$ turns through the angle $\omega_0 t$ with respect to its initial direction along the x axis. We designate by $I_x(t)$ the intensity at time t of the radiation scattered along the z axis with polarization along the x axis. It is determined by the x projection of the dipole moment, d_x. This projection is equal to $d \cos \omega_0 t$. Hence, the contribution to the intensity $I_x(t)$ will be decreased by the factor $\cos^2 \omega_0 t$, that is,

$$I_x(t) = I \cos^2 \omega_0 t,$$

where I is the intensity of scattered radiation in the absence of the magnetic field ($\omega_0 = 0$). We also take into account the exponential decrease with time of the magnitude of the dipole moment. The quantity I, which is proportional to the square of the dipole moment, also decreases exponentially, so that

$$I(t) = I_0 \exp(-t/\tau).$$

Here I_0 is the constant that determines the intensity at $t = 0$ of the scattered radiation polarized along the x axis.

We now calculate the intensity averaged over the observation time, $\overline{I_x}$, as the total radiation energy $\int_0^\infty \overline{I_x}\, dt$ for all times, divided by the damping time τ, or

$$\overline{I_x} = \frac{1}{\tau} \int_0^\infty I_x \, dt. \tag{4.7}$$

In the absence of a magnetic field ($\omega_0 = 0$), we have $I_x = I_0$. Using the above relationships to calculate the integral in Eq. (4.7), we obtain

$$\begin{aligned} \overline{I_x} &= \left(\frac{I_0}{\tau}\right) \int_0^\infty \exp\left(-\frac{t}{\tau}\right) \cos^2(\omega_0 t)\, dt \\ &= \frac{I_0}{2}\left(1 + \frac{1}{1 + 4\omega_0^2 \tau^2}\right). \end{aligned} \tag{4.8}$$

When we compare Eq. (4.8) to Eq. (4.5), we see that the time τ is the natural lifetime for the excited p state, and $\omega_H = 2\omega_0$.

In an analogous fashion, we calculate the intensity of the radiation along the direction of the z axis with polarization along the y axis. In this case, the projection $d \sin \omega_0 t$ onto the y axis replaces the projection $d \cos \omega_0 t$. Thus, instead of Eq. (4.8) we find

$$\overline{I_y} = \frac{I_0}{2}\left(1 - \frac{1}{1 + 4\omega_0^2\tau^2}\right). \tag{4.9}$$

This result vanishes when the magnetic field is zero. The scattered radiation is totally polarized along the y axis in the observation of radiation along the x axis.

Problem 4.4. Determine the directions of the polarization vector of scattered radiation for both the maximum and the minimum time-averaged intensity $\overline{I}$ under the conditions of the previous problem. Scattered radiation is observed along the direction of the magnetic field. Calculate the polarization coefficient of the scattered light

$$\overline{P} = \frac{\overline{I}_{\max} - \overline{I}_{\min}}{\overline{I}_{\max} + \overline{I}_{\min}}.$$

Intensities are averaged as in Eq. (4.7).

SOLUTION. In analogy to the preceding problem, we take the magnetic field strength vector **H** to be directed along the z axis, and the incident electromagnetic wave propagates along the y axis with its polarization along the x axis. Calculate the intensity of the light emitted in the z direction such that its polarization vector lies in the xy plane and is directed at an angle φ with respect to the x axis.

We solve this problem in a manner similar to the previous problem. Let us suppose that the dipole moment is excited at time $t = 0$, and at the time t the vector **d** rotates to the angle $\omega_0 t$ with respect to the initial direction along the x axis. Here ω_0 is the Larmor frequency of precession in the magnetic field **H**. The projection of the dipole moment onto the direction of the polarization vector of the scattered radiation at time t is given by $d \cos(\omega_0 t + \varphi)$. Hence, the averaged intensity of the scattered radiation for a fixed value of φ is

$$\overline{I_\varphi} = \left(\frac{I_0}{\tau}\right) \int_0^\infty \exp\left(-\frac{t}{\tau}\right) \cos^2(\omega_0 t + \varphi)\, dt.$$

Evaluating the integral, we obtain

$$\overline{I_\varphi} = \frac{I_0}{2}\left[1 + \frac{\cos\left[2\varphi - \arctan\left(2\omega_0\tau\right)\right]}{\left(1 + 4\omega_0^2\tau^2\right)^{1/2}}\right]. \tag{4.10}$$

As in the preceding problem, the quantity I_0 determines the radiation intensity with photons polarized along the x axis in the absence of the magnetic field ($\omega_0 = \varphi = 0$). In particular, at $\varphi = 0$ and at $\varphi = \pi/2$, Eq. (4.10) yields, respectively, the expressions

(4.8) and (4.9), as one would expect. Expression (4.10) has the maximum value

$$(I_0/2)\left[1 + \left(1 + 4\omega_0^2\tau^2\right)^{-1/2}\right]$$

at the angle

$$\varphi = \tfrac{1}{2}\arctan(2\omega_0\tau),$$

and the minimum value

$$(I_0/2)\left[1 - \left(1 + 4\omega_0^2\tau^2\right)^{-1/2}\right]$$

at the angle

$$\varphi = \tfrac{1}{2}[\pi + \arctan(2\omega_0\tau)].$$

The directions for maximum and minimum intensities are mutually perpendicular, as would be expected.

The polarization coefficient of the scattered radiation is

$$\overline{P} = \left(1 + 4\omega_0^2\tau^2\right)^{-1/2}. \tag{4.11}$$

It is seen that this value is larger than that given by Eq. (4.4) with $\omega_H = 2\omega_0$. The quantity (4.11) changes as a function of H more slowly than Eq. (4.4). Equation (4.11) is the correct polarization coefficient for scattered radiation.

It is clear that, under the conditions of Problem 4.1 for absorption of light by atoms with a resulting transition from an s state to a p state, the expression (4.11) is correct, but then τ is the lifetime of the atomic p state, and the quantity $2\omega_0$ should be replaced by ω_H. The energy $\hbar\omega_H$ is the interval between magnetic sublevels with projections $+1$ and -1 of the orbital angular momentum of the p state. This energy interval is proportional to the magnitude of H.

4.2 INTERFERENCE OF STATES DURING RADIATION

In the previous section we considered the simplest phenomena associated with coherent processes in radiative transitions. Now we wish to describe more complicated problems related to experiments where interference of atomic states during radiation is important.

Problem 4.5. Investigate the photon radiation from a damped classical dipole in a constant magnetic field, where the radiation is to be examined in the direction of the magnetic field. Radiation is incident on the system with time dependence given by

$$J(t) = \text{const}\left[1 + \varepsilon\cos(\Omega t + \psi)\right], \qquad \varepsilon \ll 1.$$

The quantity ε determines the degree of modulation of the incident radiation. The modulation frequency Ω is assumed to be small compared to the frequency ω_0 of the incident radiation.

SOLUTION. This problem is a generalization of Problem 4.4 to the case of radiation modulated in time. We suppose again that the magnetic field is directed along the z axis, and the incident radiation is polarized along x and propagates in the y direction. It should be noted (see Problem 4.1) that if the incident radiation is polarized along the z axis and propagates along y, then interference effects do not appear. We set the origin of time, $t = 0$, at the time at which the dipole is excited.

Let us consider the radiation scattered in the direction z of the magnetic field, with polarization in the xy plane. We designate by φ the angle between the polarization vector and the x axis. Repeating the procedure of Problem 4.4, we obtain for the averaged intensity

$$\overline{I_\varphi} = \frac{I_0}{\tau} \int_0^\infty \left[1 + \varepsilon \cos(\Omega t + \psi)\right] e^{-t/\tau} \cos^2(\omega_0 t + \varphi)\, dt, \tag{4.12}$$

in place of Eq. (4.10). This expression takes into account the intensity modulation of the incident radiation as given in the statement of this problem.

The first term in Eq. (4.12), independent of ε, obviously produces Eq. (4.10). Therefore only the second term, which is proportional to the modulation strength ε, is of interest here. Let us write this term in the form

$$\delta\overline{I_\varphi} = \frac{I_0 \varepsilon}{2\tau} \int_0^\infty \cos(\Omega t + \psi)\, e^{-t/\tau} \left\{1 + \cos\left[2(\omega_0 t + \varphi)\right]\right\} dt. \tag{4.13}$$

Expression (4.13) is represented again as a sum of two terms. The simplest to evaluate is the first term, since an analogous integral was calculated in the solution of Problem 4.4 [see Eq. (4.10)]. We need only make the substitutions $2\omega_0 \to \Omega$ and $2\varphi \to \psi$. Then, expressing $\delta\overline{I_\varphi}$ in the form

$$\delta\overline{I_\varphi} = \delta_1\overline{I_\varphi} + \delta_2\overline{I_\varphi},$$

we obtain for $\delta_1\overline{I_\varphi}$ the expression

$$\delta_1\overline{I_\varphi} = \frac{\varepsilon I_0}{2} \frac{\cos\left[\psi - \arctan(\Omega\tau)\right]}{\left(1 + \Omega^2\tau^2\right)^{1/2}}. \tag{4.14}$$

It should be noted that Eq. (4.14) does not depend on the magnetic field strength H, nor on the polarization angle φ of the scattered radiation.

Physically, expression (4.14) represents the intensity modulation of the scattered radiation. Indeed, if we set $\varphi = 0$ and $H = 0$, then the quantity I_0 is modified as

$$I_0 \to I_0 \left\{1 + \frac{\varepsilon}{2} \frac{\cos\left[\psi - \arctan(\Omega\tau)\right]}{\left(1 + \Omega^2\tau^2\right)^{1/2}}\right\}. \tag{4.15}$$

It follows from Eq. (4.15) that the degree of modulation of the scattered radiation is approximately $(1 + \Omega^2\tau^2)^{1/2}$ times less than the degree to which the incident radiation is modulated. In addition to this, the phase of the scattered radiation is delayed with respect to the modulation phase ψ of the incident radiation by the angle $\arctan(\Omega\tau)$, which depends on the depletion time τ of a dipole.

Now let us consider the second term in Eq. (4.13), that is, let us calculate the quantity $\delta_2\overline{I_\varphi}$. Rewriting it in the form

$$\delta_2\overline{I_\varphi} = \frac{I_0\varepsilon}{4\tau}\int_0^\infty e^{-t/\tau}\left[\cos(2\omega_0 t + 2\varphi - \Omega t - \psi) + \cos(2\omega_0 t + 2\varphi + \Omega t + \psi)\right] dt,$$

we see that it can be calculated in a fashion analogous to the way Eq. (4.14) was derived. One need only make appropriate changes in the identity of the parameters in the integrand. In this fashion we obtain

$$\delta_2\overline{I_\varphi} = \frac{\varepsilon I_0}{4}\left\{\frac{\cos\left[2\varphi + \psi - \arctan\left[(2\omega_0 + \Omega)\tau\right]\right]}{\left[1 + (2\omega_0 + \Omega)^2\tau^2\right]^{1/2}} + \frac{\cos\left[2\varphi - \psi - \arctan\left[(2\omega_0 - \Omega)\tau\right]\right]}{\left[1 + (2\omega_0 - \Omega)^2\tau^2\right]^{1/2}}\right\}. \tag{4.16}$$

Equation (4.16) describes interference oscillations that depend on the relation between the frequencies Ω and $2\omega_0$. The second term in Eq. (4.16) is resonantly large if the condition $\Omega \approx 2\omega_0$ is fulfilled. In like fashion, the first term of Eq. (4.16) increases resonantly under the condition $\Omega \approx -2\omega_0$.

We refer back to Problem 4.1, where, instead of the quantity $2\omega_0$, we had the frequency ω_H, which measures the energy splitting between Zeeman sublevels of p states with quantum numbers $+1$ and -1. We conclude that interference oscillations are important when the frequency of these oscillations is approximately equal to the transition frequency between the sublevels considered. A resonant increase of intensity is thus achieved when the photon energy $\hbar\Omega$ coincides with the energy splitting of sublevels with magnetic quantum numbers $+1$ and -1 of the atomic p state. This is the usual resonance condition in the quantum theory of atomic radiation.

We emphasize that all interference effects disappear after averaging over the phase ψ of the intensity modulation of the incident radiation.

Equation (4.16) can be simplified for a very strong magnetic field, that is, under the condition that $\omega_0\tau \gg 1$ and the resonance condition $\Omega \approx 2\omega_0$ (or $\Omega \approx -2\omega_0$). Then $\delta_1\overline{I_\varphi} \to 0$, and we obtain

$$\overline{I_\varphi} \approx \frac{1}{2}I_0\left(1 + \frac{\varepsilon}{2}\frac{\cos\left\{2\varphi - \psi - \arctan\left[(2\omega_0 - \Omega)\tau\right]\right\}}{\left[1 + (2\omega_0 - \Omega)^2\tau^2\right]^{1/2}}\right). \tag{4.17}$$

It follows from this expression that

$$\bar{I}_{\max} = \frac{1}{2}I_0\left\{1 + \frac{\varepsilon}{2\left[1 + (2\omega_0 - \Omega)^2\tau^2\right]^{1/2}}\right\},$$

$$\bar{I}_{\min} = \frac{1}{2}I_0\left\{1 - \frac{\varepsilon}{2\left[1 + (2\omega_0 - \Omega)^2\tau^2\right]^{1/2}}\right\}.$$

Finally, we find the polarization coefficient

$$\overline{P} = \frac{\overline{I}_{\max} - \overline{I}_{\min}}{\overline{I}_{\max} + \overline{I}_{\min}} = \frac{\varepsilon}{2\left[1 + (2\omega_0 - \Omega)^2\tau^2\right]^{1/2}}. \tag{4.18}$$

The polarization coefficient is largest at exact resonance. Then it is equal to $\overline{P}_{\max} = \varepsilon/2$, and does not depend on Ω, ψ, or τ.

Problem 4.6. Calculate the radiation intensity along the direction of the magnetic field for a classical dipole placed in a magnetic field H. The quantity H is subjected to a modulation in time of frequency Ω and strength ε. Investigate the phenomenon of paramagnetic resonance between the modulation frequency Ω and the Larmor precession frequency ω_0.

SOLUTION. According to the statement of this problem, we take the Larmor precession frequency in the magnetic field to be of the form

$$\frac{d\alpha}{dt} = \omega(t) = \omega_0\left[1 + \varepsilon\cos(\Omega t + \psi)\right]. \tag{4.19}$$

Here, ω_0 is the Larmor frequency in a constant magnetic field, ψ is the modulation phase angle, and $\alpha(t)$ is the angle of dipole rotation in the case of a variable rotation frequency $\omega(t)$. After performing the integration over time in Eq. (4.19), we obtain

$$\alpha(t) = \omega_0 t + \omega_0\frac{\varepsilon}{\Omega}\left[\sin(\Omega t + \psi) - \sin\psi\right].$$

We set the zero of time at the moment when the dipole is excited, so that $\alpha(0) = 0$. The initial direction of the dipole is taken to coincide with the x axis. We wish to find the radiation in the direction of the magnetic field strength, which is polarized in the xy plane. We designate by φ the angle between the polarization vector and the x axis.

In analogy to the solution of Problem 4.1, we obtain the intensity I_φ of the emitted photons as

$$\overline{I_\varphi} = \frac{I_0}{\tau}\int_0^\infty e^{-t/\tau}\cos^2\left\{\omega_0 t + \varphi + \omega_0\frac{\varepsilon}{\Omega}\left[\sin(\Omega t + \psi) - \sin\psi\right]\right\}dt. \tag{4.20}$$

In order to calculate the integral in Eq. (4.20), we introduce the generating function for the ordinary Bessel function $J_n(\xi)$ as given by

$$\exp(i\xi\sin\theta) = \sum_{n=-\infty}^{\infty} J_n(\xi)\exp(in\theta).$$

When this relation is employed in Eq. (4.20), we find

$$\overline{I_\varphi} = \frac{I_0}{2}\left\{1 + \sum_{n=-\infty}^{\infty} J_n\left(\frac{2\omega_0\varepsilon}{\Omega}\right)\frac{\cos\left[2\varphi + n\psi - (2\omega_0\varepsilon/\Omega)\sin\psi + \beta\right]}{\left[1 + (2\omega_0 + n\Omega)^2\tau^2\right]^{1/2}}\right\}. \tag{4.21}$$

Here the notation is introduced that

$$\beta = \arctan\left[(2\omega_0 + n\Omega)\,\tau\right].$$

It follows from Eq. (4.21) that, under the condition $n\Omega = \pm 2\omega_0$, one term in the sum in Eq. (4.21) will be resonantly large. Such a resonance is called a *parametric resonance*. In the vicinity of the nth resonance, which is clearly observed when $\omega_0\tau \gg 1$, we can restrict the result to a single term of the above sum. We then obtain

$$\overline{I_\varphi} = \frac{I_0}{2}\left\{1 + J_n(-n\varepsilon)\frac{\cos\left(2\varphi + n\psi + n\varepsilon\sin\psi + \beta\right)}{\left[1 + (2\omega_0 + n\Omega)^2\,\tau^2\right]^{1/2}}\right\}. \tag{4.22}$$

Since the magnitude of the second term in Eq. (4.22) is less than unity, the entire expression (4.22) is, of course, positive.

It is seen from Eq. (4.22) also that the amplitude of the resonance depends essentially on the modulation strength ε. If the quantity ε is of the order of unity, then the amplitude of the resonance is of the order of the nonresonance background, especially for small values of n. On the other hand, if $\varepsilon \ll 1$, then the amplitude of the resonance is proportional to ε^n and decreases very rapidly with an increase in the index n. Hence, parametric resonance consists of a series of resonances, and the intensity of these resonances decreases rapidly with increasing multiplicity n of the resonance.

Now we wish to determine from Eq. (4.22) the degree of polarization $\overline{P_n}$ of the emitted radiation. First we choose φ so that the cos function in the equation has the value $+1$, and then we select φ so that the cos function has the value -1. This gives us the expression

$$\overline{P_n} = \frac{|J_n(-n\varepsilon)|}{\left[1 + (2\omega_0 + n\Omega)^2\,\tau^2\right]^{1/2}} \tag{4.23}$$

for the degree of polarization. In the case of exact resonance ($n\Omega = -2\omega_0$), we find the maximum degree of polarization

$$\overline{P_{n\max}} = |J_n(-n\varepsilon)|.$$

The degree of polarization decreases as the number n grows. It is proportional to ε^n at $\varepsilon \ll 1$. If $\varepsilon \sim 1$, then the degree of polarization oscillates as a function of ε.

The condition $\omega_0\tau \gg 1$ is achieved in strong magnetic fields. In this case, the above resonances are observed clearly. From the physical point of view, a resonance takes place in a system if its inherent frequency $2\omega_0$ is equal to, or a multiple of, the "forcing" frequency Ω.

We can also conclude from Eq. (4.21) that even for very small modulation strengths $\varepsilon \ll 1$, ε still plays an important role if the inequality $\omega_0 \gg \Omega$ is fulfilled. In this case we have $(\varepsilon\omega_0/\Omega) \sim 1$, and the Bessel function cannot be expanded in a power series. Then the quantity $\overline{I_\varphi}$ oscillates but is nonresonant.

In the above, we considered the electromagnetic radiation field to be a small perturbation. The next sections will be devoted to investigation of processes in which the electric field strength $\mathbf{E}$ is sufficiently large that perturbation theory is inapplicable. We shall see later that the condition for a field to be strong is $(\mathbf{D}_{k0} \cdot \mathbf{E}\tau/\hbar) > 1$, where $\mathbf{D}_{k0}$ is the dipole matrix element of the atomic transition $k \to 0$, and the quantity τ is the broadening of this transition. We are restricted here to resonance processes where the radiation frequency is close to the frequency ω_{k0} of the atomic transition being considered. Apart from this case, problems are fundamentally complicated when the intensity of the radiation field increases into the nonperturbative domain. Nonperturbative problems will be treated in more detail in the next chapter.

4.3 RESONANCE TRANSITIONS IN TWO-LEVEL SYSTEMS

In this section, we consider problems in which the radiation frequency is approximately equal to the frequency ω_{k0} of some atomic transition. Relaxation of the excited state k is assumed to occur only by spontaneous emission to the ground state 0. We can then exclude other atomic states from consideration, which simplifies in an essential fashion the solution of problems with a strong electromagnetic field.

Problem 4.7. The radiation field with electric field strength $\mathbf{E}$ and field frequency ω excites the atomic transition $0 \leftrightarrow k$, which has the transition frequency $\omega_{k0} \approx \omega$. Relaxation of the excited state k is determined by the spontaneous transition back to the ground state 0 with the lifetime τ_k. Calculate the cross section for the absorption of radiation for an arbitrary value of $\mathbf{E}$.

SOLUTION. We shall find first the equilibrium distribution of atomic electrons in the states 0 and k in the external electromagnetic field. We consider the system of equations (H.7) and (H.9) for elements of the density matrix. Using the resonance approximation, we retain only the required exponents in the interaction potential $\mathbf{D} \cdot \mathbf{E} \cos \omega t$. We then obtain the system of equations

$$\begin{aligned} \frac{d\rho_{00}}{dt} &= \frac{1}{\tau}\rho_{kk} + \frac{i}{2\hbar}(\mathbf{D}_{k0} \cdot \mathbf{E})\left(e^{-i\omega t}\rho_{0k} - e^{i\omega t}\rho_{k0}\right), \\ \frac{d\rho_{k0}}{dt} &= -\left(i\omega_{k0} + \frac{1}{2\tau_k}\right)\rho_{k0} + \frac{i}{2\hbar}(\mathbf{D}_{k0} \cdot \mathbf{E})\, e^{-i\omega t}(\rho_{kk} - \rho_{00}); \qquad \rho_{0k} = \rho_{k0}^{*}. \end{aligned} \tag{4.24}$$

It is clear from these equations that in the stationary case, which is realized at large times, we have

$$\frac{d\rho_{00}}{dt} = 0, \qquad \frac{d\rho_{kk}}{dt} = 0,$$

and so

$$\rho_{00},\ \rho_{kk} = \text{const}, \qquad \rho_{k0} = Ae^{-i\omega t}. \tag{4.25}$$

Instead of writing a differential equation for ρ_{kk}, we can use the condition for the conservation of particle number

$$\rho_{00} + \rho_{kk} = 1. \tag{4.26}$$

When we substitute Eq. (4.25) into the second equation of the system (4.24), and take Eq. (4.26) into account, we find

$$A = \frac{G\Gamma}{\Delta - i\Gamma}\,(2\rho_{kk} - 1). \tag{4.27}$$

New quantities are introduced here for resonance detuning $\Delta = \omega_{k0} - \omega$; width of the excited state k, $\Gamma = 1/2\tau_k$; and the dimensionless parameter of the field intensity

$$G = (\mathbf{D}_{k0} \cdot \mathbf{E})\,\tau_k/\hbar.$$

Substituting Eq. (4.27) into the first equation of (4.24), we find

$$\rho_{kk} = \frac{\chi}{2}\frac{\Gamma^2}{\Delta^2 + \Gamma^2(1 + \chi)}. \tag{4.28}$$

The new dimensionless parameter introduced here is

$$\chi = 2G^2 = 2\,|(\mathbf{D}_{k0} \cdot \mathbf{E})|^2\,(\tau_k/\hbar)^2.$$

The equilibrium value obtained in Eq. (4.28) for the population of the excited level k is constant due to a balance between two processes: excitation of this state by the external radiation field and the spontaneous decay to the ground state 0. In equilibrium, both processes have the same transition rate, which is

$$g_k\frac{1}{\tau_k}\rho_{kk} = 2\Gamma\,\rho_{kk}g_k.$$

We determine the cross section for absorption of radiation as the ratio of the rate to the flux of incident photons. The flux is

$$J_\omega = \frac{1}{\hbar\omega}\frac{cE^2}{4\pi}\overline{\cos^2\omega t} = \frac{cE^2}{8\pi\hbar\omega}.$$

Then we obtain

$$\sigma_a = \frac{8\pi\hbar\omega\chi\Gamma^2 g_k}{cE^2\left[\Delta^2 + \Gamma^2(1 + \chi)\right]} \tag{4.29}$$

for the absorption cross section. We now express the square of the matrix element of the dipole transition in terms of the lifetime τ_k of the excited state k as

$$\frac{1}{\tau_k} = 2\Gamma = \frac{4\omega^3}{\hbar c^3}\left|(D_z)_{k0}\right|^2 g_0. \tag{4.30}$$

Then the substitution of Eq. (4.30) into Eq. (4.29) gives the result

$$\sigma_a = 2\pi \frac{c^2}{\omega^2} \frac{g_k}{g_0} \frac{\Gamma^2}{\Delta^2 + \Gamma^2(1+\chi)}. \tag{4.31}$$

We see that the quantity χ is the parameter that determines the influence of the radiation field on the transition cross section. The dependence of the absorption cross section on the field strength **E** in the case $\chi \gg 1$ (mentioned in the preceding section) takes place when the transition time between states k and 0 under the influence of the wave field [which is of the order of $\hbar/(D_{k0}E)$] is less than the time τ_k for spontaneous decay.

In the case $\chi \gg 1$, we have

$$\rho_{kk} = \rho_{00} = \tfrac{1}{2},$$

corresponding to equipartition between the two levels in the strong field (for resonance detunings Δ which are not extremely large). In this case we obtain

$$\sigma_a = 2\pi \frac{c^2}{\omega^2} \frac{g_k}{g_0} \frac{1}{\chi}. \tag{4.32}$$

It is seen that the absorption cross section tends to zero. This phenomenon can be referred to as a "transparency" of an atomic medium in a sufficiently strong field of resonance radiation.

In the opposite limiting case $\chi \ll 1$, Eq. (4.31) coincides with Eq. (2.89), which was obtained within the framework of perturbation theory. The corresponding statistical weights are $g_0 = 2J_0 + 1$ and $g_k = 2J_k + 1$, where J_0 and J_k are the angular momenta of the initial and final states, respectively.

Expression (4.31) can also be converted to the frequency distribution function a_ω in a strong field. After normalization, we obtain [see also Eq. (2.9)]

$$a_\omega = \frac{1}{\pi} \frac{\Gamma\sqrt{1+\chi}}{(\omega - \omega_{k0})^2 + \Gamma^2(1+\chi)}. \tag{4.33}$$

Problem 4.8. Under the conditions specified for the preceding problem, calculate the cross section for coherent scattering of the radiation; that is, the cross section for photon emission with frequency ω in a strong field.

SOLUTION. According to Eq. (1.17), the intensity of coherent radiation I_k of an atom in the classical limit is

$$I_k = 2\left(\ddot{D}_z\right)^2 / c^3,$$

where D_z is the projection of the electric dipole moment onto the direction of the electric field vector. In the case under consideration, the dipole moment is produced by an external field. According to the definition of the density matrix, we have

$$D_z = \sum_j (\widehat{D}_z\rho)_{jj} = (\widehat{D}_z)_{0k}\rho_{k0} + (\widehat{D}_z)_{k0}\rho_{0k} \tag{4.34}$$
$$= 2\,\mathrm{Re}(D_z)_{0k}\rho_{k0}.$$

Here $\widehat{D}_z$ is the quantum mechanical dipole moment operator. We find here the correspondence between the classical and quantum approaches for radiation intensity.

According to Eq. (4.25), we have $\rho_{k0} = A\exp(-i\omega t)$. Hence the dipole moment also oscillates with the frequency ω, and will therefore emit photons with this frequency. Therefore the radiation being considered is coherent, that is, the scattering is elastic.

The quantity A is determined by expression (4.27), and ρ_{kk} in turn follows from Eq. (4.28). Thus we find that

$$\rho_{k0} = \frac{G\Gamma}{\Delta - i\Gamma}\frac{\Delta^2 + \Gamma^2}{\Delta^2 + \Gamma^2(1+\chi)}e^{-i\omega t}. \tag{4.35}$$

When we substitute this result into Eq. (4.34), we obtain for the intensity of the radiation emitted with the frequency ω (averaged over a period of the field, $2\pi/\omega$), the result

$$I_k = \frac{4\omega^4}{c^3}\frac{g_k G^2\Gamma^2\left|(D_z)_{k0}\right|^2\left(\Delta^2+\Gamma^2\right)}{\left[\Delta^2+\Gamma^2(1+\chi)\right]^2}. \tag{4.36}$$

Upon division of this expression by the incident energy flux of the electromagnetic radiation,

$$\frac{cE^2}{4\pi}\overline{\cos^2\omega t} = \frac{cE^2}{8\pi},$$

we obtain the cross section of the coherent radiation

$$\sigma_r^e = 2\pi\frac{c^2}{\omega^2}\frac{g_k}{g_0}\frac{\Gamma^2\left(\Delta^2+\Gamma^2\right)}{\left[\Delta^2+\Gamma^2(1+\chi)\right]^2}. \tag{4.37}$$

We have used here the relation (4.30) as in Problem 4.7.

The cross section for incoherent radiation corresponding to the scattering of radiation with shifted frequency can be obtained as the difference between the total absorption cross section [see Eq. (4.31) in the preceding problem] and the coherent scattering cross section, Eq. (4.37). This difference is

$$\sigma_r^i = 2\pi\frac{c^2}{\omega^2}\frac{g_k}{g_0}\frac{\Gamma^2\chi}{\left[\Delta^2+\Gamma^2(1+\chi)\right]^2}. \tag{4.38}$$

Incoherent scattering can be explained physically by noting that, at high intensity, the emitted photons cover a wide frequency range, so that the energy of the emitted photons can differ significantly from the energy of the incident photon.

The ratio of the cross sections for coherent and incoherent scattering can be found from Eqs. (4.37) and (4.38) to be

$$\frac{\sigma_r^e}{\sigma_r^i} = \frac{\Delta^2 + \Gamma^2}{\Gamma^2 \chi}. \tag{4.39}$$

It is seen from Eq. (4.39) that the scattering of resonance radiation by an atom is mostly coherent at small radiation intensity, while in the high-intensity limit (with moderate resonance detunings), the scattering will be almost totally incoherent.

Problem 4.9. Calculate the absorption cross section of an atom for resonant radiation if the atom also experiences collisions with neighboring atoms.

SOLUTION. We generalize the system of equations for the density matrix given in (4.24) to the case where collisional broadening exists, making use of Eq. (H.12) to obtain

$$\begin{aligned}
\frac{d\rho_{00}}{dt} &= \frac{1}{\tau}\rho_{kk} + \frac{i}{2\hbar}(\mathbf{D}_{0k}\cdot\mathbf{E})\left(e^{-i\omega t}\rho_{0k} - e^{i\omega t}\rho_{k0}\right),\\
\frac{d\rho_{k0}}{dt} &= -\left(i\omega_{k0} + \frac{1}{2\tau_k} + \frac{\nu}{2} - i\Delta\nu\right)\rho_{k0} + \frac{i}{2\hbar}(\mathbf{D}_{k0}\cdot\mathbf{E})\,e^{-i\omega t}(\rho_{kk} - \rho_{00}),\\
\rho_{0k} &= \rho_{k0}^*.
\end{aligned} \tag{4.40}$$

As in the solution of Problem 2.9 [see Eq. (2.39)], we have employed in Eqs. (4.40) the averages over the velocities v of the colliding atoms given by

$$\nu = \overline{2Nv\sigma'}, \qquad \Delta\nu = \overline{Nv\sigma''}.$$

Here N is the density of the perturbing atoms, and the quantities σ' and σ'' are determined by expressions (2.42) and (2.43), respectively. The quantity $\delta(\rho)$ in those equations represents the difference of phases for scattering in the excited (k) and the ground (0) states.

We see that only the nondiagonal elements of the matrix element are modified. We solve the system (4.40) in a fashion analogous to that used in the solution of (4.24) in Problem 4.7. Let us find the stationary solution that does not depend on the initial conditions. The time dependence of the density matrix elements is determined by Eqs. (4.25). After substitution into the second equation of the system (4.40), we obtain in place of Eq. (4.27) the result

$$A = \frac{G\Gamma}{\widetilde{\Delta} - i\widetilde{\Gamma}}(2\rho_{kk} - 1). \tag{4.41}$$

Here, $\widetilde{\Delta} = \Delta - \Delta\nu$ is the resonance detuning taking into account the Stark shift arising from atomic collisions. The quantity $\widetilde{\Gamma} = \Gamma + \nu/2$ is the total width taking into account both spontaneous and collision broadening. Other notations are taken from Problem 4.7, to wit: $\Gamma = 1/(2\tau_k)$, $\Delta = \omega_{k0} - \omega$, and $G = (\mathbf{D}_{k0}\cdot\mathbf{E})\,\tau/\hbar$.

When we substitute ρ_{k0} as given in (4.25) into the first equation of the system (4.40), and use the value of A determined by Eq. (4.41), we obtain

$$\rho_{kk} = \frac{\widetilde{\chi}}{2} \frac{\widetilde{\Gamma}^2}{\widetilde{\Delta}^2 + \widetilde{\Gamma}^2(1 + \widetilde{\chi})}, \tag{4.42}$$

which is the required generalization of Eq. (4.28) for ρ_{kk}. We have introduced here the quantity

$$\widetilde{\chi} = 2G^2 \frac{\Gamma}{\widetilde{\widetilde{\Gamma}}} = \chi \frac{\Gamma}{\widetilde{\widetilde{\Gamma}}}.$$

Now we wish to derive the rate of decay of the excited state k, given by $\rho_{kk} g_k / \tau_k$. Due to the stationary, or equilibrium, nature of this process, the required rate is equal to the excitation rate of the ground state 0. Proceeding in analogy to Problem 4.7, we introduce the absorption cross section as a ratio of the decay rate to the flux of incident photons, or

$$J_\omega = cE^2/(8\pi\hbar\omega).$$

Then the absorption cross section σ_a can be written in the form analogous to Eq. (4.31) if the relation (4.30) is taken into account, giving

$$\sigma_a = 2\pi \frac{c^2}{\omega^2} \frac{g_k}{g_0} \frac{\Gamma\widetilde{\Gamma}}{\widetilde{\Delta}^2 + \widetilde{\Gamma}^2(1 + \widetilde{\chi})}. \tag{4.43}$$

In the limit $\nu = \Delta\nu = 0$, Eq. (4.43) reduces to Eq. (4.31). We see that collisions produce additional broadening and Stark shifting of the resonance. Expression (4.43) is in agreement with Eq. (2.89). The numerator has the product of the total width by the partial width that corresponds to the spontaneous decay of the state k.

Problem 4.10. Under the conditions of Problem 4.9, determine the cross sections for coherent and incoherent scattering of resonance radiation by an atom.

SOLUTION. This problem is the generalization of Problem 4.8 to the case where collisions are included. When Eq. (4.42) is substituted into Eq. (4.41), Eq. (4.35) is replaced by

$$\rho_{k0} = \frac{G\Gamma}{\widetilde{\Delta} - i\widetilde{\Gamma}} \frac{\widetilde{\Delta}^2 + \widetilde{\Gamma}^2}{\widetilde{\Delta}^2 + \widetilde{\Gamma}^2(1 + \widetilde{\chi})} e^{-i\omega t}. \tag{4.44}$$

We continue by substituting Eq. (4.44) into Eq. (4.34) and find the generalization of Eq. (4.36) for the intensity of coherent radiation of photons to be

$$I_k = \frac{4\omega^4 \left|(D_z)_{k0}\right|^2}{c^3} g_k \frac{G^2\Gamma^2\left(\widetilde{\Delta}^2 + \widetilde{\Gamma}^2\right)}{\left[\widetilde{\Delta}^2 + \widetilde{\Gamma}^2(1 + \widetilde{\chi})\right]^2}. \tag{4.45}$$

The cross section for coherent radiation is determined by the generalization of Eq. (4.37) as

$$\sigma_r^e = 2\pi \frac{c^2}{\omega^2} \frac{g_k}{g_0} \frac{\Gamma^2(\widetilde{\Delta}^2 + \widetilde{\Gamma}^2)}{\left[\widetilde{\Delta}^2 + \widetilde{\Gamma}^2(1 + \widetilde{\chi})\right]^2}. \tag{4.46}$$

When we subtract this from the total absorption cross section (4.43), we find the cross section for incoherent photon scattering,

$$\sigma_r^i = 2\pi \frac{c^2}{\omega^2} \frac{g_k}{g_0} \frac{(\nu/2)(\widetilde{\Delta}^2 + \widetilde{\Gamma}^2) + \chi\Gamma\widetilde{\Gamma}^2}{\left[\widetilde{\Delta}^2 + \widetilde{\Gamma}^2(1 + \widetilde{\chi})\right]^2}\Gamma. \tag{4.47}$$

Let us analyze the expression (4.47). In the weak field case when $\chi \ll 1$, the cross section (4.47) does not vanish but yields the result

$$\sigma_r^i \underset{\chi \ll 1}{=} 2\pi \frac{c^2}{\omega^2} \frac{g_k}{g_0} \frac{\nu\Gamma}{2(\widetilde{\Delta}^2 + \widetilde{\Gamma}^2)}. \tag{4.48}$$

Thus we can conclude that collisions give rise to incoherence of the radiation. In the opposite limiting case where $\chi \gg 1$, the coherent scattering cross section has the vanishing limit

$$\sigma_r^e \underset{\chi \gg 1}{=} 2\pi \frac{c^2}{\omega^2} \frac{g_k}{g_0} \frac{\widetilde{\Delta}^2 + \widetilde{\Gamma}^2}{\widetilde{\Gamma}^2 \chi^2}. \tag{4.49}$$

Hence the total cross section in this case reduces to the incoherent part

$$\sigma_a \underset{\chi \gg 1}{\approx} \sigma_r^i = 2\pi \frac{c^2}{\omega^2} \frac{g_k}{g_0} \frac{1}{\chi}. \tag{4.50}$$

This also tends to zero with increasing intensity, but more slowly than the coherent part.

Incoherent scattering is also dominant when collisional broadening is large compared to spontaneous broadening, that is, when $\nu \gg \Gamma$. In this case we get

$$\sigma_a \approx \sigma_r^i = 2\pi \frac{c^2}{\omega^2} \frac{g_k}{g_0} \frac{\nu\Gamma}{2(\widetilde{\Delta}^2 + \nu^2/4 + \nu\chi\Gamma/2)}. \tag{4.51}$$

The absorption of radiation is an incoherent process in this case, nearly independent of the intensity χ of the incident radiation. It should be noted that, if $\chi = O(1)$, then $\nu^2 \gg \nu\chi\Gamma$ in Eq. (4.51), and the cross section does not depend on χ. Intensity dependence appears only at $\chi \gg 1$, when the cross section begins to decrease.

Problem 4.11. Under the conditions of Problem 4.9, estimate the order of magnitude of the intensity of resonance radiation at which saturation appears, that is, when

$\rho_{kk} \sim 1$. Assume that broadening due to collisions is large as compared to the spontaneous width.

SOLUTION. Saturation corresponds to the case when the population ρ_{kk} of the excited state is of the same order as the population ρ_{00} of the ground state. According to Eq. (4.42) with $\nu \gg \Gamma$, we have

$$\rho_{kk} = \frac{\chi\nu\Gamma}{4\left(\widetilde{\Delta}^2 + \nu^2/4 + \chi\nu\Gamma/2\right)}, \tag{4.52}$$

where we have used the relation $\widetilde{\Gamma} \approx \nu/2$, and consequently $\widetilde{\chi} = \chi(2\Gamma/\nu)$. We can estimate the photon flux to be

$$J = cE^2/(8\pi\hbar\omega)$$

for the case $\rho_{kk} \sim 1$. It follows from Eq. (4.52) that then $2\chi\Gamma \sim \nu^2$, and

$$J_0 = \frac{\nu\omega^2}{4\pi c^2}.$$

We can now rewrite Eq. (4.52) in terms of this flux as

$$\rho_{kk} = \frac{\nu^2(J/J_0)}{8\left[\widetilde{\Delta}^2 + \nu^2(1 + J/J_0)/4\right]}. \tag{4.53}$$

It is seen that saturation is achieved on the edges of the spectral profile at much higher photon fluxes than are required near the center of the line. In fact, we find

$$J \sim \left(\widetilde{\Delta}^2/\nu^2\right) J_0.$$

Earlier we found that the ratio of the spontaneous lifetime to the transition time in the external field is

$$G = (\mathbf{D}_{k0} \cdot \mathbf{E})\, \tau_k \hbar.$$

This parameter can be written in the form

$$G = \frac{1}{2}\sqrt{\frac{J\nu}{J_0\Gamma}}.$$

Since $\nu \gg \Gamma$, we then find that $G \gg 1$ near saturation conditions ($J \sim J_0$). That is, transitions between levels take place due to induced transitions and absorption of photons only.

The rate of induced transitions between levels 0 and k is given by the relation

$$\frac{1}{\tau_a} = \frac{\mathbf{D}_{k0} \cdot \mathbf{E}}{\hbar} = \frac{G}{\tau_k} = \sqrt{\frac{J}{J_0}\Gamma\nu} = \frac{c}{\omega}\sqrt{\frac{2\pi J}{\tau_k}}. \tag{4.54}$$

This expression gives the connection between the quantity τ_a and the spontaneous lifetime of the excited state τ_k, as well as with the photon flux J.

Problem 4.12. Obtain Eqs. (4.42) and (4.43) from solutions of the approximate rate equations that take into account only the variations in populations of atomic levels. Discuss the applicability of these equations to the derivation of the coherent scattering cross section.

SOLUTION. Take w_{k0} to be the transition rate between the states 0 and k due to an external electric field $\mathbf{E}$. Then we designate by $1/\tau_k$ the spontaneous rate for the transition $k \to 0$. The quantity w_{k0} is determined by Eq. (1.7), where the δ function should be replaced by the distribution function a_ω for the emitted photons [see Eq. (2.38)]

$$a_\omega = \frac{1}{\pi} \frac{\widetilde{\Gamma}}{\widetilde{\Delta}^2 + \widetilde{\Gamma}^2}. \tag{4.55}$$

This distribution function takes into account both spontaneous and collision broadening of levels. Hence, according to Eq. (1.17), we have

$$w_{k0} = \frac{\pi}{2\hbar^2} |(\mathbf{D}_{k0} \cdot \mathbf{E})|^2 a_\omega. \tag{4.56}$$

For the sake of simplicity, we set $g_0 = g_k = 1$.

The rate equation for the population of level k is of the obvious form

$$\frac{d\rho_{kk}}{dt} = -\frac{1}{\tau_k}\rho_{kk} + w_{k0}(\rho_{kk} - \rho_{00}). \tag{4.57}$$

From conservation of the number of particles, it follows that

$$\rho_{kk} + \rho_{00} = 1.$$

It should be noted that Eq. (4.55) does not take into account the influence of a strong field on the distribution function a_ω. Indeed, the quantity w_{k0} determines transition rates on short time scales so that an atom cannot transit several times between states 0 and k. However, we are interested in obtaining steady-state solutions of equations over such long time intervals that many transitions between states 0 and k can take place. The steady-state solution of Eq. (4.57) has the simple form

$$\rho_{kk} = \frac{1}{2 + (w_{k0}\tau_k)^{-1}}. \tag{4.58}$$

We have taken into account the conservation of the number of particles. Substituting Eq. (4.56) into (4.58) and again introducing the notation $G = (\mathbf{D}_{k0} \cdot \mathbf{E})\, \tau_k/\hbar$, we find

$$\rho_{kk} = \frac{1}{2 + \left(\widetilde{\Delta}^2 + \widetilde{\Gamma}^2\right)/\left(G^2\widetilde{\Gamma}\Gamma\right)}.$$

This coincides with Eq. (4.42) if the notations $\Gamma = 1/2\tau_k$ and $\widetilde{\chi} = 2G^2\Gamma/\widetilde{\Gamma}$ are introduced again. Hence, Eq. (4.43) based on Eq. (4.42) correctly describes the photon absorption cross section within the framework of the rate equation method.

This means that, for the calculation of the photon absorption cross section, we can use the simple rate equation method instead of the more cumbersome density matrix method. However, the rate equation method does not permit the correct calculation of coherent scattering cross sections for photons since, according to the solution of Problem 4.8, nondiagonal elements of the density matrix ρ_{k0} are required. The rate equation method applies only to diagonal matrix elements of the density matrix.

The solution of Problem 4.10 has shown that coherent scattering is negligibly small when $\nu \gg \Gamma$, and all scattering is inelastic. Thus, we can conclude that rate equations give correct descriptions of physical processes when collision broadening is predominant, that is, when $\nu \gg \Gamma$. We see in the system of equations (4.40) that the condition $\nu \gg \Gamma$ causes nondiagonal matrix elements ρ_{k0} to vanish much more quickly than diagonal matrix elements, since we have the behavior $\rho_{k0} \sim \exp(-\nu t)$ in contrast to $\rho_{kk}, \rho_{00} \sim \exp(-\Gamma t)$.

4.4 RESONANCE TRANSITIONS IN MULTILEVEL ATOMS

The approach of the preceding section is inapplicable if the initial state 0 is not a ground state or if the excited state k can experience spontaneous transitions to other atomic levels. The reason is that the previous approach was based on a conservation law for the number of particles. Here we consider a situation where the spontaneous transition between resonant levels k and 0 is negligibly small. After the external radiation field is turned on, more and more electrons will disappear from the initially presumed two-level system. This changes the nature of the analysis. The following discussion is concerned with this modified physical situation. Problem 4.13 is simplified in that it does not contain depletion. It is the so-called classical Rabi problem.

Problem 4.13. Obtain an expression for the wave function of an atom perturbed by monochromatic resonance radiation. Relaxation of the levels is assumed not to occur.

SOLUTION. We start from Eq. (1.4) for the amplitudes c_0 and c_k for an atomic electron with a lower state 0 and an upper state k available to it. We shall write these equations in resonance approximation, assuming that the resonance detuning $\Delta = \omega_{k0} - \omega$ is small compared to the field frequency ω. We thus obtain

$$
\begin{aligned}
i\hbar\frac{dc_0}{dt} &= -\tfrac{1}{2}\left(\mathbf{D}_{0k}\cdot\mathbf{E}\right)\exp(-i\Delta t)c_k, \\
i\hbar\frac{dc_k}{dt} &= \tfrac{1}{2}\left(\mathbf{D}_{k0}\cdot\mathbf{E}\right)\exp(i\Delta t)c_0.
\end{aligned}
\tag{4.59}
$$

After the substitution

$$\widetilde{c}_0 = \exp(i\Delta t)c_0,$$

system (4.59) reduces to the system of differential equations with constant coefficients

$$i\hbar \frac{dc_k}{dt} = \tfrac{1}{2}\left(\mathbf{D}_{k0} \cdot \mathbf{E}\right) \widetilde{c}_0, \tag{4.60}$$
$$i\hbar \frac{d\widetilde{c}_0}{dt} = -\hbar\Delta \widetilde{c}_0 + \tfrac{1}{2}\left(\mathbf{D}_{0k} \cdot \mathbf{E}\right) c_k.$$

Upon eliminating c_k, we obtain the second-order equation for $\widetilde{c}_0$,

$$\frac{d^2\widetilde{c}_0}{dt^2} - i\Delta \frac{d\widetilde{c}_0}{dt} + \frac{1}{4\hbar^2}\left|\left(\mathbf{D}_{k0} \cdot \mathbf{E}\right)\right|^2 \widetilde{c}_0 = 0. \tag{4.61}$$

It has the two linearly independent solutions

$$\widetilde{c}_0^{\pm} = B^{\pm} \exp\left[i\left(\tfrac{1}{2}\Delta \pm \Omega\right) t\right]. \tag{4.62}$$

We have introduced here the quantity

$$\Omega = \sqrt{\frac{1}{4}\Delta^2 + \frac{1}{4\hbar^2}\left|\left(\mathbf{D}_{k0} \cdot \mathbf{E}\right)\right|^2},$$

which is known as the *Rabi frequency*.

Substituting Eq. (4.62) into the second of the Eqs. (4.60), we obtain results for c_k. The coefficients $B^{\pm}$ are determined by the initial conditions. We suppose that at $t = 0$ an atom is found in the state 0, so that

$$c_0(0) = 1, \qquad c_k(0) = 0.$$

Then the wave function at time t is given as

$$\Psi(t) = \left[\cos\Omega t + (i\Delta/2\Omega)\sin\Omega t\right]\exp\left(-i\Delta t/2 - i\varepsilon_0 t/\hbar\right)\Psi_0(\mathbf{r}) \tag{4.63}$$
$$- i\left[\left(\mathbf{D}_{k0} \cdot \mathbf{E}\right)/2\hbar\Omega\right]\sin\Omega t \exp\left(i\Delta t/2 - i\varepsilon_k t/\hbar\right)\Psi_k(\mathbf{r}).$$

We find that the population of the excited state k is

$$\rho_{kk}(t) = \frac{\left|\left(\mathbf{D}_{k0} \cdot \mathbf{E}\right)\right|^2}{4\hbar^2\Omega^2}\sin^2\Omega t. \tag{4.64}$$

This oscillates monochromatically in time from zero to the maximum value

$$\left(\rho_{kk}\right)_{\max} = \frac{\left|\left(\mathbf{D}_{k0} \cdot \mathbf{E}\right)\right|^2}{4\hbar^2\Omega^2}, \tag{4.65}$$

with the frequency of oscillation 2Ω, that is, with twice the Rabi frequency.

If $\Delta = 0$ (exact resonance), then Eq. (4.64) takes the simpler form

$$\rho_{kk}(t) \underset{\Delta=0}{=} \frac{1}{2}\left\{1 - \cos\left[\frac{1}{\hbar}\left(\mathbf{D}_{k0} \cdot \mathbf{E}\right)\right]\right\}. \tag{4.66}$$

This probability changes periodically from zero to unity and back. Thus the atomic system oscillates between being entirely in state 0 and entirely in state k. Such oscillations of population are explained by the suddenness of the turn on of the perturbation at time $t = 0$. They decrease with a more gradual turn on of the perturbation, and they vanish for a turn on that is adiabatically slow.

Problem 4.14. Under the conditions of Problem 4.13, take account of the broadening of states 0 and k due to transitions to other states. Calculate the photon absorption cross section and the absorption coefficient. Assume that the states have the same lifetime for spontaneous decay.

SOLUTION. We denote by τ the spontaneous lifetimes of each of the states. Then, according to the theory of spontaneous radiation considered in Section 2.1, one should change the energy of each of the states by the replacement

$$\varepsilon_0 \rightarrow \varepsilon_0 - i/(2\tau), \qquad \varepsilon_k \rightarrow \varepsilon_k - i/(2\tau) \tag{4.67}$$

in all formulas of the preceding problem. Since the resonance detuning Δ and Rabi frequency Ω depend only on the difference of energies of the states 0 and k, the modification (4.67) leaves the values of Δ and Ω unaltered.

Thus all of the modification of the wave function (4.63) consists of attaching the factors $\exp(-t/2\tau)$ to each of the functions $\Psi_0(\mathbf{r})$ and $\Psi_k(\mathbf{r})$. Hence, the population of state k at time t is given by the generalization of Eq. (4.64),

$$\rho_{kk}(t) = \frac{|(\mathbf{D}_{k0} \cdot \mathbf{E})|^2}{4\hbar^2\Omega^2} \sin^2 \Omega t \exp\left(-\frac{t}{\tau}\right). \tag{4.68}$$

The rate of decay of this state is given by $\rho_{kk}(t)/\tau$. When we integrate this over the entire time the perturbation is applied, we obtain the total probability of decay of the state k,

$$W_k = \int_0^\infty \rho_{kk}(t)\frac{dt}{\tau} = \frac{1}{2}\frac{|(\mathbf{D}_{k0} \cdot \mathbf{E})|^2/\hbar^2}{\Delta^2 + |(\mathbf{D}_{k0} \cdot \mathbf{E})|^2/\hbar^2 + 1/\tau^2}. \tag{4.69}$$

The quantity W_k is also the photon absorption probability. That is, it is equivalent to the stationary population $\rho_{kk}(\infty)$ of the state k found in Problem 4.7.

We now introduce the notations

$$\Gamma = 1/\tau, \qquad G = (\mathbf{D}_{k0} \cdot \mathbf{E})\,\tau/\hbar, \qquad \chi = G^2. \tag{4.70}$$

With this terminology, Eq. (4.69) becomes

$$W_k = \frac{\chi}{2}\frac{\Gamma^2}{\Delta^2 + \Gamma^2(1+\chi)}. \tag{4.71}$$

Since, at $t = 0$, we have one electron in state 0, then the quantity W_0 can be found from the obvious conservation law

$$W_0 + W_k = 1. \tag{4.72}$$

The parameter χ defined in Eq. (4.70) is called the *saturation parameter*. It gives the population of the excited state k as a function of the intensity of the radiation field.

In a very strong field, when $\chi \gg 1$, the quantities W_k and W_0 achieve their limiting values, which is $\frac{1}{2}$ for both of them. In the opposite case, when $\chi \ll 1$, we obtain the perturbation theory result

$$W_k = \frac{\chi}{2}\frac{\Gamma^2}{\Delta^2 + \Gamma^2}. \tag{4.73}$$

In this case, the photon absorption probability W_k is proportional to the intensity of the radiation field. A continuing increase of the field leads to a slowing of the growth of W_k. Finally, at $\chi \gg 1$ we obtain saturation, that is, $W_k = \frac{1}{2}$.

The close correspondence between Eqs. (4.71) and (4.28) should be noted. The difference is that, in Eq. (4.71), both k and 0 states have spontaneous widths, while in Eq. (4.28) only the excited state k has a spontaneous width.

We shall now calculate the photon absorption cross section. The quantity ΓW_k determines the rate at which state k loses population. In the steady-state case, this quantity is also the photon absorption rate. When we divide it by the flux of incident photons, $J = cE^2/(8\pi\hbar\omega)$, we obtain the photon absorption cross section

$$\sigma_a = \pi\frac{c^2}{\omega^2}\frac{g_k}{g_0}\frac{\Gamma^2}{\Delta^2 + \Gamma^2(1+\chi)}. \tag{4.74}$$

We have introduced in this expression the statistical weights g_0 and g_k of the initial and final states, respectively. The close correspondence of Eqs. (4.74) and (4.31) should be noted.

According to Eqs. (2.87) and (2.88), the absorption coefficient is given by the relation

$$k_\omega = N\sigma_a(W_0 - W_k), \tag{4.75}$$

where N is the density of atoms. Substituting Eq. (4.74) for the absorption cross section and Eq. (4.71) for W_k into Eq. (4.75), and taking account of Eq. (4.72), we find

$$k_\omega = N\frac{\pi c^2}{\omega^2}\frac{g_k}{g_0}\frac{\Gamma^2\left(\Delta^2 + \Gamma^2\right)}{\left[\Delta^2 + \Gamma^2(1+\chi)\right]^2}. \tag{4.76}$$

It is seen from Eq. (4.76) that the absorption cross section diminishes as the field intensity parameter χ increases and tends toward zero for $\chi \gg 1$.

It should be observed that the width Γ in this problem can be produced, for example, by the ionization of the excited state by the given radiation field (or other radiation fields), instead of arising from spontaneous relaxation of the state. In this case, the solution of this problem includes the solution of the problem of resonance two-photon ionization of an atom in a strong field.

Problem 4.15. Solve the preceding problem when states 0 and k have different spontaneous lifetimes. Determine how the population of the excited state varies with time.

SOLUTION. We refer again to the wave function of Eq. (2.63) and make the change

$$\varepsilon_0 \to \varepsilon_0 - i/(2\tau_0), \qquad \varepsilon_k \to \varepsilon_k - i/(2\tau_k). \tag{4.77}$$

With this modification, the population (4.64) of the state k takes the form

$$\rho_{kk}(t) = \frac{|(\mathbf{D}_{k0}\cdot\mathbf{E})|^2}{4\hbar^2\widetilde{\Omega}^2}\sin^2\left(\widetilde{\Omega}t\right)\exp\left(-\frac{t}{\tau}\right). \tag{4.78}$$

We have introduced here the notation

$$\widetilde{\Omega} = \frac{1}{2}\sqrt{\left(\Delta - \frac{i}{2\widetilde{\tau}}\right)^2 + \frac{1}{\hbar^2}|(\mathbf{D}_{k0}\cdot\mathbf{E})|^2}, \tag{4.79}$$

where $\widetilde{\tau}$ is given by

$$\widetilde{\tau}^{-1} = \tau_k^{-1} - \tau_0^{-1}.$$

In addition, the quantity τ in Eq. (4.78) is now determined by the relation

$$\tau^{-1} = \left(\tau_k^{-1} + \tau_0^{-1}\right)/2. \tag{4.80}$$

The probability for photon absorption through the action of the external field, integrated over all time, is

$$W_k = \frac{1}{\tau_k}\int_0^\infty \rho_{kk}(t)\,dt. \tag{4.81}$$

Substituting Eq. (4.78) into Eq. (4.81), and carrying out an elementary but cumbersome integration, we find

$$W_k = \frac{\chi}{2\tau_k}\frac{\Gamma}{\Delta^2 + \Gamma^2(1+\chi)}. \tag{4.82}$$

The terminology we employ here has the meaning

$$\chi = \frac{\tau_0\tau_k}{\hbar^2}|(\mathbf{D}_{k0}\cdot\mathbf{E})|^2, \qquad \Gamma = \frac{1}{\tau} = \frac{1}{2}\left(\frac{1}{\tau_k} + \frac{1}{\tau_0}\right). \tag{4.83}$$

If the two spontaneous lifetimes are the same, then Eq. (4.82) coincides, as it should, with Eq. (4.71).

When $\chi \ll 1$, Eq. (4.82) gives the perturbation theory result

$$W_k \underset{\chi\ll 1}{=} \frac{\chi}{2\tau_k}\frac{\Gamma}{\Delta^2 + \Gamma^2}. \tag{4.84}$$

As χ increases, the growth of W_k slows, and when $\chi \gg 1$, we obtain the saturated probability

$$W_k = \frac{1}{1 + \tau_k/\tau_0}. \tag{4.85}$$

Therefore we can call the parameter χ of Eq. (4.83) the saturation parameter. The result (4.85) can be obtained more directly if we make use of the facts that, at $\chi \gg 1$, the ratio W_k/W_0 is simply τ_0/τ_k, and that $W_k + W_0 = 1$ according to Eq. (4.72).

It follows from Eq. (4.83) that saturation can be approached not only in a very strong field but even in weak fields if the inequalities $\tau_k \gg \tau_0$ or $\tau_k \ll \tau_0$ are satisfied, that is, if one of the spontaneous widths is much less than the other.

We now derive the absorption cross section. By proceeding in a fashion parallel to the derivations in the preceding problem, we obtain

$$\sigma_a = \pi \frac{c^2}{\omega^2} \frac{g_k}{g_0} \frac{\tau_k}{\tau_{k0}} \frac{\Gamma^2}{\Delta^2 + \Gamma^2(1 + \chi)}. \tag{4.86}$$

Here the quantity τ_{k0} refers to the partial lifetime of state k with respect to state 0.

In conclusion, we wish to establish the time dependence of $\rho_{kk}(t)$, the population of the state k. For simplicity, we restrict ourselves to the case of zero resonance detuning, $\Delta = 0$. It follows from Eq. (4.78) that for moderate fields, where Eq. (4.79) shows that

$$\frac{1}{\hbar} |(\mathbf{D}_{k0} \cdot \mathbf{E})| < \frac{1}{2\widetilde{\tau}}, \tag{4.87}$$

then the quantity $\widetilde{\Omega}$ takes on imaginary values. The population $\rho_{kk}(t)$ is then of the form

$$\rho_{kk}(t) \sim \sinh^2\left(\left|\widetilde{\Omega}\right|t\right) \exp(-t/\tau), \tag{4.88}$$

that is, the population of state k decreases aperiodically. If, on the other hand, the inequality is reversed, so that

$$\frac{1}{\hbar} |(\mathbf{D}_{k0} \cdot \mathbf{E})| > \frac{1}{2\widetilde{\tau}},$$

then we have from Eq. (4.78)

$$\rho_{kk}(t) \sim \sin^2\left(\left|\widetilde{\Omega}\right|t\right) \exp(-t/\tau),$$

and the population of the state oscillates with a decaying amplitude. Finally, in the intermediate case

$$\frac{1}{\hbar} |(\mathbf{D}_{k0} \cdot \mathbf{E})| = \frac{1}{2\widetilde{\tau}},$$

the population of the state behaves as

$$\rho_{kk}(t) \sim t^2 \exp(-t/\tau).$$

We have emphasized in the foregoing problem that the results are applicable when the broadening of the states is produced by mechanisms other than spontaneous relaxation. In particular, they can be used for the solution of the problem of resonance ionization of an atom in a strong field. Then the quantity τ_k^{-1} will be the ionization rate of the excited resonance state k.

As in the preceding problem, we can also derive the absorption coefficient.

5

ATOMS IN STRONG FIELDS

The physical phenomena that can appear under strong-field circumstances are often qualitatively different from those familiar in the perturbative domain. This subject has recently received so much theoretical and experimental attention that it has now become a recognized independent area of research within the larger field of atomic physics. Although this chapter is not intended to present a complete treatment of strong-field atomic physics, a modern discussion of radiative processes in atoms would not be complete without an introduction to intense-field phenomena.

When the interaction energy of an atom with an externally applied field is sufficiently large, then a perturbation expansion to describe this interaction will not be convergent. It is thus necessary to seek some method other than perturbation theory to describe the behavior of the atom in the field. Some problems employing nonperturbative methods have appeared earlier in this book, but the great majority of the problems are solved using perturbation theory, as has been true in atomic physics until very recently. We shall give here a brief survey of the physical environments that demand intense-field treatment and of some of the methods adapted to the strong-field regime in atomic physics. The emphasis here will be on methods that attain an analytical simplicity through the restriction that field strengths are well beyond the perturbation theory limits. This is in contrast to perturbation theory, which achieves simplicity through the assumption of small field intensity. The intensity domain that marks the transition between perturbation methods and strong-field methods is characterized by considerable analytical complexity, and will not be treated here.

5.1 PROPERTIES USEFUL FOR STRONG-FIELD METHODS

We shall begin this chapter with some basic results that will set the stage for other problems exploring the methods of strong-field physics and their physical consequences. The problems in this section have universal applicability, but they are especially relevant to the strong-field domain.

Problem 5.1. Find general criteria to judge when the limits of perturbation theory will be exceeded.

SOLUTION. Several criteria for the applicability of perturbation theory can be stated on very general grounds. The violation of any one of these restrictions is adequate grounds for the conclusion that perturbation theory gives incorrect predictions for the strong-field environment.

A perturbation theory of atomic phenomena is normally couched in terms of an interaction Hamiltonian H_I, whose magnitude is small as compared to that Hamiltonian H_0 that describes the unperturbed system. In the case of an atom subjected to a strong field, H_0 includes the effects of the atomic binding potential, and H_I is the additional part of the total Hamiltonian arising from the externally applied electromagnetic field. To associate a magnitude with the ratio $|H_I|/|H_0|$, we may presume that the transition caused by H_I is ionization of an atom from its ground state, so that the transition energy is of the order of E_i, the ionization potential of the atom. We take the interaction Hamiltonian to be of the form

$$H_I = -\frac{e}{mc}\mathbf{A}\cdot\mathbf{p} + \frac{e^2}{2mc^2}\mathbf{A}^2, \tag{5.1}$$

where $\mathbf{A}$ is the vector potential of the externally applied field (which we shall, for convenience, often refer to as the "laser field"). To assign a magnitude to the first term in Eq. (5.1), we replace the $\mathbf{p}$ operator by the commutator

$$\frac{\mathbf{p}}{m} = \frac{i}{\hbar}\left[H_0, \mathbf{r}\right], \tag{5.2}$$

which will be valid when H_0 contains the kinetic energy operator and potentials that depend only on spatial coordinates. The magnitude of the $\mathbf{A}\cdot\mathbf{p}$ term in Eq. (5.1) is then $eA_0E_iR_0/(\hbar c)$, where A_0 is the amplitude of $\mathbf{A}$, and R_0 is a radius typical of the bound system. Furthermore, if we take the magnitude of H_0 to be E_i, and replace the radius R_0 by the general expression $R_0 \sim \hbar/(mE_i)^{1/2}$, then we find the ratio of the $\mathbf{A}\cdot\mathbf{p}$ term to the magnitude of H_0 to be

$$\frac{1}{|H_0|}\left|\frac{e}{mc}\mathbf{A}\cdot\mathbf{p}\right| \sim \frac{eA_0R_0}{\hbar c} = \frac{eE_0R_0}{\hbar\omega}, \tag{5.3}$$

where the last result in this equation follows from the replacement of the vector potential A_0 of the laser field of frequency ω by E_0c/ω, where E_0 is the amplitude of the electric field. The quantity in Eq. (5.3) appears in the literature in two different guises. The inverse of this quantity is widely known as the Keldysh parameter γ,

where

$$\gamma \equiv \frac{\hbar\omega}{eE_0R_0}. \tag{5.4}$$

The Keldysh parameter is also known as the *adiabaticity parameter*. If the quantity in Eq. (5.3) is squared, and the generic replacement is made that $R_0^2 = \hbar^2/(mE_i)$, then we obtain the intensity parameter often designated z_1, where

$$z_1 \equiv \frac{2U_p}{E_i}, \tag{5.5}$$

and U_p is the ponderomotive potential

$$U_p \equiv \frac{e^2}{2m\omega^2}\left\langle \mathbf{E}^2 \right\rangle. \tag{5.6}$$

The angle bracket in Eq. (5.6) refers to a time average over a period of the applied laser field. The name *bound-state intensity parameter* is sometimes applied to z_1, since it relates an energy characterizing the electromagnetic field to an energy characterizing a bound state.

Other implications of the significance of z_1 come from a consideration of the "power broadening" of atomic energy levels, that is, the broadening due to the shortened lifetime of a level in the presence of a strong field. This broadening is approximately $|e\mathbf{E}\cdot\mathbf{r}| \approx eEa_0$, where a_0 is the Bohr radius. When this broadening approaches the energy of a single photon of the field, it then becomes indefinite as to the number of photons required to connect that level to others. This is nonperturbative behavior, and so perturbation theory requires $eEa_0 < \hbar\omega$. If we square the ratio of eEa_0 to $\hbar\omega$, and replace a_0^2 by $\hbar^2/(2mR_\infty)$, (where R_∞ is the Rydberg constant, equal to the binding energy of hydrogen), we find that we have $(eEa_0/\hbar\omega)^2 \approx z_1 < 1$.

Both from the need to have the interaction Hamiltonian of smaller magnitude than the unperturbed Hamiltonian, and to have the power broadening of atomic energy levels smaller than a photon energy, we conclude that a necessary (but not sufficient) condition for perturbation theory to be valid is that

$$\gamma > 1 \qquad \text{or} \qquad z_1 < 1. \tag{5.7}$$

The ponderomotive potential of Eq. (5.6) is a quantity that appears frequently in strong-field atomic problems. It can be viewed as the minimum kinetic energy a free electron possesses as a result of its oscillations in a laser field, or equivalently, as the potential energy of a free electron due to its interaction with the field.

Another intensity parameter that measures the need for intense-field treatment is

$$z \equiv \frac{U_p}{\hbar\omega} = \frac{e^2}{2m\hbar\omega^3}\left\langle \mathbf{E}^2 \right\rangle. \tag{5.8}$$

The transition of an ionized electron to a detached state in which the motion of the electron in the laser field is fully developed (as opposed to the case in which the electron emerges into a simple continuum Coulomb state) is an explicitly nonperturbative effect. That is, no small number of photons can place an ionized electron into the state of oscillatory motion characteristic of classical electron motion in the external field when the energy of that motion is of the order of the photon energy. Perturbative treatment of ionization thus requires the condition

$$z < 1. \tag{5.9}$$

The quantity z has been called the *continuum-state intensity parameter* to contrast it with the bound-state intensity parameter z_1. It has also been called the *nonperturbative intensity parameter*, since, as will be seen shortly, it is usually the factor that is critical in determining the limits of perturbation theory.

Yet another limitation on the application of perturbation theory comes from the requirement that the internal electric field in the atom should exceed the externally applied electric field. That is, in atomic units, the external electric field should be less than unity. In Gaussian units, this means that $E_0 < 2R_\infty/(ea_0)$, where R_∞ is the Rydberg unit of energy, and a_0 is the Bohr radius. This yields the upper limit of perturbation theory as

$$E_0 < \frac{m^2e^5}{\hbar^4} \quad \text{or} \quad U_p < \frac{e^2}{2m\omega^2}\left(\frac{m^2e^5}{\hbar^4}\right)^2, \tag{5.10}$$

where the value of E_0 is interpreted as a root mean square for purposes of the limit on the ponderomotive potential. Expressed in atomic units $\hbar = m = e^2 = 1$, Eq. (5.10) becomes $E_0 < 1$.

The upper bounds on field intensity in Eqs. (5.7), (5.9), and (5.10) are all necessary, but not sufficient, conditions for the convergence of perturbation theory. At the very least, all of these conditions must be satisfied simultaneously. Each of these conditions has a field frequency dependence different from the others. The resulting limitation as a function of frequency is illustrated in Figure 5.1. The range of laser frequencies

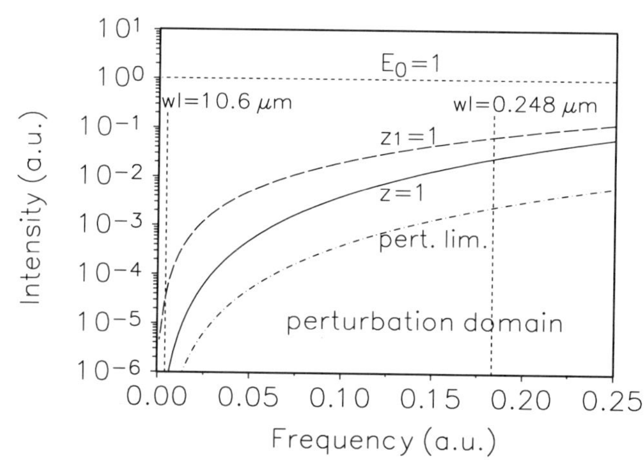

Figure 5.1. Limits on perturbation theory as a function of frequency. The vertical lines corresponding to wavelengths of 10.6 and 0.248 μm encompass the full range of frequencies for which intense-field lasers are available. Perturbation theory is valid below the curve at $z = 0.1$, labeled "Pert. lim."

for which nonperturbative intense fields can be attained extends from about 1 μm to 248 nm (excimer laser). As shown in the figure, it is the intensity parameter z that measures the effective limit of perturbation theory. This conclusion remains true even if the wavelength is extended to 10 μm, representing the CO_2 laser.

Problem 5.2. Suppose that a particle of mass m_1 and charge q_1 is bound by a central potential to a particle of mass m_2 and charge q_2. Show how the two-particle Schrödinger equation separates into a center-of-mass equation and a relative-coordinate equation when an external electromagnetic field is present. This field is described by a vector potential $\mathbf{A}(t)$ and/or a scalar potential of the form $-q\mathbf{r}\cdot\mathbf{E}(t)$, where $\mathbf{E}(t)$ is the electric field vector.

SOLUTION. The procedure we follow is the same as that normally employed to show how the two-particle Schrödinger equation separates, except that we include the presence of external fields. This will lead to the appearance of a reduced charge in addition to the familiar reduced mass.

The new coordinates $\mathbf{R}$ and $\mathbf{r}$ for the location of the center of mass and for the relative coordinate are given in terms of the coordinates of the two separate particles by

$$\mathbf{R} = \frac{m_1\mathbf{r}_1 + m_2\mathbf{r}_2}{m_1 + m_2}, \qquad \mathbf{r} = \mathbf{r}_1 - \mathbf{r}_2, \tag{5.11}$$

with inverse transformations

$$\mathbf{r}_1 = \mathbf{R} + \frac{m_2}{m_1 + m_2}\mathbf{r}, \qquad \mathbf{r}_2 = \mathbf{R} - \frac{m_1}{m_1 + m_2}\mathbf{r}. \tag{5.12}$$

It is useful to define symbols for the total mass m_t and total charge q_t,

$$m_t = m_1 + m_2, \qquad q_t = q_1 + q_2. \tag{5.13}$$

The gradient operators in the $\mathbf{R}$, $\mathbf{r}$ coordinates are related to those in the $\mathbf{r}_1$, $\mathbf{r}_2$ coordinates by

$$\boldsymbol{\nabla}_1 = \boldsymbol{\nabla}_r + \frac{m_1}{m_t}\boldsymbol{\nabla}_R, \qquad \boldsymbol{\nabla}_2 = -\boldsymbol{\nabla}_r + \frac{m_2}{m_t}\boldsymbol{\nabla}_R. \tag{5.14}$$

Direct substitution of the transformation Eqs. (5.11)–(5.14) into the two-body Schrödinger equation

$$\begin{aligned} i\hbar\frac{\partial\psi(\mathbf{r}_1,\mathbf{r}_2,t)}{\partial t} = &\left\{\frac{1}{2m_1}\left[-i\hbar\boldsymbol{\nabla}_1 - \frac{q_1}{c}\mathbf{A}(t)\right]^2 - q_1\mathbf{r}_1\cdot\mathbf{E}(t) + V\left(|\mathbf{r}_1 - \mathbf{r}_2|\right)\right. \\ &\left.+ \frac{1}{2m_2}\left[-i\hbar\boldsymbol{\nabla}_2 - \frac{q_2}{c}\mathbf{A}(t)\right]^2 - q_2\mathbf{r}_2\cdot\mathbf{E}(t)\right\}\partial\psi(\mathbf{r}_1,\mathbf{r}_2,t) \end{aligned} \tag{5.15}$$

leads to the new Schrödinger equation

$$i\hbar\frac{\partial\psi(\mathbf{R},\mathbf{r},t)}{\partial t} = \left\{\frac{1}{2m_t}\left[-i\hbar\boldsymbol{\nabla}_R - \frac{q_t}{c}\mathbf{A}(t)\right]^2 - q_t\mathbf{R}\cdot\mathbf{E}(t) + \frac{1}{2m_r}\left[-i\hbar\boldsymbol{\nabla}_r - \frac{q_r}{c}\mathbf{A}(t)\right]^2 - q_r\mathbf{r}\cdot\mathbf{E}(t) + V(r)\right\}\partial\psi(\mathbf{R},\mathbf{r},t). \tag{5.16}$$

In addition to the total mass m_t and the total charge q_t, Eq. (5.16) now contains also the reduced mass and reduced charge

$$m_r = \frac{m_1 m_2}{m_t}, \qquad q_r = \frac{q_1 m_2 - q_2 m_1}{m_t}. \tag{5.17}$$

Equation (5.16) is plainly separable. We introduce the product solution

$$\psi(\mathbf{R},\mathbf{r},t) = \psi_R(\mathbf{R},t)\psi_r(\mathbf{r},t) \tag{5.18}$$

into Eq. (5.16), and divide by ψ in the form of Eq. (5.18) to obtain

$$\frac{1}{\psi_R}\left\{i\hbar\frac{\partial\psi_R}{\partial t} - \left[\frac{1}{2m_t}\left(-i\hbar\boldsymbol{\nabla}_R - \frac{q_t}{c}\mathbf{A}\right)^2 - q_t\mathbf{R}\cdot\mathbf{E}\right]\psi_R\right\} = -\frac{1}{\psi_r}\left\{i\hbar\frac{\partial\psi_r}{\partial t} - \left[\frac{1}{2m_r}\left(-i\hbar\boldsymbol{\nabla}_r - \frac{q_r}{c}\mathbf{A}\right)^2 - q_r\mathbf{r}\cdot\mathbf{E}+V(r)\right]\psi_r\right\}. \tag{5.19}$$

The left side of this equation is a function of $\mathbf{R}$, t only. The right side is a function of $\mathbf{r}$, t only. Hence each side of Eq. (5.19) must be independent of either $\mathbf{R}$ or $\mathbf{r}$ and can be set equal to some arbitrary function of time, $f(t)$. Our final result is then the two separated equations:

$$i\hbar\frac{\partial\psi_R}{\partial t} = \left[\frac{1}{2m_t}\left(-i\hbar\boldsymbol{\nabla}_R - \frac{q_t}{c}\mathbf{A}\right)^2 - q_t\mathbf{R}\cdot\mathbf{E} + f(t)\right]\psi_R \tag{5.20}$$

for the center-of-mass equation and

$$i\hbar\frac{\partial\psi_r}{\partial t} = \left[\frac{1}{2m_r}\left(-i\hbar\boldsymbol{\nabla}_r - \frac{q_r}{c}\mathbf{A}\right)^2 - q_r\mathbf{r}\cdot\mathbf{E}+V(r) - f(t)\right]\psi_r \tag{5.21}$$

for the relative-coordinate equation.

These equations have a number of interesting features. The c.m. (center-of-mass) equation depends entirely on the total mass and charge, as one would expect. If we are concerned with a system such as a neutral atom, we have $q_t = 0$, and Eq. (5.20) reduces to the free-particle equation

$$i\hbar\frac{\partial\psi_R}{\partial t} = \left[\frac{1}{2m_t}(-i\hbar\boldsymbol{\nabla}_R)^2 + f(t)\right]\psi_R. \tag{5.22}$$

For the neutral atom considered as a two-body system, we can set q_1 equal to the electron charge, $q_1 = -e$, and so the nuclear charge is $q_2 = e$. Equation (5.17) for the reduced charge then gives identically $q_r = -e$. If, however, we consider a positive ion like He^+, the single electron again has $q_1 = -e$, but the nuclear charge is $q_2 = 2e$, so the reduced charge is

$$q_r = -e\left(\frac{2m_1 + m_2}{m_1 + m_2}\right), \tag{5.23}$$

and the electron in the ion has a charge slightly modified from the usual $-e$. For a negative ion such as H^-, considering the two-body problem to consist of a neutral core atom and a single electron, the reduced charge will be

$$q_r = -e\left(\frac{m_1}{m_1 + m_2}\right). \tag{5.24}$$

Ions will thus exhibit reduced charge effects in addition to reduced mass effects. There will be, for example, an isotopic effect on the apparent electron charge as well as on the apparent electron mass.

The arbitrary function $f(t)$ in Eqs. (5.20) and (5.21) may be viewed as the generating function for a gauge change corresponding to an alteration of the scalar potential. A change in electromagnetic gauge accomplished by the generating function $\chi(t)$ will not affect the vector potential but will change the scalar potential ϕ according to the prescription

$$\phi \to \phi' = \phi - \frac{1}{c}\frac{\partial}{\partial t}\chi(t). \tag{5.25}$$

When changing gauge, the wave function has its phase altered such that

$$\psi' = \exp\left[i\chi(t)/\hbar\right]\psi.$$

If we now associate $\chi(t)$ with the $f(t)$ in Eqs. (5.20) and (5.21), we can set

$$\psi_R = \exp\left[if(t)/\hbar\right]\psi_R', \qquad \psi_r = \exp\left[-if(t)/\hbar\right]\psi_r', \tag{5.26}$$

so that $\psi_R\psi_r = \psi_R'\psi_r'$. Equations (5.20) and (5.21) then become

$$i\hbar\frac{\partial\psi_R'}{\partial t} = \left[\frac{1}{2m_t}\left(-i\hbar\nabla_R - \frac{q_t}{c}\mathbf{A}\right)^2 - q_t\mathbf{R}\cdot\mathbf{E}\right]\psi_R', \tag{5.27}$$

and

$$i\hbar\frac{\partial\psi_r'}{\partial t} = \left[\frac{1}{2m_r}\left(-i\hbar\nabla_r - \frac{q_r}{c}\mathbf{A}\right)^2 - q_r\mathbf{r}\cdot\mathbf{E} + V(r)\right]\psi_r'. \tag{5.28}$$

Problem 5.3. Repeat Problem 5.2 when the field is described by a vector potential only, and this potential has the specific form of a plane wave, where $\mathbf{A} = \mathbf{a}\cos(\omega t - \mathbf{k}\cdot\mathbf{r})$. That is, the dipole approximation is not used.

SOLUTION. It will be found that the two-particle Schrödinger equation does not separate into uncoupled c.m. and relative-coordinate equations, unlike Problem 5.2.

It will be instructive to examine the nature and extent of the approximation made in presuming that separation does occur.

The Schrödinger equation for the two particles labeled by subscripts 1 and 2 is

$$i\hbar\frac{\partial\psi(\mathbf{r}_1,\mathbf{r}_2,t)}{\partial t} = \left\{\frac{1}{2m_1}\left[-i\hbar\boldsymbol{\nabla}_1 - \frac{q_1}{c}\mathbf{A}(\mathbf{r}_1,t)\right]^2 + V(|\mathbf{r}_1-\mathbf{r}_2|)\right.$$
$$\left. + \frac{1}{2m_2}\left[-i\hbar\boldsymbol{\nabla}_2 - \frac{q_2}{c}\mathbf{A}(\mathbf{r}_2.t)\right]^2\right\}\partial\psi(\mathbf{r}_1,\mathbf{r}_2,t), \quad (5.29)$$

which differs from Eq. (5.15) in that the Coulomb gauge is specified (so there is no scalar potential associated with the laser field), and there is no longer a dipole approximation employed for the vector potential of the electromagnetic field. Some assumptions must be introduced in Eq. (5.29) in order to proceed further.

The vector potential will be taken to have the explicit form

$$\mathbf{A} = \mathbf{a}\cos(\omega t - \mathbf{k}\cdot\mathbf{r}),$$

and it will be assumed that $|\mathbf{k}\cdot\mathbf{r}| \ll 1$ and $|\mathbf{k}\cdot\mathbf{R}| \approx 0$. The change of variables expressed in Eqs. (5.11)–(5.14) puts Eq. (5.29) in the form

$$i\hbar\frac{\partial\psi(\mathbf{R},\mathbf{r},t)}{\partial t} = \left\{\frac{1}{2m_t}\left[-i\hbar\boldsymbol{\nabla}_{\mathbf{R}} - q_t\frac{\mathbf{a}}{c}\cos\omega t - q_r\frac{\mathbf{a}}{c}\mathbf{k}\cdot\mathbf{r}\sin\omega t\right]^2\right.$$
$$+ \frac{1}{2m_r}\left[-i\hbar\boldsymbol{\nabla}_{\mathbf{r}} - q_r\frac{\mathbf{a}}{c}\cos\omega t - q_e\frac{\mathbf{a}}{c}\mathbf{k}\cdot\mathbf{r}\sin\omega t\right]^2$$
$$\left. + V(r)\right\}\psi(\mathbf{R},\mathbf{r},t). \quad (5.30)$$

In addition to the total mass and charge of Eq. (5.23) and the reduced mass and charge of Eq. (5.27), there is now also an effective charge q_e given by

$$q_e = \frac{1}{m_t^2}\left(q_1m_2^2 + q_2m_1^2\right). \quad (5.31)$$

It is plain that Eq. (5.30) cannot be separated. The term $q_r\left(\mathbf{a}/c\right)(\mathbf{k}\cdot\mathbf{r})\sin\omega t$ contained in the same square bracket as $-i\hbar\boldsymbol{\nabla}_{\mathbf{R}}$ makes this impossible. We can estimate the importance of this term by comparing its energy to that associated with the atom, which we take to be the atomic unit of energy, which is $2R_\infty$. The notation is that R_∞ is the Rydberg unit, which is approximately 13.6 eV. That is, we wish to estimate the dimensionless ratio

$$\mathcal{R} = \frac{1}{2R_\infty}\frac{1}{2m_t}\left|q_r\frac{\mathbf{a}}{c}\mathbf{k}\cdot\mathbf{r}\right|^2.$$

Since $|\mathbf{k}| = \omega/c$, and $|\mathbf{r}| \approx a_0 =$ *Bohr radius*, we find that

$$\mathcal{R} \sim \frac{1}{2R_\infty}\frac{1}{2m_tc^2}(q_rEa_0)^2, \quad (5.32)$$

where $E = a\omega/c$ is the amplitude of the electric field associated with the vector potential of amplitude a. From Eq. (5.32), we see that the coupling between c.m. and relative motions in the two-body system is proportional to the intensity of the laser field. If we suppose that the laser is focused to an energy flux of 3.5×10^{16} W/cm^2, this corresponds to the applied electric field equal to the internal Coulomb field in a hydrogen atom. Equation (5.32) is then $\mathcal{R} \sim R_\infty/m_t c^2$. For a hydrogen atom, this is $\mathcal{R} \sim 10^{-8}$, and can be discarded as negligible. However, laser energy fluxes of as much as 3×10^{21} W/cm^2 are anticipated for the near future, which would make $\mathcal{R} \sim 10^{-3}$, which is still small but might be observable. With laser energy fluxes this high, the present nonrelativistic treatment is inadequate. Nevertheless, one would still expect this coupling of c.m. and relative coordinates to occur at these extreme laser intensities.

Finally, we note the term $q_e\left(\mathbf{a}/c\right)(\mathbf{k}\cdot\mathbf{r})\sin\omega t$ contained in the same square bracket as $-i\hbar\nabla_r$ in Eq. (5.30). This is of the form to be expected as a first correction to the leading dipole approximation term. The only unusual feature is that it occurs with the effective charge q_e rather than the reduced charge q_r. As long as one of the masses in the two-body system is of the order of a nuclear mass, with the other being the electron mass, then $q_e \approx q_r \approx -e$, and the difference in charge is unimportant.

Problem 5.4. Consider an atom subjected to an interaction Hamiltonian that is periodic in time. That is, the total Hamiltonian is of the form $H(\mathbf{r},t) = H_0(\mathbf{r}) + V(\mathbf{r},t)$, and V is periodic with period T, so that $V(\mathbf{r},t+T) = V(\mathbf{r},t)$. Show that the complete set of solutions of the Schrödinger equation with such a Hamiltonian may be divided into subsets of solutions, each of which is physically equivalent to all the others. This is called the *Floquet property*.

SOLUTION. The Schrödinger equation is

$$i\hbar\frac{\partial}{\partial t}\Psi(\mathbf{r},t) = H(\mathbf{r},t)\Psi(\mathbf{r},t) = \left[H_0(\mathbf{r}) + V(\mathbf{r},t)\right]\Psi(\mathbf{r},t). \tag{5.33}$$

From general symmetry principles, if H is a periodic function of time,

$$H(t+T) = H(t), \tag{5.34}$$

one should be able to select a set of solutions of the Schrödinger equation that are also periodic, with the same period T. We associate a frequency ω with this periodicity in the usual fashion, with $\omega = 2\pi/T$. Periodicity demands only that the value of the wave function be repeated after a period to within an arbitrary constant phase, so we require

$$\Psi_\epsilon(\mathbf{r},t+T) = \exp\left(-i\epsilon T/\hbar\right)\Psi_\epsilon(\mathbf{r},t), \tag{5.35}$$

where ϵ is a real parameter defined so as to have units of energy. It is called the quasi energy.

We can introduce functions $\Phi(\mathbf{r}, t)$, which are exactly periodic, by setting

$$\Psi_\epsilon(\mathbf{r}, t) = \exp\left(-i\epsilon t/\hbar\right)\Phi_\epsilon(\mathbf{r}, t), \qquad \Phi_\epsilon(\mathbf{r}, t+T) = \Phi_\epsilon(\mathbf{r}, t). \tag{5.36}$$

Suppose now that we have states of different quasi energies ϵ_1 and ϵ_2. Hermiticity of the Hamiltonian requires that the inner product of any two states $\left(\Psi_{\epsilon_1}, \Psi_{\epsilon_2}\right)$ be independent of time. Therefore, if $\epsilon_2 - \epsilon_1 \neq n\hbar\omega$, where n is an integer, then the states must be orthogonal at any given time, $(\Psi_{\epsilon_1}(t), \Psi_{\epsilon_2}(t)) = 0$. We also conclude that the quasi energy is defined only up to an additive constant $n\hbar\omega$.

The conclusion from the above is that states $\Phi_{\epsilon_1}(t)$ and $\Phi_{\epsilon_2}(t)$, which differ only in that $\epsilon_2 = \epsilon_1 \pm n\hbar\omega$, are physically indistinguishable. Nevertheless, they represent different states in the complete set of $\Psi_\epsilon(t)$ states.

In particular, in an external monochromatic electromagnetic field, the wave function can be represented in the form

$$\Psi = \exp\left(-i\frac{Et}{\hbar}\right)\sum_n \exp(-in\omega t)\,\phi_n. \tag{5.37}$$

Determination of the expansion coefficients ϕ_n constitutes the content of the Floquet method for the solution of strong-field problems in atomic physics.

Problem 5.5. Consider a system bound by a time-independent potential and subjected to a laser field that may be treated in the dipole approximation. Find a unitary transformation to a moving coordinate system such that the transformation removes the $\mathbf{A}\cdot\mathbf{p}$ and $\mathbf{A}^2$ terms from the Schrödinger equation. The electromagnetic field after the transformation will be manifested as a field-induced modification in the central potential term.

SOLUTION. The Schrödinger equation is

$$\begin{aligned} i\hbar\frac{\partial}{\partial t}\Psi(\mathbf{r}, t) &= \left\{\frac{1}{2m}\left[-i\hbar\nabla - \frac{e}{c}\mathbf{A}(t)\right]^2 + V(\mathbf{r})\right\}\Psi(\mathbf{r}, t) \\ &= \left[H_0 + H_I(t)\right]\Psi(\mathbf{r}, t), \end{aligned} \tag{5.38}$$

where the unperturbed Hamiltonian H_0 and the interaction Hamiltonian H_I are given by

$$H_0 = \frac{1}{2m}(-i\hbar\nabla)^2 + V(\mathbf{r}), \tag{5.39}$$

$$H_I(t) = -\frac{e}{c}\mathbf{A}(t)\cdot\frac{-i\hbar\nabla}{m} + \frac{e^2\mathbf{A}^2}{2mc^2}. \tag{5.40}$$

Because the vector potential $\mathbf{A}$ is a function only of the time, the term $e^2\mathbf{A}^2/\left(2mc^2\right)$ can be removed by an energy shift operator we shall designate by Ω_E. The other term in H_I contains the momentum operator $-i\hbar\nabla$ multiplied by a quantity that is a function of time only. This suggests removal of this term by the translation operator $\Omega_{\mathbf{r}}$. That is, we set

$$\Psi = \Omega_E \Omega_{\mathbf{r}} \Phi,$$
$$\Omega_E = \exp\left[-\frac{i}{\hbar}\frac{e^2}{2mc^2}\int_{-\infty}^{t} \mathbf{A}^2(\tau)\,d\tau\right], \tag{5.41}$$
$$\Omega_{\mathbf{r}} = \exp\left[\frac{i}{\hbar}(-i\hbar\boldsymbol{\nabla})\cdot\frac{e}{mc}\int_{-\infty}^{t} \mathbf{A}(\tau)\,d\tau\right].$$

Since Ω_E commutes with all the terms on the right-hand side of Eq. (5.38), substitution of Eq. (5.41) into Eq. (5.38) yields the equation for Φ,

$$i\hbar\frac{\partial}{\partial t}\Phi(\mathbf{r},t) = \frac{1}{2m}(-i\hbar\boldsymbol{\nabla})^2\Phi(\mathbf{r},t) + \Omega_{\mathbf{r}}^{-1}V(\mathbf{r})\,\Omega_{\mathbf{r}}\Phi(\mathbf{r},t). \tag{5.42}$$

To simplify Eq. (5.42), we define

$$\boldsymbol{\alpha} \equiv -\frac{e}{mc}\int_{-\infty}^{t}\mathbf{A}(\tau)\,d\tau, \tag{5.43}$$

so that

$$\Omega_{\mathbf{r}} = \exp(-\boldsymbol{\alpha}\cdot\boldsymbol{\nabla}). \tag{5.44}$$

We now employ the theorem, often called the Baker–Hausdorff theorem, that

$$e^B C e^{-B} = C + [B, C] + \frac{1}{2!}\big[B, [B, C]\big] + \frac{1}{3!}\big[B, [B, [B, C]]\big] + \ldots, \tag{5.45}$$

where the square bracket indicates the commutator of the generally noncommuting quantities B and C. We associate B with $\Omega_{\mathbf{r}}^{-1}$ and C with $V(\mathbf{r})$, so that Eqs. (5.43) and (5.44) give

$$\begin{aligned}\Omega_{\mathbf{r}}^{-1}V(\mathbf{r})\,\Omega_{\mathbf{r}} &= V(\mathbf{r}) + \boldsymbol{\alpha}\cdot\boldsymbol{\nabla} + \frac{1}{2!}(\boldsymbol{\alpha}\cdot\boldsymbol{\nabla})^2 V(\mathbf{r}) + \frac{1}{3!}(\boldsymbol{\alpha}\cdot\boldsymbol{\nabla})^3 V(\mathbf{r}) + \ldots \\ &= V(\mathbf{r}+\boldsymbol{\alpha}).\end{aligned} \tag{5.46}$$

Finally, the equation satisfied by Φ is

$$i\hbar\frac{\partial}{\partial t}\Phi(\mathbf{r},t) = \left[\frac{1}{2m}(-i\hbar\boldsymbol{\nabla})^2 + V(\mathbf{r}+\boldsymbol{\alpha})\right]\Phi(\mathbf{r},t). \tag{5.47}$$

The transformation given by Eq. (5.41), leading to the result (5.47), is generally called the Kramers–Henneberger transformation. It has been known since the early days of quantum mechanics.

To interpret the result physically, we observe that the $\boldsymbol{\alpha}$ quantity defined in Eq. (5.43) satisfies the equation

$$m\frac{d^2}{dt^2}\boldsymbol{\alpha} = -\frac{e}{c}\frac{\partial}{\partial t}\mathbf{A}(t) = e\mathbf{E}(t), \tag{5.48}$$

where $\mathbf{E}(t)$ is the electric field. This is the classical equation of motion for the displacement $\boldsymbol{\alpha}$ of a free electron from its position of equilibrium. Thus, $V[\mathbf{r} + \boldsymbol{\alpha}(t)]$ is an expression of the potential experienced by the electron as measured in a system of coordinates oscillating with the electron, and $\Omega_{\mathbf{r}}$ gives the transformation to that system. The quantity

$$U_E = \frac{e^2 \mathbf{A}^2}{2mc^2} \tag{5.49}$$

is the energy of the oscillatory motion of a free electron in an electromagnetic field defined by the vector potential $\mathbf{A}$. That is, it is the potential energy of the motion of a charged particle in this field. The transformation Ω_E displaces the energy in the problem by the amount U_E. The time average of U_E over a period of the field gives the ponderomotive energy U_p of the electron in the field, encountered above in Eq. (5.6). That is, we have the connection

$$U_p = \langle U_E \rangle = \frac{e^2 \langle \mathbf{A}^2 \rangle}{2mc^2}. \tag{5.50}$$

Only for the special case of circularly polarized laser light do we have $U_p = U_E$.

Problem 5.6. Find an exact solution (within the nonrelativistic dipole approximation) for the motion of a free, charged particle interacting with an electromagnetic field.

SOLUTION. The solution can be found directly from the results of the preceding problem. For a free particle in a laser field, the equation of motion is just Eq. (5.38) without the $V(\mathbf{r})$ term. The transformation (5.41) leads to, from Eq. (5.47), simply

$$i\hbar \frac{\partial}{\partial t} \Phi(\mathbf{r}, t) = \frac{1}{2m} (-i\hbar \boldsymbol{\nabla})^2 \Phi(\mathbf{r}, t),$$

which has the solution

$$\Phi(\mathbf{r}, t) = C \exp\left[\frac{i}{\hbar}\left(\mathbf{p} \cdot \mathbf{r} - \frac{p^2}{2m} t\right)\right], \tag{5.51}$$

where C is a normalization constant. This free-particle solution is an eigenfunction of the operator $\Omega_{\mathbf{r}}$, and so the Kramers–Henneberger transformation (5.41) applied to Eq. (5.51) gives the required solution

$$\Psi(\mathbf{r}, t) = C \exp\left\{\frac{i}{\hbar}\left[\mathbf{p} \cdot \mathbf{r} - \frac{p^2}{2m} t - \int_{-\infty}^{t} \left(-\frac{e}{mc} \mathbf{A}(\tau) \cdot \mathbf{p} + \frac{e^2}{2mc^2} \mathbf{A}^2(\tau)\right) d\tau\right]\right\}. \tag{5.52}$$

Equation (5.52) is known as the Volkov solution, or the Gordon–Volkov solution. Though derived here for the case of the nonrelativistic dipole approximation case, the exact solution can also be stated for a relativistic Dirac or Klein–Gordon particle.

Problem 5.7. Devise a general system of calculating quantum transition rates, suited to a nonperturbative environment associated with very intense laser pulses. A characteristic of such problems is that the laser field (i.e., the "perturbing" field that causes the transitions) is present only during a short period of time. The system is initially prepared, and final measurements are made, in space and/or time domains in which the laser field is absent. The calculational scheme should be such that there is no need to rely on sharp turn-on or turn-off of the field, nor need there be any resort to "adiabatic decoupling" schemes. S-matrix methods, originally introduced in connection with scattering problems, are particularly adapted to this purpose.

SOLUTION. The initial and final states of the atomic system are states in which no electromagnetic interaction exists. These states form the complete set $\{\Phi_n\}$ of solutions of the Schrödinger equation

$$i\hbar\partial_t\Phi = H_0\Phi, \tag{5.53}$$

where H_0 is given in Eq. (5.39), and the shorthand notation is introduced that $\partial_t \equiv \partial/\partial t$. The full Schrödinger equation containing all external field effects is

$$i\hbar\partial_t\Psi = (H_0 + H_I)\Psi, \tag{5.54}$$

where H_I is the interaction Hamiltonian given in Eq. (5.40), and it is understood that the vector potential $\mathbf{A}$ satisfies the asymptotic condition that

$$\lim_{t\to\pm\infty} \mathbf{A}(t) = 0, \qquad \Rightarrow \lim_{t\to\pm\infty} H_I = 0. \tag{5.55}$$

We remark that the way in which this limit occurs is unimportant for the formulation, and also that the adjective "asymptotic" simply means before the laser pulse is on and after it is off. Thus $t \to \pm\infty$ might refer to femtoseconds in some modern laser systems.

The fully interacting atomic system is prepared initially in a well-defined noninteracting state Φ_i, so that

$$\lim_{t\to-\infty} \Psi_i^{(+)} = \Phi_i, \tag{5.56}$$

where the superscript $(+)$ designates a Ψ state that satisfies the condition (5.56). The results of an experiment are evaluated by examining the end products, which is to say that one finds the relative amplitude that the fully interacting state Ψ_i is to be found in a noninteracting state Φ_f. This gives a transition amplitude that, to follow the convention established in the scattering community, will be called an S matrix. The probability amplitude that the Ψ state starting as Φ_i ends as Φ_f is

$$S_{fi} = \lim_{t\to+\infty} \left(\Phi_f, \Psi_i^{(+)}\right). \tag{5.57}$$

The two statements (5.56) and (5.57) can be combined as

$$
\begin{aligned}
(S-1)_{fi} = S_{fi} - \delta_{fi} &= \lim_{t\to+\infty}\left(\Phi_f, \Psi_i^{(+)}\right) - \lim_{t\to-\infty}\left(\Phi_f, \Psi_i^{(+)}\right) \\
&= \int_{-\infty}^{\infty} dt \frac{\partial}{\partial t}\left(\Phi_f, \Psi_i^{(+)}\right) \\
&= -\frac{i}{\hbar}\int_{-\infty}^{\infty} dt \left[\left(-i\hbar\partial_t\Phi_f, \Psi_i^{(+)}\right) + \left(\Phi_f, i\hbar\partial_t\Psi_i^{(+)}\right)\right].
\end{aligned} \tag{5.58}
$$

The equations of motion (5.53) and (5.54), when employed in Eq. (5.58), give the result

$$
(S-1)_{fi} = -\frac{i}{\hbar}\int_{-\infty}^{\infty} dt\left(\Phi_f, H_I\Psi_i^{(+)}\right). \tag{5.59}
$$

The equations of motion in quantum mechanics are invariant under time reversal, so a fully equivalent transition amplitude comes from inquiring of the relative probability amplitude that a well-defined final state Φ_f can arise from some initial state. That is, Eqs. (5.56) and (5.57) are replaced by

$$
\lim_{t\to+\infty}\Psi_f^{(-)} = \Phi_f, \tag{5.60}
$$

$$
S_{fi} = \lim_{t\to-\infty}\left(\Psi_f^{(-)}, \Phi_i\right). \tag{5.61}
$$

These equations lead to

$$
(S-1)_{fi} = -\frac{i}{\hbar}\int_{-\infty}^{\infty} dt\left(\Psi_f^{(-)}, H_I\Phi_i\right), \tag{5.62}
$$

which is fully equivalent to Eq. (5.59).

Equations (5.59) and (5.62) are exact as long as all their component quantities are exactly stated. Since one normally does not know an exact expression for $\Psi_i^{(+)}$ or $\Psi_f^{(-)}$, some approximation must be employed for this purpose, as will be seen in later problems.

Despite the use of the historically motivated terminology of calling the above transition amplitudes S matrices, they are in no sense confined to free-free (or scattering) problems. Either the initial or final state can be either free or bound.

5.2 QUALITATIVE STRONG-FIELD PROPERTIES

There have been essentially five means of treating the response of an atom to the application of a laser field too strong to permit a perturbative treatment. One method is purely numerical. The Schrödinger equation is solved directly on a computer. This method has yielded many useful results, but it is inevitably limited by computer capacity despite major recent advances in the speed of computers. For example, it has not been found possible to cope with low-frequency problems (even the commonly used laser wavelength of 1.06 μm presents serious difficulties), only limited information can be gleaned about photoelectron spectra, the so-called stabilization regime can be treated only for very high frequencies, and there seems to be little realistic hope for the treatment of problems in the relativistic domain. The main focus of this book is analytical, and so no further mention of purely numerical methods will be made.

Methods based on the Floquet property have been developed, but they have proven to be at least as computer-intensive as direct numerical treatment of the equation of motion. A third approach is the "high-frequency approximation," which follows from approximating the atomic potential $V(\mathbf{r} + \boldsymbol{\alpha})$ [see Eq. (5.47)] by the first term in its Fourier expansion. This limits the method to high frequencies well beyond the presently available range of laser frequencies. This leaves two methods which are analytical and with a wide domain of applicability: tunneling methods and the *strong-field approximation*, or SFA.

The SFA is similar to one of the tunneling methods, known as the Keldysh approximation. Both the SFA and the Keldysh method use Eq. (5.62) as a starting point, the difference being that the SFA uses (in atomic units) the interaction Hamiltonian $H_I = -\mathbf{A} \cdot \mathbf{p}/c + \mathbf{A}^2/\left(2c^2\right)$, while the Keldysh approximation employs the gauge where $H_I = -\mathbf{E} \cdot \mathbf{r}$. Exact gauge invariance does not hold in these strong-field approximations, and so the two results are not identical. The analytical form of the transition amplitude and transition rate turns out to be significantly simpler in the SFA. Furthermore, the choice of $-\mathbf{A} \cdot \mathbf{p}/c + \mathbf{A}^2/\left(2c^2\right)$ for the interaction Hamiltonian blends smoothly into the corresponding relativistic theory.

The property of continuous connection to a relativistic treatment is an important advantage in a strong-field theory, since for sufficiently strong fields the quiver motion of a free electron in the laser field acquires relativistic velocities, and the nonrelativistic theory becomes inadequate purely as a result of high field intensity. Another way to view this physical situation is to note that a free electron in a linearly polarized plane-wave electromagnetic field acquires a figure-8 motion with the long axis of the figure along the polarization direction and the short axis along the direction of propagation of the laser field. When the amplitude of this figure-8 motion in the direction of the field propagation approaches the size of the atom, then the dipole approximation is no longer valid. This occurs at field intensities well below those found in Problem 5.3 to impede the separation of the Schrödinger equation into c.m. and relative coordinates. The impossibility of using the dipole approximation represents a problem most easily solved by employing a relativistic theory. The connection of a theory with $H_I = -\mathbf{E} \cdot \mathbf{r}$ to non-dipole-approximation and/or relativistic theories is very difficult to establish.

The analytical difficulty of the Keldysh method has confined its application to the limiting case of large photon numbers only. This leads to tunneling results. Since the tunneling case can also be obtained as a limit of much more general SFA results, it is the SFA that will be treated here. That is, the last two of the five theoretical approaches to strong-field theories as listed above are both accessible from the SFA.

To illustrate the principal properties of strong-field results with the minimum of complexity, we shall examine one-dimensional problems.

As is the general practice in strong-field work, atomic units shall be used in the remainder of this chapter.

Problem 5.8. Find the general expression for the ionization rate of a one-dimensional "atom" when the interaction Hamiltonian is $H_I = -pA/c + A^2/(2c^2)$ and the laser vector potential A is so intense that the ponderomotive energy of the ionized electron in the laser field is much larger than the initial field-free atomic binding energy.

SOLUTION. The transition amplitude is given either by Eq. (5.59) or (5.62). The state $\Psi_i^{(+)}$ required for Eq. (5.59) is, however, very difficult to approximate when the laser field is strong. Although strong fields suggest an approximation in which we take the laser field to be more important than the atomic binding potential, it is never appropriate to simply neglect the effect of the atomic potential in the initial bound state $\Psi_i^{(+)}$. No laser field can be so strong that it can be dominant over the effect of the atomic field in the immediate vicinity of the center of the binding potential. Yet, by hypothesis, the laser electric field is so strong that it can be larger than the atomic field at a Bohr radius from the atomic center, and so we cannot presume the laser field to be dominated by the atomic field throughout the atom.

The above dilemma does not arise for the $\Psi_f^{(-)}$ state required for Eq. (5.62). Since the electron in the state $\Psi_f^{(-)}$ is ionized, it is consistent to assume that a sufficiently strong laser field will so dominate the effect of the atomic potential on the electron in the continuum that we can neglect the atomic potential altogether in the ionized state. The state $\Psi_f^{(-)}$ will then be replaced by the Volkov state $\Psi_f^{(-)V}$, which follows from Eq. (5.52). As written, Eq. (5.52) refers to a $\Psi_f^{(+)V}$ state. The corresponding expression for a one-dimensional $\Psi_f^{(-)V}$ state is, in atomic units,

$$\Psi^{(-)V} = \frac{1}{L^{1/2}} \exp\left\{ i \left[px - \frac{p^2}{2}t + \int_t^{+\infty} \left(-\frac{pA(\tau)}{c} + \frac{A^2(\tau)}{2c^2} \right) d\tau \right] \right\}, \tag{5.63}$$

where the normalization constant C of Eq. (5.52) is replaced by $L^{-1/2}$, where L is the length of a "box normalization" volume as often employed for continuum electron states. We shall introduce a monochromatic field given by the one-dimensional vector potential

$$A = a \cos \omega t \tag{5.64}$$

associated with the electric field

$$E = E_0 \sin \omega t; \qquad E_0 = a\omega/c. \tag{5.65}$$

In this case, the Volkov solution is simply

$$\Psi^V = \frac{1}{L^{1/2}} \exp\left[i \left(px - \frac{p^2}{2}t - z\omega t + \zeta \sin \omega t - \frac{z}{2} \sin 2\omega t \right) \right], \tag{5.66}$$

where

$$\zeta \equiv \frac{ap}{\omega c} = \frac{E_0 p}{\omega^2}, \qquad z \equiv \frac{a^2}{4\omega c^2} = \frac{E_0^2}{4\omega^3}. \tag{5.67}$$

The quantity z in Eq. (5.67) is the one-dimensional version of the intensity parameter z introduced in Eq. (5.8).

With the interaction Hamiltonian

$$H_I = i\frac{A}{c}\frac{\partial}{\partial x} + \frac{A^2}{2c^2}, \tag{5.68}$$

and with Eq. (5.63) replacing $\Psi_f^{(-)}$, the transition amplitude (5.62) becomes the strong-field approximation, which we shall identify by a superscript SFA. We note that the Volkov solution in Eq. (5.66) is an eigenvalue of the H_I operator of Eq. (5.68). It is thus permissible to employ H_I with the $-i(\partial/\partial x)$ operator replaced by the scalar quantity p.

The combination of Eqs. (5.62), (5.64), (5.66), (5.67), and (5.68) gives the transition amplitude

$$\begin{aligned}(S-1)_{fi}^{\mathrm{SFA}} &= -i\int_{-\infty}^{\infty} dt\,\left(\Psi_f^V(x,t), H_I\Phi_i(x,t)\right)\\ &= \frac{i}{L^{1/2}}\left(e^{ipx}, (-\omega\zeta\cos\omega t + 2\omega z\cos^2\omega t)\phi_i(x)\right)\\ &\quad\times\int_{-\infty}^{\infty} dt\,\exp\left[i\left(\frac{p^2}{2}t + z\omega t - \zeta\omega t + \frac{z}{2}\sin 2\omega t\right)\right]e^{iE_Bt}.\end{aligned} \tag{5.69}$$

We have used the usual stationary-state relation

$$\Phi_i(x,t) = \phi_i(x)\,e^{-iE_it} = \phi(x)\,e^{iE_Bt}, \tag{5.70}$$

recognizing that the initial state is bound, and so

$$E_B \equiv |E_i| = -E_i. \tag{5.71}$$

We identify the Fourier transform of the spatial part of the initial state wave function (i.e., the momentum space wave function) as

$$\widehat{\phi}_i(p) = \left(e^{ipx}, \phi_i(x)\right) = \int_{-\infty}^{\infty} e^{-ipx}\phi_i(x)\,dx. \tag{5.72}$$

Equation (5.69) becomes

$$\begin{aligned}(S-1)_{fi}^{\mathrm{SFA}} &= -\frac{i}{L^{1/2}}\widehat{\phi}_i(p)\\ &\quad\times\int_{-\infty}^{\infty} dt\,\exp\left[i\left(\frac{p^2}{2} + z\omega + E_B\right)t - i\zeta\sin\omega t + i\frac{z}{2}\sin 2\omega t\right]\\ &\quad\times(-\omega\zeta\cos\omega t + \omega z + \omega z\cos wt).\end{aligned} \tag{5.73}$$

The integral over time in Eq. (5.73) suggests an integration by parts. When this is done, the *surface terms* at $t = \pm\infty$ are simple harmonic terms with phases linear in t. They are thus representations of the zero distribution (generalized function) and can be set to zero. We are left with

$$(S-1)_{fi}^{\mathrm{SFA}} = \frac{i}{L^{1/2}}\widehat{\phi}_i(p)\left(\frac{p^2}{2}+E_B\right)\int_{-\infty}^{\infty}dt\exp\left[i\left(\frac{p^2}{2}+z\omega+E_B\right)t\right.$$
$$\left.-i\zeta\sin\omega t+i\frac{z}{2}\sin 2\omega t\right]. \tag{5.74}$$

This quantity can be expressed in terms of a generalized Bessel function, whose principal properties are summarized in Appendix J. The generalized Bessel function can be defined as

$$J_n(u,v) = \frac{1}{2\pi}\int_{-\pi}^{\pi}d\theta\exp\left[i\left(u\sin\theta+v\sin 2\theta-n\theta\right)\right]. \tag{5.75}$$

It enters into Eq. (5.74) because of the generating function for $J_n(u,v)$, which is

$$\exp\left[-i\left(u\sin x+v\sin 2x\right)\right] = \sum_{n=-\infty}^{\infty}J_n(u,v)\exp(-inx). \tag{5.76}$$

Equation (5.74) then becomes

$$(S-1)_{fi}^{\mathrm{SFA}} = \frac{i}{L^{1/2}}\sum_n J_n\left(\zeta,-\frac{z}{2}\right)\widehat{\phi}_i(p)\left(\frac{p^2}{2}+E_B\right)$$
$$\times\int_{-\infty}^{\infty}dt\exp\left[i\left(\frac{p^2}{2}+z\omega+E_B-n\omega\right)t\right],$$

which yields the delta-function expression

$$(S-1)_{fi}^{\mathrm{SFA}} = \frac{2\pi i}{L^{1/2}}\sum_{n=n_0}^{\infty}J_n\left(\zeta,-\frac{z}{2}\right)\widehat{\phi}_i(p)\left(\frac{p^2}{2}+E_B\right)\delta\left(\mathcal{E}-n\omega\right), \tag{5.77}$$

in which we have introduced the definitions

$$\mathcal{E} \equiv \frac{p^2}{2}+z\omega+E_B \tag{5.78}$$

and

$$n_0 \equiv \left\{z+E_B/\omega\right\}, \tag{5.79}$$

where the brace $\{\}$ in Eq. (5.79) means the smallest integer containing the quantity within the brace.

The delta function in Eq. (5.77) conveys important physical information. From the meaning of $\mathcal{E}$ given in Eq. (5.78), $\delta(\mathcal{E}-n\omega)$ signifies that n photons from the laser field supply the atomic binding energy E_B, the kinetic energy $p^2/2$ of the emitted photoelectron, and the ponderomotive energy $z\omega = U_p$ of the free electron in interaction with the laser field. This last requirement is an important feature unique to strong-field physics. It does not arise in perturbation theories.

The delta function also leads to the imposition of a minimum order n_0 in the sum over n, to make it possible to achieve a zero in the argument of the delta function. The fact that the sum over n continues to larger values means that the formalism accounts for contributions of n_0, $n_0 + 1$, $n_0 + 2, \ldots$ photons. We shall see that contributions to the transition amplitude by greater than the minimum number of photons can be very important, in sharp contrast to perturbation theory, where only the lowest possible photon order is of any significance. The phenomenon where more than n_0 photons contributes to the ionization process has come to be called *above-threshold ionization*, or ATI. Furthermore, we see from Eq. (5.79) that the minimum number of photons required for ionization will index upward as the value of the intensity parameter z increases. This occurs because the laser field must supply enough energy to sustain the field-induced *quiver energy* of the detached electron in the presence of the laser field. Each time the value of n_0 indexes upward to the next integer because of an increase in field intensity, it is referred to as a *channel closing*.

To form a transition rate from the transition amplitude $(S-1)_{fi}^{\mathrm{SFA}}$, we must evaluate

$$w = \lim_{t\to\infty} \frac{1}{t} \left|(S-1)_{fi}^{\mathrm{SFA}}\right|^2 . \tag{5.80}$$

In doing this, a square of the delta function arises, which can be viewed as

$$\left[\delta\left(\mathcal{E}-n\omega\right)\right]^2 = \delta\left(\mathcal{E}-n\omega\right)\delta\left(0\right). \tag{5.81}$$

The delta function of zero argument can be replaced by the limiting form of the integral representation of the delta function,

$$\delta\left(0\right) = \lim_{t\to\infty}\frac{1}{2\pi}\int_{-t/2}^{t/2}\exp\left(i0\right)dt = \lim_{t\to\infty}\frac{1}{2\pi}t, \tag{5.82}$$

which yields

$$w = \frac{2\pi}{L}\sum_{n=n_0}^{\infty}\delta\left(\mathcal{E}-n\omega\right)\left[\widehat{\phi}_i(p)\left(\frac{p^2}{2}+E_B\right)J_n\left(\frac{E_0 p}{\omega^2},-\frac{z}{2}\right)\right]^2 . \tag{5.83}$$

The total differential transition rate is then found by integration over the phase space available to the emerging photoelectrons, which, in this one-dimensional problem, gives

$$\frac{dW}{d\Omega} = \int w\frac{Ldp}{2\pi}, \tag{5.84}$$

where the 2π in the denominator is the volume (in atomic units) of a unit cell in the phase space. Integration over solid angle in one dimension is simply a matter of summing over backward and forward directions, but the general notation $dW/d\Omega$ is employed in Eq. (5.84) to guide the three-dimensional procedure.

The delta function in Eq. (5.83) will be used to accomplish the integral in Eq. (5.84). In terms of the quantities

$$\mathsf{E} = \omega\left(n-z-\epsilon_B\right); \qquad \epsilon_B = E_B/\omega, \tag{5.85}$$

we can transform the delta function by the following steps:

$$\begin{aligned}\delta(\mathcal{E}-n\omega) &= \delta\left(\frac{p^2}{2}-\mathsf{E}\right) = \delta\left[\left(\frac{p}{2^{1/2}}-\mathsf{E}^{1/2}\right)\left(\frac{p}{2^{1/2}}+\mathsf{E}^{1/2}\right)\right] \\ &= \frac{1}{2\mathsf{E}^{1/2}}\left[\delta\left(\frac{p}{2^{1/2}}-\mathsf{E}^{1/2}\right)+\delta\left(\frac{p}{2^{1/2}}+\mathsf{E}^{1/2}\right)\right] \\ &= \frac{1}{(2\mathsf{E})^{1/2}}\left\{\delta\left[p-(2\mathsf{E})^{1/2}\right]+\delta\left[p+(2\mathsf{E})^{1/2}\right]\right\}. \end{aligned} \tag{5.86}$$

The two terms in Eq. (5.86) represent the two possible directions of emission in one dimension. Each of the factors $\widehat{\phi}_i(p)$ and $J_n(E_0 p/\omega^2, -z/2)$ in Eq. (5.83) has definite parity with respect to the replacement $p \to -p$. It follows that each of the two emission directions gives an equal contribution to the differential transition rate in Eq. (5.83), so the result for the total transition rate is

$$W = 2\sum_{n=n_0}^{\infty}\frac{1}{p}\left[\widehat{\phi}_i(p)\left(\frac{p^2}{2}+E_B\right)J_n\left(\frac{E_0 p}{\omega^2},-\frac{z}{2}\right)\right]^2; \qquad p=(2\mathsf{E})^{1/2}, \tag{5.87}$$

with $\mathsf{E} = \omega(n - z - \epsilon_B)$ as given in Eq. (5.85).

An alternative expression for Eq. (5.87) is to substitute the explicit value for p as given by the definition of E to obtain

$$W = \omega(2\omega)^{1/2}\sum_{n=n_0}^{\infty}\frac{(n-z)^2}{(n-z-\epsilon_B)^{1/2}}\left[\widehat{\phi}_i(p)J_n\left(\frac{E_0 p}{\omega^2},-\frac{z}{2}\right)\right]^2; \qquad p=(2\mathsf{E})^{1/2}. \tag{5.88}$$

Problem 5.9. Find the low-intensity, single-photon limit of the SFA, and compare with the first-order perturbation theory result.

SOLUTION. It is sufficient to examine only the transition amplitude, since passage from the amplitude to the total transition rate would follow the same series of steps in perturbation theory as in the SFA.

The low-intensity limit allows us to retain only terms of first order in E_0 or A, and the single-photon condition means that $\omega > E_B$ and $n_0 = 1$. The generalized Bessel function in Eq. (5.77) can be approximated as

$$J_1\left(\frac{E_0 p}{\omega^2},-\frac{z}{2}\right) \approx J_1\left(\frac{E_0 p}{\omega^2}\right) \approx \frac{E_0 p}{2\omega^2}.$$

Equation (5.77) then approaches the expression

$$(S-1)_{fi}^{\mathrm{SFA}} \underset{A\to 0, n_0=1}{\longrightarrow} \frac{\pi i}{L^{1/2}}\frac{E_0 p}{\omega}\delta\left(\frac{p^2}{2}+E_B-\omega\right)\widehat{\phi}_i(p). \tag{5.89}$$

For the perturbation theory result, we can start with either of the general amplitudes (5.59) or (5.62), with Ψ replaced by Φ. The two terms in the expression for the interaction Hamiltonian H_I are replaced by only the first-order term. That is, the first-order time-dependent perturbation theory (*PT*) transition amplitude is

$$
\begin{aligned}
(S-1)^{\mathrm{PT}}_{fi} &= -i\int_{-\infty}^{\infty} dt\left(\Phi_f^{\mathrm{free}},\left(-\frac{A}{c}\right)\left(-i\frac{\partial}{\partial x}\right)\Phi_i^{\mathrm{bound}}\right)\\
&= -i\int_{-\infty}^{\infty} dt\left(\left(i\frac{A}{c}\frac{\partial}{\partial x}\right)\Phi_f,\Phi_i\right)\\
&= -i\int_{-\infty}^{\infty} dt\left(-\frac{Ap}{c}\Phi_f,\Phi_i\right),
\end{aligned}
\tag{5.90}
$$

where the p in the last line of Eq. (5.90) is the scalar p, and no longer the operator. When we substitute $A(t) = a\cos\omega t$, as in Eq. (5.64), use the ordinary free-particle plane wave solution for Φ_f and the stationary state (5.70) for Φ_i, we find

$$
\begin{aligned}
(S-1)^{\mathrm{PT}}_{fi} &= -\pi i\frac{1}{L^{1/2}}\frac{ap}{c}\left(e^{ipx},\phi_i(x)\right)\left[\delta\left(\frac{p^2}{2}+E_B+\omega\right)\right.\\
&\quad\left.+\,\delta\left(\frac{p^2}{2}+E_B-\omega\right)\right].
\end{aligned}
\tag{5.91}
$$

Of the delta functions in Eq. (5.91), the first can never have a zero in the argument, and so we need retain only the latter. We have then

$$
(S-1)^{\mathrm{PT}}_{fi} = -\frac{\pi i}{L^{1/2}}\frac{E_0 p}{\omega}\delta\left(\frac{p^2}{2}+E_B-\omega\right)\widehat{\phi}_i(p),
\tag{5.92}
$$

differing from Eq. (5.89) only in an irrelevant overall sign. (The change in sign is an artifact of the integration by parts that was done in deriving the SFA result.) Thus the SFA reduces to exactly the perturbation limit for large ω and low intensity.

Expressed as a transition rate, Eq. (5.89) or (5.92) gives

$$
W^{\mathrm{SFA}} \to W^{\mathrm{PT}} = (E_0/\omega)^2\left[(\omega-E_B)/2\right]^{1/2}\left|\widehat{\phi}_i(p)\right|^2, \qquad p=\pm\left[2(\omega-E_B)\right]^2.
$$

Problem 5.10. Find the asymptotic form taken by the SFA transition amplitude in the limiting case where the laser field is of very low frequency. This means that $\epsilon_B \gg 1$. For ionization to occur at all, the field must be very intense, or $z \gg 1$. The minimum photon order has $n_0 > z$, so $n_0 \gg 1$. It will also be assumed that $z \gg \epsilon_B$.

SOLUTION. From Eqs. (5.5) and (5.8), it is seen that $z_1 = 2z/\epsilon_B$. The statement that $z \gg \epsilon_B$ means that $z_1 \gg 1$.

To evaluate the asymptotic behavior of the SFA transition amplitude in Eq. (5.77), it is necessary to examine the generalized Bessel function. From the integral representation in Eq. (5.75), we can write

$$J_n\left(\zeta, -\frac{z}{2}\right) = \frac{1}{2\pi}\int_{-\pi}^{\pi} d\theta \exp\left[ing(\theta)\right], \tag{5.93}$$

$$g(\theta) = \frac{\zeta}{n}\sin\theta - \frac{z}{2n}\sin 2\theta - \theta, \tag{5.94}$$

where the n has been extracted as the large parameter necessary for the application of asymptotic methods. We shall here take $n_0 = z + \epsilon_B$, which is equivalent to Eq. (5.79) for large orders. When the value of the momentum p allowed by the delta function is inserted into the definition of ζ in Eq. (5.67), we find

$$\zeta = (8z)^{1/2}(n - z - \epsilon_B)^{1/2} = (8z)^{1/2}(n - n_0)^{1/2}. \tag{5.95}$$

The ratio ζ/n in Eq. (5.94) will then always be extremely small, since the spread of photon orders will be much less than the lowest order. Since the other two terms in Eq. (5.94) are of order unity, the first term will be neglected, and we shall henceforth not distinguish between n_0 and n. The saddle points in $g(\theta)$ are at

$$\frac{dg}{d\theta} = -\frac{z}{n}\cos 2\theta - 1 = 0.$$

The saddle points so located are designated θ_0 and are at

$$\cos 2\theta_0 = -\frac{n}{z} = -\frac{z + \epsilon_B}{z}. \tag{5.96}$$

This means that the saddle points cannot be on the real axis. If we set $\theta_0 = x + iy$, then the saddle points are at

$$x = \pm\frac{\pi}{2}, \qquad y = -\operatorname{arcsinh}\left(1/z_1^{1/2}\right). \tag{5.97}$$

Saddle points at a corresponding location in the upper half of the complex θ plane are not considered because one cannot deform the initial path of integration along the real axis into a path of steepest descent through the upper-half-plane points.

To evaluate the leading behavior of $J_n(\zeta, -z/2)$, we must find the value of $\exp[ing(\theta_0)]$, that is, we wish to find

$$ing(\theta_0) = -i\frac{z}{2}\sin 2\theta_0 - in\theta_0, \tag{5.98}$$

where, for the saddle point at $x = \pi/2$ we have

$$\cos\theta_0 = -i\frac{1}{z_1^{1/2}}, \qquad \sin\theta_0 = -\left(1 + \frac{1}{z_1}\right)^{1/2}, \qquad \theta_0 = \frac{\pi}{2} - i\operatorname{arcsinh}\left(\frac{1}{z_1^{1/2}}\right),$$

and for the saddle point at $x = -\pi/2$ we have

$$\cos\theta_0 = -i\frac{1}{z_1^{1/2}}, \qquad \sin\theta_0 = \left(1 + \frac{1}{z_1}\right)^{1/2}, \qquad \theta_0 = -\frac{\pi}{2} - i\operatorname{arcsinh}\left(\frac{1}{z_1^{1/2}}\right).$$

Since $\left(1/z_1^{1/2}\right) \ll 1$, we can expand the inverse hyperbolic function as

$$\operatorname{arcsinh}\left(\frac{1}{z_1^{1/2}}\right) \approx \frac{1}{z_1^{1/2}} - \frac{1}{6}\frac{1}{z_1^{3/2}}.$$

Both saddle points give the same real part in Eq. (5.98), and we obtain the asymptotic result that

$$J_n\left(\zeta, -\frac{z}{2}\right) \sim \exp\left(-\frac{4}{3}\frac{z}{z_1^{3/2}}\right). \tag{5.99}$$

The transition rate is proportional to the square of this quantity, so when the argument in Eq. (5.99) is doubled and converted to the directly physical parameters, the asymptotic result for the transition rate behaves as

$$W \sim \exp\left[-\frac{2\,(2E_B)^{3/2}}{3E_0}\right]. \tag{5.100}$$

Equation (5.100) has long been known as the behavior of the transition rate for tunneling of a bound particle through a potential barrier for both constant and slowly varying electric fields of amplitude E_0.

Problem 5.11. Apply the SFA formalism to find the total transition rate and photoelectron energy spectrum for ionization from a one-dimensional system bound by the delta function potential $V(x) = -(2E_B)^{1/2}\,\delta(x)$, where E_B is the binding energy of the single bound state possessed by this system.

SOLUTION. In terms of the notation $\eta^2 = 2E_B$, the normalized solution of the Schrödinger equation for the delta function potential is well known to be

$$\Phi(x,t) = \eta^{1/2}\exp(-\eta\,|x|)\exp(iE_B t). \tag{5.101}$$

The Fourier transform of the space part of (5.101) is

$$\widehat{\phi}(p) = \frac{2\eta^{3/2}}{\eta^2 + p^2}. \tag{5.102}$$

The momentum conservation condition $p = (2\mathsf{E})^{1/2}$ is associated with the SFA transition amplitude of Eq. (5.88), where E is defined in Eq. (5.85). When this is employed in the denominator of Eq. (5.102), one finds that $\eta^2 + p^2 = (2\omega)(n - z)$. The SFA transition rate is thus

$$W = 4E_B\epsilon_B^{1/2}\sum_{n=n_0}^{\infty}\frac{1}{(n - z - \epsilon_B)^{1/2}}\left[J_n\left(\zeta, -\frac{z}{2}\right)\right], \tag{5.103}$$
$$\zeta = (8z)^{1/2}(n - z - \epsilon_B)^{1/2}.$$

The rate in (5.103) is for a monochromatic wave. It is a per-atom transition rate. In practical application to the description of realistic laser experiments, this elementary rate is employed in the solution of a rate equation applied to the problem of finding the number density of ions formed from a collection of initially neutral atoms. The atomic rate enters into this procedure as the elementary atomic rate $W[I(x,t)]$ dependent on the field intensity distribution $I(x,t)$ describing the space and time profile of an actual laser pulse. When so employed, depletion effects in the initial collection of atoms are accounted for. This is important, since the large transition rates typical of very intense field laser experiments means that depletion of the initial population of atoms plays a significant role in the measurements of ion yield and photoelectron spectra. Nevertheless, the monochromatic rate W given above is sufficient in itself to describe most of the qualitative features of real experiments.

Expression (5.103) gives the total transition rate for the formation of ions, but it also contains within it the information necessary to find the energy spectrum of emitted photoelectrons. Each of the terms in the sum over the index n represents an electron with a different energy of emission. As we have seen, the kinetic energy of the emitted electron is $p^2/2 = \omega(n - z - \epsilon_B)$, so each successively larger value of n corresponds to the addition of another increment of energy ω to the photoelectron. This spectrum can be very extensive in strong fields, and for really extended ATI spectra, the individual peaks for given n values can no longer be distinguished. The spectrum appears to be continuous.

Figure 5.2 gives the spectrum predicted by Eq. (5.103) in a laser pulse of Gaussian shape in time, but without the spatial variation to be found in a focused laser in a laboratory experiment. No depletion effects are included. The Gaussian distribution has a peak intensity of $I = 1$, E_B has been set to $\frac{1}{2}$ (as in the hydrogen atom),

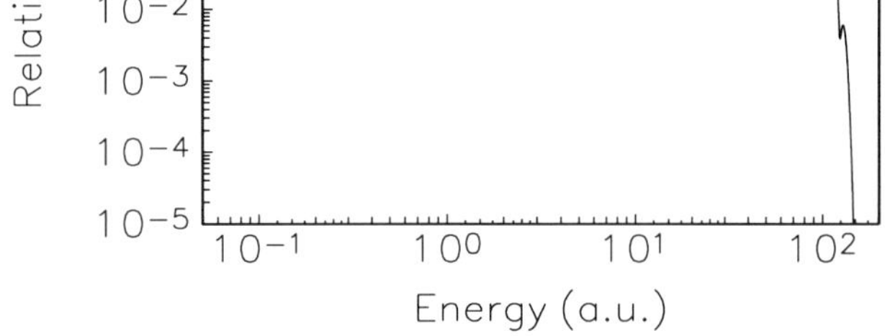

Figure 5.2. Spectrum of photoelectrons emitted from a one-dimensional "atom" bound by a delta function potential with a binding energy of 0.5 a.u., when the laser causing the ionization has a frequency of $\frac{1}{16}$ a.u. and a pulse shape Gaussian in time with a peak intensity of 1 a.u.

and the wavelength is taken to be $\frac{1}{16}$, a typical value for a laser. It is seen that the spectrum extends to an energy of about 100 a.u. Since each photon has an energy of $\frac{1}{16}$ a.u., the spectrum spans about 1600 photon orders, which is an extreme example of the ATI phenomenon. Although this example refers to a very high intensity, real laboratory experiments have yielded spectra of extent nearly that shown in the figure. This phenomenon plainly bears no resemblance to traditional atomic physics and has no explanation within perturbation theory.

APPENDIX A

ANGULAR MOMENTUM

The angular momentum of a classical particle is given by

$$\mathbf{L} = \mathbf{r} \times \mathbf{p},$$

where $\mathbf{r}$ and $\mathbf{p}$ are the radius vector and the linear momentum of the particle. If the motion is in a central field, $\mathbf{L}$ is a conserved quantity. In quantum mechanics, it is no longer true that each component of the angular momentum vector is separately conserved. A more limited statement of conservation of angular momentum in quantum mechanics can be achieved, however, as a result of the separation of radial and angular variables in the equation of motion, made possible by the central force property. The angular part of the Schrödinger equation is independent of the form of the central-field potential, and so the angular part of the equation of motion is connected only with angular momentum properties. The basic properties of the angular momentum of a particle in quantum mechanics will be summarized below. (See Ref. 8 for further details.)

A.1 PROJECTION OF ANGULAR MOMENTUM

The angular momentum operator in quantum mechanics has the same expression as in classical physics,

$$\widehat{\mathbf{L}} = \widehat{\mathbf{r}} \times \widehat{\mathbf{p}},$$

except for the operator nature of the physical variables. In the configuration representation, the linear momentum operator is

$$\widehat{\mathbf{p}} = \frac{\hbar}{i}\frac{\partial}{\partial \mathbf{r}}.$$

The component of the angular momentum operator in the z direction is

$$\widehat{L}_z = \frac{\hbar}{i}\left(x\frac{\partial}{\partial y} - y\frac{\partial}{\partial x}\right).$$

We now introduce a spherical coordinate system (r, θ, φ) with the polar direction along the z axis. The angle θ is the polar angle, that is, the angle between the radius vector $\mathbf{r}$ and the z axis. The azimuthal angle φ is measured from the x axis to the projection of the radius vector $\mathbf{r}$ onto the xy plane. In this coordinate system, we have

$$\widehat{L}_z = \frac{\hbar}{i}\frac{\partial}{\partial\varphi}. \tag{A.1}$$

The eigenfunctions $\phi(\varphi)$ of this operator satisfy the equation

$$\widehat{L}_z\phi = L_z\phi,$$

and are of the form

$$\phi(\varphi) = C\exp(iL_z\varphi).$$

For this function to be single valued in coordinate space, we must have

$$\phi(\varphi) = \phi(\varphi + 2\pi),$$

so that

$$L_z/\hbar = m,$$

where m is an integer. That is, the eigenvalue of the z projection of the angular momentum operator is

$$L_z = \hbar m. \tag{A.2}$$

A.2 SQUARE OF THE ANGULAR MOMENTUM

We transform the square of the angular momentum operator,

$$\widehat{\mathbf{L}}^2 = (\mathbf{r}\times\widehat{\mathbf{p}})\cdot(\mathbf{r}\times\widehat{\mathbf{p}})$$

so that the linear momentum operators are on the right. To do this, we must use the commutation properties of the linear momentum operators with the coordinates,

$$\widehat{p}_k r_j - r_j\widehat{p}_k = \frac{\hbar}{i}\delta_{jk},$$

where j, k are rectangular components of the vectors. Using the properties of the triple scalar product of three vectors, we obtain

$$\widehat{\mathbf{L}}^2 = \mathbf{r}^2\widehat{\mathbf{p}}^2 - r^2\widehat{p}_r^2.$$

The $\widehat{\mathbf{p}}^2$ operator can be written in terms of the Laplacian $\boldsymbol{\nabla}^2$ as

$$\widehat{\mathbf{p}}^2 = -\hbar^2 \boldsymbol{\nabla}^2,$$

and so, using the expression for the Laplacian in spherical coordinates, we find

$$\begin{aligned} \widehat{\mathbf{L}}^2 &= -\hbar^2 r^2 \boldsymbol{\nabla}^2_{\theta,\varphi} \\ &= -\hbar^2 \left[\frac{\partial}{\partial(\cos\theta)} \left(1 - \cos^2\theta\right) \frac{\partial}{\partial(\cos\theta)} + \frac{1}{\sin^2\theta} \frac{\partial^2}{\partial\varphi^2} \right]. \end{aligned} \tag{A.3}$$

The Hamiltonian for a particle in a central field is

$$\begin{aligned} \widehat{H} &= \frac{1}{2m}\widehat{\mathbf{p}}^2 + U(r) \\ &= \frac{1}{2m}\widehat{p}_r^2 + \frac{1}{2mr^2}\widehat{\mathbf{L}}^2 + U(r). \end{aligned} \tag{A.4}$$

The squared angular momentum operator is seen to be the only part of the Hamiltonian that relates to angular coordinates, and so it determines the angular state of the particle.

A.3 COMMUTATION PROPERTIES OF ANGULAR MOMENTUM OPERATORS

From the commutation properties of the linear momentum and the coordinates,

$$\widehat{p}_k r_j - r_j \widehat{p}_k = \frac{\hbar}{i}\delta_{jk},$$

where the indices refer to projections onto x, y, z axes, we obtain the commutation relations for the components of angular momentum,

$$\begin{aligned} \left[\widehat{L}_x, \widehat{L}_y\right] &\equiv \widehat{L}_x\widehat{L}_y - \widehat{L}_y\widehat{L}_x = i\hbar\widehat{L}_z, \\ \left[\widehat{L}_y, \widehat{L}_z\right] &= i\hbar\widehat{L}_x, \\ \left[\widehat{L}_z, \widehat{L}_x\right] &= i\hbar\widehat{L}_y. \end{aligned}$$

We now introduce the operators

$$\widehat{L}_+ = \widehat{L}_x + i\widehat{L}_y, \qquad \widehat{L}_- = \widehat{L}_x - i\widehat{L}_y.$$

The commutation relations for these operators are

$$\left[\widehat{L}_z, \widehat{L}_+\right] = \hbar\widehat{L}_+, \qquad \left[\widehat{L}_z, \widehat{L}_-\right] = -\hbar\widehat{L}_-.$$

We also find the commutators of the rectangular components with $\widehat{L}^2$ to be

$$\left[\widehat{\mathbf{L}}^2, \widehat{L}_x\right] = \left[\widehat{\mathbf{L}}^2, \widehat{L}_y\right] = \left[\widehat{\mathbf{L}}^2, \widehat{L}_z\right] = 0.$$

The quantum state of a particle can be determined by the eigenvalues of a set of mutually commuting operators. In the case of angular momentum, the square of the angular momentum and one of the *x, y, x* components of angular momentum form such a set. These operators commute with the Hamiltonian in Eq. (A.4) for a particle in a central field, as well as commuting between themselves. Therefore the state of the particle can be characterized by the eigenvalues of the angular momentum squared and of one of the projections onto a fixed axis of the angular momentum.

A.4 EIGENVALUE OF THE SQUARED ANGULAR MOMENTUM OPERATOR

The value of an operator averaged over its values in a given state will be indicated by a superposed bar. With that notation, we can write

$$\overline{\widehat{\mathbf{L}}^2} = \overline{\widehat{L}_x^2} + \overline{\widehat{L}_y^2} + \overline{\widehat{L}_z^2}, \qquad \overline{\widehat{L}_x^2} \geq 0, \qquad \overline{\widehat{L}_y^2} \geq 0.$$

From this and Eq. (A.2), we conclude that

$$\overline{\widehat{\mathbf{L}}^2} - \hbar^2 m^2 \geq 0.$$

The eigenvalue of one of the projections of the angular momentum operator has a range of possible values, but they are bounded from above and below by the last inequality. We introduce a quantum number l associated with the squared angular momentum operator, which we can specify as

$$l = \max(|m|).$$

We shall be able to express the eigenvalue of the squared angular momentum operator explicitly in terms of l.

The wave function that is a simultaneous eigenfunction of the operators $\widehat{\mathbf{L}}^2$ and $\widehat{L}_z$ will be written in terms of the quantum numbers as Ψ_{lm}. From the properties of the operator $\widehat{L}_+$, it follows that

$$\left(\widehat{L}_z\widehat{L}_+ - \widehat{L}_+\widehat{L}_z\right)\Psi_{lm} = \hbar\widehat{L}_+\Psi_{lm},$$

and, since we know that

$$\widehat{L}_z\Psi_{lm} = \hbar m\Psi_{lm},$$

we can conclude that

$$\widehat{L}_z\left(\widehat{L}_+\Psi_{lm}\right) = \hbar\,(m+1)\left(\widehat{L}_+\Psi_{lm}\right).$$

The functions $\widehat{L}_+\Psi_{lm}$ are seen to satisfy the same equation as would $\Psi_{l,m+1}$. We can thus write

$$\widehat{L}_+\Psi_{lm} = \text{const} \times \Psi_{l,m+1}.$$

Since m cannot exceed l, we have

$$\widehat{L}_{+}\Psi_{ll} = 0. \tag{A.5}$$

We now apply these results to the determination of the eigenvalues of the squared angular momentum operator. We find

$$\widehat{\mathbf{L}}^2 = \widehat{L}_x^2 + \widehat{L}_y^2 + \widehat{L}_z^2 = \widehat{L}_{-}\widehat{L}_{+} + \widehat{L}_z^2 + \hbar\widehat{L}_z. \tag{A.6}$$

If we operate with this expression on the state Ψ_{ll} and use Eq. (A.5), we obtain

$$\widehat{\mathbf{L}}^2\Psi_{ll} = \hbar^2 l\,(l+1)\,\Psi_{ll}.$$

The eigenvalue of the squared angular momentum operator does not depend on the eigenvalues of a projection of the angular momentum. We then conclude that

$$\widehat{\mathbf{L}}^2\Psi_{lm} = \hbar^2 l\,(l+1)\,\Psi_{lm}. \tag{A.7}$$

We now calculate matrix elements of the angular momentum operators. Since the angular momentum is a Hermitian operator, we have

$$\left\langle l, m+1 \left| \widehat{L}_{+} \right| lm \right\rangle = \left\langle lm \left| \widehat{L}_{-} \right| l, m+1 \right\rangle^{*}. \tag{A.8}$$

In view of the raising operator property of $\widehat{L}_{+}$, this is the only nonzero matrix element of this operator for the given values of l and m. Equations (A.6) and (A.7) then lead to

$$\left| \left\langle l, m+1 \left| \widehat{L}_{+} \right| lm \right\rangle \right|^2 = \hbar^2 \left[l\,(l+1) - m\,(m+1) \right].$$

We obtain, finally, the relations

$$\begin{aligned} \widehat{L}_{+}\Psi_{lm} &= \hbar\sqrt{(l-m)\,(l+m+1)}\Psi_{l,m+1}, \\ \widehat{L}_{-}\Psi_{lm} &= \hbar\sqrt{(l+m)\,(l-m+1)}\Psi_{l,m-1}. \end{aligned} \tag{A.9}$$

APPENDIX B

CLEBSCH–GORDAN COEFFICIENTS

Clebsch–Gordan coefficients arise when two angular momenta are combined into a total angular momentum. This will occur when the angular momentum of a system is found as the combination of the angular momenta of two subsystems or when two types of angular momenta relating to the same particle are combined to find the total angular momentum for that particle, as in the addition of orbital and spin angular momenta to obtain a total angular momentum for the particle.

Let us find the total angular momentum $\mathbf{j}$ as a sum of the momenta $\mathbf{j}_1$ and $\mathbf{j}_2$. The wave function for the total momentum can be written as

$$\Psi_{jm} = \sum_{m_1,m_2} \langle j_1 j_2, m_1 m_2 \,|\, jm \rangle \, \psi_{j_1 m_1} \psi_{j_2 m_2}, \tag{B.1}$$

where the indices on the wave functions characterize the angular momentum and its projection onto a fixed axis. The coefficients in this expansion are called Clebsch–Gordan coefficients. We now examine their properties.

B.1 PROPERTIES OF CLEBSCH–GORDAN COEFFICIENTS

B.1.1 Condition for Addition of Angular Momentum Projections

It follows from the conservation law for the sum of angular momentum projections that

$$\langle j_1 j_2, m_1 m_2 \,|\, jm \rangle = 0 \qquad \text{if} \qquad m_1 + m_2 \neq m. \tag{B.2}$$

B.1.2 Orthogonality Condition

Orthogonality properties of the wave functions are expressed as

$$\langle \Psi_{jm} | \Psi_{jm'} \rangle = \delta_{mm'},$$
$$\langle \psi_{j_1 m_1} | \psi_{j_1 m_1'} \rangle = \delta_{m_1 m_1'},$$

This leads to the orthogonality condition for the Clebsch–Gordan coefficients, which is

$$\sum_{m_1 m_2} \langle j_1 j_2, m_1 m_2 \,|\, jm \rangle \left\langle j_1 j_2, m_1 m_2 \,\middle|\, jm' \right\rangle = \delta_{mm'}, \tag{B.3}$$

where $\delta_{mm'}$ is the Kronecker delta symbol defined by

$$\delta_{mm'} = \begin{cases} 1, & m = m' \\ 0, & m \neq m' \end{cases} .$$

B.1.3 Inversion Property

In the inversion transformation of the radius vector, $\mathbf{r} \to -\mathbf{r}$, the wave functions will change sign or not, depending on their parity. That is, the wave functions transform as

$$\Psi_{jm} \to (-1)^{j-m} \Psi_{j,-m}.$$

This leads to

$$\begin{aligned}(-1)^{j-m} \Psi_{j,-m} &= (-1)^{j-m} \sum_{m_1 m_2} \langle j_1 j_2, -m_1, -m_2 \,|\, j, -m \rangle \, \psi_{j_1,-m_1} \psi_{j_2,-m_2} \\ &= \sum_{m_1 m_2} \langle j_1 j_2, m_1 m_2 \,|\, jm \rangle \, \psi_{j_1 m_1} \psi_{j_2 m_2},\end{aligned}$$

so that we obtain

$$\langle j_1 j_2, -m_1, -m_2 \,|\, j, -m \rangle = (-1)^{j-j_1-j_2} \langle j_1 j_2, m_1 m_2 \,|\, jm \rangle. \tag{B.4}$$

B.1.4 Permutation Properties

It follows from the rules for the construction of the Clebsch–Gordan coefficients and the rules for the addition of two angular momenta into a zero total angular momentum that

$$\begin{aligned}\langle j_1 j_2, m_1 m_2 \,|\, jm \rangle &= (-1)^{j_1-m_1} \sqrt{\frac{2j+1}{2j_2+1}} \langle j_1 j, m_1, -m \,|\, j_2 - m_2 \rangle \qquad \text{(B.5)} \\ &= (-1)^{j_2+m_2} \sqrt{\frac{2j+1}{2j_1+1}} \langle j j_2, -m, m_2 \,|\, j_1 - m_1 \rangle.\end{aligned}$$

B.2 EVALUATION OF CERTAIN CLEBSCH–GORDAN COEFFICIENTS

We consider first the evaluation of Clebsch–Gordan coefficients for the frequently occurring case $j_2 = \frac{1}{2}$. We begin by obtaining a relation for Clebsch–Gordan coefficients that is valid for any angular momenta. From the addition properties used to form $j = j_1 + j_2$, we obtain the condition

$$\widehat{j}^2 - \widehat{j_1}^2 - \widehat{j_2}^2 = 2\widehat{\mathbf{j}_1} \cdot \widehat{\mathbf{j}_2} = 2\widehat{j_{1z}}\widehat{j_{2z}} + \widehat{j_{1+}}\widehat{j_{2-}} + \widehat{j_{1-}}\widehat{j_{2+}}.$$

When this operator acts on the wave function in Eq. (B.1), the result is

$$\begin{aligned} j(j+1) - j_1(j_1+1) - j_2(j_2+1) = \sum_{m_2 m_2'} &\langle j_1 j_2, m_1 m_2 | jm\rangle \langle j_1 j_2, m_1' m_2' | jm\rangle \\ &\times \left\langle \psi_{j_1 m_1'} \psi_{j_2 m_2'} | 2 j_{1z} j_{2z} + j_{1+} j_{2-} \right. \\ &\left. + j_{1-} j_{2+} | \psi_{j_1 m_1} \psi_{j_2 m_2} \right\rangle. \end{aligned}$$

For the given value of m, the conservation rule for angular momentum projections gives

$$m_1 = m - m_2, \qquad m_1' = m - m_2'.$$

With the help of the momentum operator eigenvalues in Eqs. (A.2) and (A.9), we find that

$$\begin{aligned} &j(j+1) - j_1(j_1+1) - j_2(j_2+1) \\ &\quad = 2\sum_{m_2} m_1 m_2 \langle j_1 j_2, m_1 m_2 | jm\rangle^2 \\ &\qquad + \sum_{m_2} \sqrt{(j_1 - m_1)(j_1 + m_1 + 1)(j_2 - m_2 + 1)(j_2 + m_2)} \\ &\qquad \times \langle j_1 j_2, m_1 m_2 | jm\rangle \langle j_1 j_2, m_1 + 1, m_2 - 1 | jm\rangle \\ &\qquad + \sum_{m_2} \sqrt{(j_1 - m_1 + 1)(j_1 + m_1)(j_2 - m_2)(j_2 + m_2 + 1)} \\ &\qquad \times \langle j_1 j_2, m_1 m_2 | jm\rangle \langle j_1 j_2, m_1 - 1, m_2 + 1 | jm\rangle. \end{aligned} \tag{B.6}$$

Now we apply the above results to the particular case where $j_2 = \frac{1}{2}$. For the given values j, j_1, there are two nonzero Clebsch–Gordan coefficients that we shall denote as

$$X = \left\langle j_1 \tfrac{1}{2}, m - \tfrac{1}{2}, \tfrac{1}{2} \,|jm \right\rangle,$$
$$Y = \left\langle j_1 \tfrac{1}{2}, m + \tfrac{1}{2}, -\tfrac{1}{2} \,|jm \right\rangle.$$

Using this notation, Eq. (B.6) for $j_2 = \frac{1}{2}$ becomes

$$j(j+1) - j_1(j_1+1) - \tfrac{3}{4} = \left(m - \tfrac{1}{2}\right) X^2 - \left(m + \tfrac{1}{2}\right) Y^2 + 2\sqrt{\left(j_1 + \tfrac{1}{2}\right)^2 - m^2}\, XY.$$

The normalization condition for the Clebsch–Gordan coefficients given in Eq. (B.3) is of the form

$$X^2 + Y^2 = 1.$$

When this is inserted into the preceding expression, we obtain

$$(j - j_1)(j + j_1 + 1) - \tfrac{1}{4} = m\left(X^2 - Y^2\right) + 2\sqrt{\left(j_1 + \tfrac{1}{2}\right)^2 - m^2} XY.$$

With the notation

$$t = \frac{m}{j + \frac{1}{2}} \leq 1,$$

we can rewrite our results as the system of equations

$$t(X^2 - Y^2) + 2\sqrt{1 - t^2} XY = \pm 1, \qquad X^2 + Y^2 = 1.$$

The ambiguous sign $\pm$ is such that the upper sign corresponds to $j = j_1 + \frac{1}{2}$, while the lower sign is associated with $j = j_1 - \frac{1}{2}$.

The solution of the equations we have obtained is

$$X = \sqrt{\frac{1+t}{2}}, \qquad Y = \sqrt{\frac{1-t}{2}}$$

for $j = j_1 + \frac{1}{2}$, and for $j = j_1 - \frac{1}{2}$ it is

$$X = \sqrt{\frac{1-t}{2}}, \qquad Y = -\sqrt{\frac{1+t}{2}}.$$

These results are summarized in Table B.1.

TABLE B.1. $\left\langle j_1 \tfrac{1}{2}, m - \sigma, \sigma \mid jm \right\rangle$

	σ	
$j - j_1$	$\frac{1}{2}$	$-\frac{1}{2}$
$\frac{1}{2}$	$\sqrt{\frac{j_1 + m + \frac{1}{2}}{2j_1 + 1}}$	$\sqrt{\frac{j_1 - m + \frac{1}{2}}{2j_1 + 1}}$
$-\frac{1}{2}$	$\sqrt{\frac{j_1 - m + \frac{1}{2}}{2j_1 + 1}}$	$-\sqrt{\frac{j_1 + m + \frac{1}{2}}{2j_1 + 1}}$

Other values for Clebsch–Gordan coefficients are generally more complicated to express. Thorough treatments of the relations between Clebsch–Gordan coefficients and tables of values are available in the literature. See, for example, Refs. 9 or 10. Another relatively simple form taken by the Clebsch–Gordan coefficients occurs when the projection of the angular momentum coincides with this momentum. In that case we have

$$\langle j_1 j_2, j_1 m_2 \,|\, jm\rangle = \frac{\delta_{j_1+m_2,m}}{\sqrt{(j_1+j_2+j+1)!}} \times \sqrt{\frac{(j_2-m_2)!\,(j+m)!\,(2j+1)\,(2j_1)!\,(j-j_1+j_2)!}{(j_2+m_2)!\,(j-m)!\,(j_1-j_2+j)!\,(j_1+j_2-j)!}}, \tag{B.7}$$

$$\langle j_1 j_2, m_1 j_2 \,|\, jm\rangle = \frac{(-1)^{j_1+j_2-j}\,\delta_{m_1+j_2,m}}{\sqrt{(j_1+j_2+j+1)!}} \times \sqrt{\frac{(j_1-m_1)!\,(j+m)!\,(2j+1)\,(2j_2)!\,(j_1-j_2+j)!}{(j_1+m_1)!\,(j-m)!\,(j-j_1+j_2)!\,(j_1+j_2-j)!}}, \tag{B.8}$$

$$\langle j_1 j_2, -j_1 m_2 \,|\, jm\rangle = (-1)^{j_1+j_2-j}\langle j_1 j_2, j_1-m_2 \,|\, j-m\rangle, \tag{B.9}$$

$$\langle j_1 j_2, m_1-j_2 \,|\, jm\rangle = (-1)^{j_1+j_2-j+1}\langle j_1 j_2, -m_1 j_2 \,|\, j-m\rangle. \tag{B.10}$$

Values of the Clebsch–Gordan coefficients are given in Table B.2 for the case $j_2 = 1$. If $m_2 = 1$ or -1, then the coefficients can be calculated using the connections in Eqs. (B.9) and (B.10). The third possibility, $m_2 = 0$, can be calculated using the values obtained for $m_2 = 1$ and -1, and the normalization condition (B.3) for the Clebsch–Gordan coefficients.

TABLE B.2. $\langle j_1 1; m-m_2, m_2 \mid jm\rangle$

	m_2		
$j-j_1$	1	0	-1
1	$\sqrt{\dfrac{(j_1+m+1)(j_1+m)}{(2j_1+2)(2j_1+1)}}$	$\sqrt{\dfrac{(j_1-m+1)(j_1+m+1)}{(2j_1+1)(j_1+1)}}$	$\sqrt{\dfrac{(j_1-m+1)(j_1-m)}{(2j_1+2)(2j_1+1)}}$
0	$-\sqrt{\dfrac{(j_1-m+1)(j_1+m)}{2j_1(2j_1+1)}}$	$\dfrac{m}{\sqrt{j_1(j_1+1)}}$	$\sqrt{\dfrac{(j_1+m+1)(j_1-m)}{2j_1(j_1+1)}}$
-1	$\sqrt{\dfrac{(j_1-m+1)(j_1-m)}{2j_1(2j_1+1)}}$	$-\sqrt{\dfrac{(j_1-m)(j_1+m)}{j_1(2j_1+1)}}$	$\sqrt{\dfrac{(j_1+m+1)(j_1+m)}{2j_1(2j_1+1)}}$

B.3 WIGNER 3j SYMBOLS

A quantity closely related to the Clebsch–Gordan coefficient is the Wigner 3j coefficient, designed to achieve maximum symmetry. It can be defined as

$$\begin{pmatrix} j_1 & j_2 & j_3 \\ m_1 & m_2 & m_3 \end{pmatrix} = (-1)^{j_1-j_2-m_3} \langle j_1 j_2, m_1 m_2 \, | j_3, -m_3 \rangle . \tag{B.11}$$

The 3j symbol has the property that

$$\begin{pmatrix} j_1 & j_2 & j_3 \\ m_1 & m_2 & m_3 \end{pmatrix} = 0 \qquad \text{if} \qquad m_1 + m_2 + m_3 \neq 0,$$

in place of Eq. (B.2).

We list the principal symmetry and orthogonality properties. Even permutation of the columns leaves the 3j symbol unchanged, or

$$\begin{pmatrix} j_1 & j_2 & j_3 \\ m_1 & m_2 & m_3 \end{pmatrix} = \begin{pmatrix} j_2 & j_3 & j_1 \\ m_2 & m_3 & m_1 \end{pmatrix} = \begin{pmatrix} j_3 & j_1 & j_2 \\ m_3 & m_1 & m_2 \end{pmatrix} .$$

Odd permutation of the columns, on the other hand, is equivalent to multiplication by $(-1)^{j_1+j_2+j_3}$, so that

$$\begin{aligned} (-1)^{j_1+j_2+j_3} \begin{pmatrix} j_1 & j_2 & j_3 \\ m_1 & m_2 & m_3 \end{pmatrix} &= \begin{pmatrix} j_2 & j_1 & j_3 \\ m_2 & m_1 & m_3 \end{pmatrix} \\ &= \begin{pmatrix} j_1 & j_3 & j_2 \\ m_1 & m_3 & m_2 \end{pmatrix} = \begin{pmatrix} j_3 & j_2 & j_1 \\ m_3 & m_2 & m_1 \end{pmatrix} . \end{aligned}$$

Orthogonality properties of the Wigner 3j symbols are

$$\begin{aligned} \sum_{j_3 m_3} \begin{pmatrix} j_1 & j_2 & j_3 \\ m_1 & m_2 & m_3 \end{pmatrix} \begin{pmatrix} j_1 & j_2 & j_3 \\ m_1' & m_2' & m_3 \end{pmatrix} &= \delta_{m_1 m_1'} \delta_{m_2 m_2'}, \\ \sum_{m_1 m_2} \begin{pmatrix} j_1 & j_2 & j_3 \\ m_1 & m_2 & m_3 \end{pmatrix} \begin{pmatrix} j_1 & j_2 & j_3' \\ m_1 & m_2 & m_3' \end{pmatrix} &= \frac{\delta_{j_3 j_3'} \delta_{m_3 m_3'} \delta\,(j_1 j_2 j_3)}{2 j_3 + 1}, \end{aligned} \tag{B.12}$$

where $\delta\,(j_1 j_2 j_3)$ in Eq. (B.12) is a quantity defined as

$$\delta\,(j_1 j_2 j_3) = \begin{cases} 1 & \text{if } |j_1 - j_2| \leq j_3 \leq j_1 + j_2 \\ 0 & \text{otherwise} \end{cases} . \tag{B.13}$$

The statement in Eq. (B.13) is called the *triangular condition*.

APPENDIX C

ROTATION FUNCTIONS

Rotation functions can be introduced as the coefficients of the transformation of a wave function from one set of quantization axes to another:

$$\psi_{jm} = \sum_K D^j_{mK}(\vartheta)\, \psi_{jK}. \tag{C.1}$$

The transformed wave function ψ_{jm} is for the state with angular momentum j and angular momentum projection m in the new coordinate system, ψ_{jK} is a wave function describing the state with quantum numbers j and K in the old coordinate system, D^j_{mK} is the rotation function, and θ represents the set of angles that defines the transformation from the old to the new coordinate system. The rotation functions are also referred to as generalized spherical functions, rotation matrices, D functions, or D matrices.

We shall now explore the basic properties of the rotation functions. From orthonormalization of the wave functions and the relation (C.1), we find that the rotation functions satisfy the relation

$$\sum_m \left(D^j_{mK}\right)^* D^j_{mK'} = \sum_m \left(D^j_{Km}\right)^* D^j_{K'm} = \delta_{KK'}. \tag{C.2}$$

Another property of these functions is connected with the parity of the wave functions, that is, with the transformation of the wave functions upon inversion of the coordinate axes. When we also incorporate the relation between the rotation functions in direct and reverse transformations, we find

$$D^j_{mK}(\vartheta) = (-1)^{K-m}\left[D^j_{-m,-K}(-\vartheta)\right]^*. \tag{C.3}$$

Other important properties of the rotation functions follow from the action of the angular momentum operator on these functions. The rotation operator can be described either in terms of the angular momentum operator or in terms of the rotation functions. A simple relation exists, therefore, between these two quantities. We denote by $x,\ y,\ x$ the axes in a stationary coordinate system, while the axes $\xi,\ \eta,\ \varsigma$ refer to a moving coordinate system. (See Fig. C.1.) The rotation functions establish the connection between these two coordinate systems. These functions satisfy the eigenvalue relations

$$\begin{aligned} \widehat{L}_z D^j_{mK}(\vartheta) &= \hbar m D^j_{mK}(\vartheta), \\ \widehat{L}_\zeta D^j_{mK}(\vartheta) &= \hbar K D^j_{mK}(\vartheta), \\ \widehat{\mathbf{L}}^2 D^j_{mK}(\vartheta) &= \hbar^2 j(j+1) D^j_{mK}(\vartheta). \end{aligned} \tag{C.4}$$

Equations (C.4) allow us to give an alternative interpretation to the rotation functions. They are eigenfunctions of the operators corresponding to simultaneous rotations with respect to both the stationary and moving coordinate systems. In particular, a rotation function represents an eigenfunction of the rotational state of a multiatomic molecule with two equal moments of inertia. If the ζ axis is the third direction, then we can describe the rotation of a molecule by the set of quantum numbers j, K, and m. These quantum numbers represent, respectively, the total rotational momentum, the projection of this momentum on the moving axis ζ, and the projection of the total momentum on the fixed axis z. The motion of a molecule can be represented as successive rotations of the molecule with respect to its axis ζ, and the rotation of this axis in the stationary coordinate system. The rotation function is the eigenfunction describing such a motion of the molecule.

Another example illustrating an application of the rotation function relates to a linear molecule rotated around an axis normal to the molecular axis. We assume that the interaction is weak between the rotation of the molecule and the motions of the electrons in the molecule. Then the rotation function D^j_{mK} represents an eigenfunction

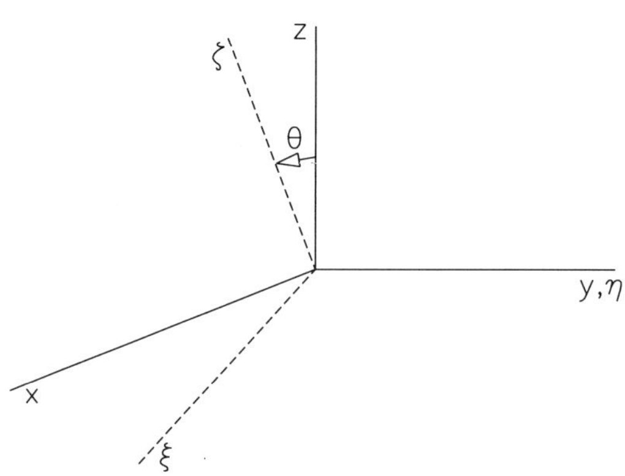

Figure C.1. Relation between the fixed x, y, z axes and the rotating ξ, η, ζ axes.

of the state that includes both the rotation of the molecular axis and the rotation of the electrons in the molecule. The quantum number j is the total angular momentum composed of the sum of the rotational molecular and electron angular momenta, m is the projection of the rotational momentum on the stationary axis z, and K is the projection of the electron momentum onto the molecular axis.

In both examples, when $K = 0$, the rotation function coincides with the wave function describing the rotation of a linear molecule and is of the form

$$D^j_{m0} = \sqrt{\frac{4\pi}{2j+1}} Y_{jm}(\theta, \varphi),$$
$$Y_{jm}(\theta, \varphi) = \frac{1}{\sqrt{2\pi}} e^{im\varphi} \sqrt{\frac{(2j+1)(j-m)!}{2(j+m)!}} P^{|m|}_j(\cos\theta). \tag{C.5}$$

Here $Y_{jm}(\theta, \varphi)$ is the normalized spherical harmonic corresponding to rotational motion with angular momentum j and projection m of this momentum on the fixed reference axis; θ and φ are the polar and azimuthal angles, respectively; and $P^{|m|}_j$ is the associated Legendre function.

Since the rotation functions describe the motion of a symmetrical rotator, they satisfy the addition rules

$$D^j_{m_1K_1}(\vartheta) D^j_{m_2K_2}(\vartheta) = \sum_{j=|j_1-j_2|}^{j_1+j_2} \langle j_1 j_2, m_1 m_2 \,|\, jm \rangle \times \langle j_1 j_2, K_1 K_2 \,|\, jK \rangle D^j_{mK}(\vartheta), \tag{C.6}$$

and

$$D^j_{mK}(\vartheta) = \sum_{m_1 K_1} \langle j_1 j_2, m_1 m_2 \,|\, jm \rangle \times \langle j_1 j_2, K_1 K_2 \,|\, jK \rangle D^{j_1}_{m_1K_1}(\vartheta) D^{j_2}_{m_2K_2}(\vartheta). \tag{C.7}$$

Here ϑ represents the set of angles that determine the position of the new coordinate system as a rotation with respect to the stationary coordinate system; $\langle j_1 j_2, m_1 m_2 \,|\, jm \rangle$ are Clebsch–Gordan coefficients; and, in accordance with the properties of these coefficients, we have $m = m_1 + m_2$ and $K = K_1 + K_2$.

C.1 MATRIX ELEMENTS OF ROTATION FUNCTIONS

The next important property of the rotation functions is the integral of the triple product of these functions,

$$\int \left[D^j_{mK}(\vartheta)\right]^* D^{j_1}_{m_1K_1}(\vartheta) D^{j_2}_{m_2K_2}(\vartheta)\, d\Omega$$
$$= \frac{8\pi^2}{2j+1} \langle j_1 j_2, m_1 m_2 \,|\, jm \rangle \langle j_1 j_2, K_1 K_2 \,|\, jK \rangle, \tag{C.8}$$

where $d\Omega$ is the element of solid angle and integration is over all angles.

Using Eq. (C.8), we can now calculate matrix elements of the rotation functions taken on rotational states of the symmetrical rotator. With $j_2 = m_2 = K_2 = 0$, we find that orthonormalized wave functions for the symmetrical rotator are

$$|jmK\rangle = \sqrt{\frac{2j+1}{8\pi^2}} D^j_{mK}(\vartheta),$$

where j, m, and K are, respectively, the angular momentum, its projection on the fixed axis, and its projection on the molecular axis. With the use of Eq. (C.8), we find that the matrix elements of the rotation function are

$$\left\langle j_1 m_1 K_1 \left| D^{j_2}_{m_2 K_2}(\vartheta) \right| jmK \right\rangle = \sqrt{\frac{2j_1+1}{2j+1}} \langle j_1 j_2, m_1 m_2 | jm \rangle \langle j_1 j_2, K_1 K_2 | jK \rangle . \tag{C.9}$$

Using Eq. (C.9), we can relate matrix elements of a vector in the stationary and in the rotating coordinate systems. If we express a matrix element in terms of its value in the rotating coordinate system, we can thereby exclude the rotation of the axis of the symmetrical rotator and reduce the problem to the calculation of quantities that depend only on the internal states of the rotator. We consider a vector **A**, with projections A_x, A_y, A_z on the stationary set of axes, and projections A_ξ, A_η, A_ζ on the axes of the rotating coordinate system. Our goal is to connect the matrix elements of these two sets of vector projections. It is simpler to find first the relationships between the sets of quantities $A_x + iA_y$, A_z, $A_x - iA_y$ and $A_\xi + iA_\eta$, A_ζ, $A_\xi - iA_\eta$. We shall denote the first set by A_q, where $q = 1, 0, -1$, and the second set by A_μ, where $\mu = 1, 0, -1$. From the transformation rules for vectors, we have

$$A_q = \sum_\mu D^1_{q\mu}(\vartheta) A_\mu, \tag{C.10}$$

where ϑ is the set of angles defining the transformation from the stationary coordinates to the rotating coordinates.

From Eqs. (C.9) and (C.10) we find the matrix elements of the vector components in the stationary system to be

$$\begin{aligned} \left\langle j'm'K'\alpha' \left| A_q \right| jmK\alpha \right\rangle &= \sum_\mu \left\langle j'm'K' \left| D^1_{q\mu} \right| jmK \right\rangle \left\langle K'\alpha' \left| A_\mu \right| K\alpha \right\rangle \\ &= \sum_\mu \sqrt{\frac{2j'+1}{2j+1}} \left\langle j'1, m'q | jm \right\rangle \left\langle j'1, K'\mu | jK \right\rangle \\ &\quad \times \left\langle K'\alpha' \left| A_\mu \right| K\alpha \right\rangle . \end{aligned} \tag{C.11}$$

Here j, j' are total angular momentum quantum numbers in initial and final states, m, m' are the angular momentum projections onto the stationary axes, K, K' are projections of the angular momenta onto the molecular axes, and α, α' are all the other quantum numbers necessary to define the initial and final states, respectively.

In conclusion, we give explicit results for the rotation functions for the small values of angular momentum $j = \frac{1}{2}$ and $j = 1$. We take the transformation between the stationary and rotating systems to be such that the η axis coincides with the y axis. That is, the transformation corresponds to a rotation around the η axis. The angle between the ς and z axes is labeled ϑ (as in Fig. C.1). The value of the rotation functions are given in Tables C.1 and C.2.

TABLE C.1. $D^{1/2}_{mK}(\vartheta)$

	m	
K	$\frac{1}{2}$	$-\frac{1}{2}$
$\frac{1}{2}$	$\cos(\vartheta/2)$	$-\sin(\vartheta/2)$
$-\frac{1}{2}$	$\sin(\vartheta/2)$	$\cos(\vartheta/2)$

TABLE C.2. $D^{1}_{mK}(\vartheta)$

		m	
K	1	0	-1
1	$\frac{1+\cos\vartheta}{2}$	$-\frac{\sin\vartheta}{\sqrt{2}}$	$\frac{1-\cos\vartheta}{2}$
0	$\frac{\sin\vartheta}{\sqrt{2}}$	$\cos\vartheta$	$-\frac{\sin\vartheta}{\sqrt{2}}$
-1	$\frac{1-\cos\vartheta}{2}$	$\frac{\sin\vartheta}{\sqrt{2}}$	$\frac{1+\cos\vartheta}{2}$

APPENDIX D

WIGNER $6j$ SYMBOLS

In this appendix we consider the addition of three angular momenta: j_1, j_2, j_3. The total angular momentum is J, with projection M onto a fixed axis. Addition can be carried out in two ways. In the first, we add the angular momenta j_1 and j_2 to form an angular momentum j_4, and then add j_4 and j_3 to obtain J. See Figure D.1. We represent by $\Psi_{JM}^{j_4}$ a wave function formed in this manner. The second approach is to add the angular momenta j_1 and j_3 to form j_5, and then combine the angular momenta j_5 and j_2 into the total angular momentum J. This wave function is denoted by $\Psi_{JM}^{j_5}$.

The overlap between these functions can be written as

$$\left\langle \Psi_{JM}^{j_4} | \Psi_{JM}^{j_5} \right\rangle = (-1)^{j_1+j_2+j_3+J} \sqrt{(2j_4+1)(2j_5+1)} \begin{Bmatrix} j_1 & j_2 & j_4 \\ J & j_3 & j_5 \end{Bmatrix}. \tag{D.1}$$

The quantity contained in the brace is called the *Wigner 6j symbol*, and the quantity

$$W(j_1, j_2, j_3, J, j_4, j_5) = (-1)^{j_1+j_2+j_3+J} \begin{Bmatrix} j_1 & j_2 & j_4 \\ J & j_3 & j_5 \end{Bmatrix}$$

is called the *Racah coefficient*.

Successive applications of Eq. (B.1) to write a wave function for added angular momenta gives the Wigner $6j$ symbol in the form

$$\begin{Bmatrix} j_1 & j_2 & j_4 \\ J & j_3 & j_5 \end{Bmatrix} = \frac{(-1)^{j_1+j_2+j_3+J}}{\sqrt{(2j_4+1)(2j_5+1)}} \sum_{m_1 m_2 m_3 m_4 m_5} \langle j_1 j_2, m_1 m_2 \,|\, j_4 m_4 \rangle \times \langle j_3 j_4, m_3 m_4 \,|\, JM \rangle \langle j_1 j_3, m_1 m_3 \,|\, j_5 m_5 \rangle \langle j_2 j_5, m_2 m_5 \,|\, JM \rangle. \tag{D.2}$$

From the definition of the Wigner $6j$ symbols, we can obtain the sum rule for the product of three Clebsch–Gordan coefficients,

$$\sum_{m_1 m_3 m_4} \langle j_1 j_2, m_1 m_2 \,|j_4 m_4\rangle \langle j_3 j_4, m_3 m_4 \,|JM\rangle \langle j_1 j_3, m_1 m_3 \,|j_5 m_5\rangle$$

$$= \frac{(-1)^{j_1+j_2+j_3+J}}{\sqrt{(2j_4+1)(2j_5+1)}} \begin{Bmatrix} j_1 & j_2 & j_4 \\ J & j_3 & j_5 \end{Bmatrix} \langle j_2 j_5, m_2 m_5 \,|JM\rangle, \qquad \text{(D.3)}$$

which follows from Eqs. (D.2) and (B.3).

From the definition of the Wigner $6j$ symbols, we find the sum of binary products of the $6j$ symbols to be

$$\sum_{j_5} (2j_5+1) \begin{Bmatrix} j_1 & j_2 & j_4 \\ j & j_3 & j_5 \end{Bmatrix} \begin{Bmatrix} j_1 & j_2 & j_4' \\ j & j_3 & j_5 \end{Bmatrix}$$

$$= \frac{\delta_{j_4 j_4'}}{2j_4+1} \sum_{j_4} (-1)^{j_2+j_3+j_5} (2j_4+1) \begin{Bmatrix} j_1 & j_2 & j_4 \\ j & j_3 & j_5 \end{Bmatrix} \begin{Bmatrix} j_2 & j_3 & j_4 \\ j_1 & j & j_6 \end{Bmatrix} \qquad \text{(D.4)}$$

$$= \begin{Bmatrix} j_1 & j_2 & j_5 \\ j_3 & j & j_6 \end{Bmatrix}. \qquad \text{(D.5)}$$

D.1 PROPERTIES OF $6j$ SYMBOLS

The principal symmetry property of the $6j$ symbols is that the permutation of any pairs

$$\left.\begin{matrix} j_1 \\ j_2 \end{matrix}\right\} \rightleftarrows \left\{\begin{matrix} j_3 \\ j_4 \end{matrix}\right. \qquad \text{or} \qquad \left.\begin{matrix} j_1 \\ j_2 \end{matrix}\right\} \rightleftarrows \left\{\begin{matrix} j_2 \\ j_1 \end{matrix}\right.$$

leaves the value of the $6j$ symbol unaltered. In all, there are 48 such permutations possible for each Wigner $6j$ symbol.

This property can be understood geometrically from the diagram in Figure D.1. The momentum represented by one side of a triangle is the vector sum of the other two sides. Choosing some sets of sides for the momentum triangles that define the tetrahedron leads to equivalences with other schemes for defining the same tetrahedron. According to this property, the $6j$ symbol is the same for any possible way to add the angular momentum vectors. We can calculate the number of such combinations by noting that we have $4! = 24$ possibilities for the composition of the four triangles in the given scheme of addition, and two ways in each variation for choosing the first angular momenta that are added. We thus have 48 different ways to add the angular momenta, which therefore leads to 48 different Wigner $6j$ symbols that have the same value but differ from each other by permutation of several of the elements.

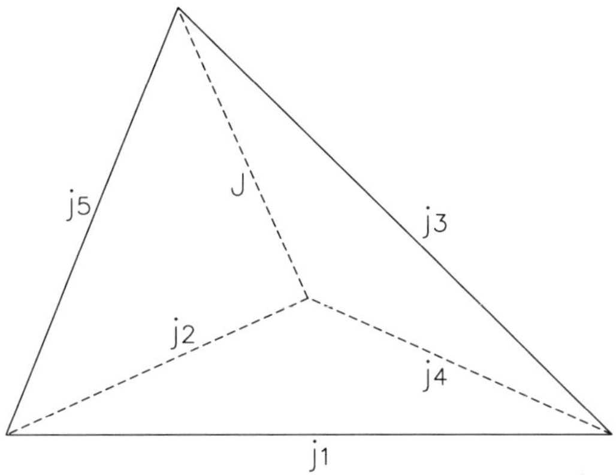

Figure D.1. Diagram showing alternative schemes for the coupling of three angular momenta.

We exhibit here another property of the Wigner $6j$ symbol:

$$\begin{Bmatrix} j_1 & j_2 & j_4 \\ j & j_3 & j_5 \end{Bmatrix} = \begin{Bmatrix} j_1 & \frac{1}{2}(j_2 + j_3 + j_4 - j_5) & \frac{1}{2}(j_2 - j_3 + j_4 + j_5) \\ j & \frac{1}{2}(j_2 + j_3 - j_4 + j_5) & \frac{1}{2}(-j_2 + j_3 + j_4 + j_5) \end{Bmatrix}. \tag{D.6}$$

Next we consider the simplest analytical expressions for the $6j$ symbols. An analytical form exists for the case when one of the angular momenta is an arithmetic sum of two of the others. In this case, the tetrahedron in Figure D.1 degenerates into a triangle in a plane. The expression for this case is

$$\begin{aligned} &\begin{Bmatrix} j_1 & j_2 & j_1 + j_2 \\ j & j_3 & j_5 \end{Bmatrix} \\ &\quad = (-1)^{j_1+j_2+j_3+j} \left\{ \frac{(2j_1)!\,(2j_2)!\,(j_1 + j_2 + j_3 + 1)!}{(2j_1 + 2j_2 + 1)!\,(j_1 + j_3 + j_5 + 1)!} \right. \\ &\qquad \times \frac{(j_1 + j_2 + j_3 - j)!\,(j_1 + j_2 - j_3 + j)!}{(j_2 + j_5 + j + 1)!\,(-j_1 - j_2 + j_3 + j)!\,(j_1 + j_3 - j_5)!} \\ &\qquad \times \left. \frac{(-j_1 + j_3 + j)!\,(-j_2 + j_3 + j)!}{(j_1 - j_3 + j_5)!\,(j_2 + j - j_5)!\,(j_2 - j + j_5)!} \right\}^{1/2}. \end{aligned} \tag{D.7}$$

D.2 PARTIAL VALUES OF $6j$ SYMBOLS

If one of the angular momenta is zero, then the tetrahedron collapses to a triangle, with only three different nonzero angular momentum values. If we set $j_1 = 0$, then Eq. (D.7) gives the Wigner $6j$ symbol in the form

$$\begin{Bmatrix} 0 & j_2 & j_2 \\ j & j_3 & j_3 \end{Bmatrix} = \frac{(-1)^{j_2+j_3+j}}{\sqrt{(2j_2 + 1)(2j_3 + 1)}}. \tag{D.8}$$

If one of the angular momenta has the value $\frac{1}{2}$, then the Wigner $6j$ symbol can be expressed as

$$\begin{Bmatrix} j_1 & j_2 & j_4 \\ \frac{1}{2} & j_3 & j_5 \end{Bmatrix} = \frac{(-1)^{j_1+j_2+j_4}}{\sqrt{(2j_2+1)(2j_3+1)(2j_4+1)(2j_5+1)}} \times \left\{ \frac{(j_1+j_2+j_4+1)!}{(j_1+j_3+j_5+1)!} \times \frac{(j_1-j_3+j_5)!\,(j_1-j_2+j_4)!\,(-j_1+j_2+j_4)!}{(j_1+j_3-j_5)!\,(j_1+j_2-j_4)!\,(-j_1+j_3+j_5)!} \right\}^{1/2}. \tag{D.9}$$

D.3 MATRIX ELEMENTS IN ADDITION OF MOMENTA

Since the Wigner $6j$ symbol expressed in Eq. (D.2) is a sum of quaternary products of Clebsch–Gordan coefficients, these symbols will appear in matrix element calculations for a system consisting of two subsystems. We illustrate this assertion with an example of such a matrix element. Take J to be a total angular momentum arising from the angular momenta j_1 and j_2 of two subsystems. We wish to calculate the matrix element of a quantity that depends on variables of the first subsystem. We assume that this quantity is a spherical tensor, that is, its dependence on angles is of the form

$$A_{lq} = AD^l_{q0}(\vartheta), \tag{D.10}$$

where A is independent of the angles, D^l_{q0} is the rotation function, and ϑ is the set of angles. Our goal is to calculate the matrix element

$$\left\langle J'M'\alpha' \left| A_{lq} \right| JM\alpha \right\rangle, \tag{D.11}$$

where $J,\ J'$ are the angular momenta of the initial and final states, respectively; $M,\ M'$ are their projections on the quantization axis; and $\alpha,\ \alpha'$ are the sets of other quantum numbers necessary to fully define the states.

We now calculate the matrix element (D.11) using the fact that the operator (D.10) depends only on variables of the first subsystem. Using Eq. (B.1) for the wave function of the system, we can write the matrix element in the form

$$\left\langle J'M'\alpha' \left| A_{lq} \right| JM\alpha \right\rangle = \sum_{m_1 m_1' m_2} \langle j_1 j_2, m_1 m_2 | JM \rangle \times \left\langle j_1' j_2, m_1' m_2 \left| J'M' \right.\right\rangle \left\langle j_1' m_1', J'\alpha' \left| A_{lq} \right| j_1 m_1 J\alpha \right\rangle.$$

We calculate the matrix element of the operator (D.10) taken over states of the first subsystem, employ Eq. (C.9), and obtain

$$\begin{aligned}\left\langle J'M'\alpha' \left| A_{lq} \right| JM\alpha \right\rangle = \sum_{m_1 m_1' m_2} & \langle j_1 j_2, m_1 m_2 \,| JM \rangle \langle j_1' j_2, m_1' m_2 \left| J'M' \right\rangle \\ & \times \left\langle j_1' l, m_1' q \,| j_1 m_1 \right\rangle \sqrt{\frac{2j_1' + 1}{2j_1 + 1}} \left\langle j_1' l, 00 \,| j_1 0 \right\rangle \\ & \times \left\langle j_1' J' \alpha' \,| A |\, j_1 J \alpha \right\rangle .\end{aligned}$$

The sums over projections of the angular momenta are accomplished using Eq. (D.3). The final result is

$$\begin{aligned}\left\langle J'M'\alpha' \left| A_{lq} \right| JM\alpha \right\rangle = (-1)^{j_1 + j_1' + 2j_2 + l} & \sqrt{(2j_1' + 1)(2j_1 + 1)} \left\langle j_1' l, 00 \,| j_1 0 \right\rangle \\ & \times \left\langle lJ', qM' \,| JM \right\rangle \begin{Bmatrix} j_1' & l & j_1 \\ J & j_2 & J' \end{Bmatrix} \left\langle j_1' \alpha' \,| A |\, j_1 \alpha \right\rangle .\end{aligned} \tag{D.12}$$

We have thus explicitly determined the dependence of the matrix element on the magnetic quantum numbers. Equation (D.12) is called the Wigner–Eckart theorem.

APPENDIX E

FRACTIONAL PARENTAGE COEFFICIENTS

We consider the wave function Ψ of N valence electrons with the same principal quantum number, bound in an atom. This wave function can be represented as a product of the wave function φ of one of the valence electrons, and a wave function ϕ for the other valence electrons. We denote by l_e the orbital angular momentum of the electron thus singled out, with its angular momentum projection on a fixed direction called μ, and with σ as the electron spin projection in the same direction. We adopt the notation

$$\varphi = \varphi_{l_e \mu \frac{1}{2} \sigma}(N)$$

for the wave function of this electron. In analogous fashion, we can describe the remaining valence electrons by the wave function

$$\phi = \phi_{l m_l s m_s}(1, 2, \ldots, N-1),$$

where l, s are the orbital and spin momenta of the other $N-1$ valence electrons, with projections m_l, m_s on the fixed direction.

We now introduce the Clebsch–Gordan coefficients

$$\left\langle \tfrac{1}{2} s, \sigma m_s \left| S M_S \right. \right\rangle, \qquad \left\langle l_e l, \mu m_l \left| L M_L \right. \right\rangle$$

describing the addition of the electron spins into the total spin S of the atom, and also the addition of the orbital angular momenta of the electrons into a total orbital angular momentum L of the atom, with projections on the fixed direction of M_S and M_L, respectively.

Multielectron configurations are generally in several levels. We classify each of these levels by definition of the initial level. The initial level, in turn, is identified as a level that produces the given atomic state after addition of an electron. The specification of the initial level can be regarded as a definition of the parentage of the level. The total wave function $\Psi_{LM_LSM_S}(1, 2, \ldots, N)$ is represented as a linear combination of products φ and ϕ corresponding to various initial levels. The coefficients in this combination are denoted as $G_{ls}^{LS}(l_e, N)$. They are called fractional parentage coefficients. That is, we have the total wave function expressed as

$$\Psi_{LM_LSM_S}(1, 2, \ldots, N) = \widehat{P} \sum_{lm_lsm_s} G_{ls}^{LS}(l_e, N) \left\langle \tfrac{1}{2}s, \sigma m_s \,|SM_S \right\rangle \times \langle l_e l, \mu m_l \,|LM_L\rangle \, \phi_{lm_lsm_s}(1, 2, \ldots, N-1) \, \varphi_{l_e\mu\frac{1}{2}\sigma}(N),$$

where $\widehat{P}$ is the operator for permutation of identical electrons.

TABLE E.1. Fractional parentage coefficients for the case $L = 1$

p-Electron State	Atomic State	G_{ls}^{LS}
	$p^2(^3P)$	1
$p(^2P)$	$p^2(^1D)$	1
	$p^2(^1S)$	1
	$p^3(^4S)$	1
$p^2(^3P)$	$p^3(^2D)$	$1/\sqrt{2}$
	$p^3(^2P)$	$-1/\sqrt{2}$
	$p^3(^4S)$	0
$p^2(^1D)$	$p^3(^2D)$	$-1/\sqrt{2}$
	$p^3(^2P)$	$-\sqrt{\frac{5}{18}}$
	$p^3(^4S)$	0
$p^2(^1S)$	$p^3(^2D)$	0
	$p^3(^2P)$	$\sqrt{\frac{2}{3}}$
	$p^4(^3P)$	$\sqrt{\frac{5}{12}}$
$p^3(^2D)$	$p^4(^1D)$	$\sqrt{\frac{3}{4}}$
	$p^4(^3S)$	0
	$p^4(^3P)$	$-1/\sqrt{3}$
$p^3(^4S)$	$p^4(^1D)$	0
	$p^4(^1S)$	0
	$p^4(^3P)$	$-\frac{1}{2}$
$p^3(^2P)$	$p^4(^1D)$	$-\frac{1}{2}$
	$p^4(^1S)$	1
$p^4(^3P)$	$p^5(^2P)$	$\sqrt{\frac{3}{5}}$
$p^4(^1D)$	$p^5(^2P)$	$1/\sqrt{3}$
$p^4(^1S)$	$p^5(^2P)$	$1/\sqrt{15}$

Fractional parentage coefficients are calculated by making use of the fact that among those states that may be constructed from φ and ϕ by straightforward addition of angular momenta, there will be some that are excluded by the Pauli principle. That is, only some of the combinations of φ and ϕ will satisfy the Pauli principle.

Fractional parentage coefficients are determined by incremental construction starting with $N = 3$. For example, the $N = 3$ wave function begins with a pair of electrons antisymmetrized in accordance with the Pauli principle. This is then combined with the wave function of the third electron. The wave function thus obtained will change sign upon permutation of the first two electrons but will not have this property in terms of permutations of each of the first two electrons with the third. If we alter the scheme by which the angular momenta are combined, we can obtain a wave function that will change sign upon permutation of the second and third electrons. A linear combination of the original and permuted wave functions can be required to change sign upon the permutations $1 \rightleftarrows 2$ and $2 \rightleftarrows 3$. This is then the required fully antisymmetrized wave function Ψ for the three valence electrons. By analogous procedures, we can obtain systems of equations to determine fractional parentage coefficients in the case of four or more equivalent electrons.

Fractional parentage coefficients must also permit the normalization of the wave function Ψ, which means they must have the property

$$\sum_{ls} \left[G_{ls}^{LS}(l_e, N)\right]^2 = 1.$$

For the case of valence s electrons, the fractional parentage coefficient is unity. Fractional parentage coefficients for an electron system in a valence p shell are given in Table E.1.

APPENDIX F

ATOMIC PROPERTIES

We shall treat an atom as a system comprised of electrons and a Coulomb central potential. Since an ion can be described in the same terms, the results obtained below are valid both for neutral atoms and for ions.

Relativistic effects will be neglected. This is a valid assumption for atomic nuclei of small and moderate charge numbers. The Hamiltonian of the electron system is of the form

$$H = -\sum_i \frac{\hbar^2}{2m}\nabla_i^2 - \sum_i \frac{Ze^2}{r_i} + \sum_{i>k} \frac{e^2}{|\mathbf{r}_i - \mathbf{r}_k|}, \tag{F.1}$$

where i and k are indices enumerating the electrons, $\mathbf{r}_i$ and $\mathbf{r}_k$ are position vectors of the corresponding electrons, and Z is the nuclear charge number. The wave function of the electron system changes sign upon the permutation of any pair of electrons.

It is easily shown that the orbital angular momentum operator of the electrons,

$$\widehat{\mathbf{L}} = \sum_i \widehat{\mathbf{l}}_i = \sum_i \mathbf{r}_i \times \widehat{\mathbf{p}}_i,$$

commutes with the Hamiltonian. In this expression, $\widehat{\mathbf{l}}_i$ is the orbital angular momentum operator of the ith electron, with position vector $\mathbf{r}_i$ and linear momentum operator $\widehat{\mathbf{p}}_i$. Hence, the good quantum numbers describing a state of the atom are:

L, the orbital angular momentum of the atom

M_L, the projection of this momentum on a fixed direction

S, the spin angular momentum of the atom, since the Hamiltonian of Eq. (F.1) does not depend on spin, and therefore commutes with the spin operator of the atom

M_S, the projection of the spin angular momentum on the same fixed direction relating to M_L

The spin state of an atom is determined by S and M_S only, although any arbitrary superposition of spins of various electrons commutes with the Hamiltonian. However, only that spin operator for an atom that sums the spins of all the electrons will have the necessary antisymmetry property when acting on the atomic wave function. That is, this operator must produce a change in sign of the wave function upon the permutation of any two electrons.

We shall now replace the field acting on all the individual electrons by a self-consistent field depending on the coordinates of only that one electron being studied. This field is spherically symmetrical, so the introduction of a self-consistent field is equivalent to a replacement of the Hamiltonian (F.1) by the model Hamiltonian

$$\widehat{H_0} = \sum_i \widehat{h_i}, \qquad \widehat{h_i} = -\frac{\hbar^2}{2m}\nabla_i^2 + V(r_i), \tag{F.2}$$

where $\widehat{h_i}$ is the Hamiltonian of a single electron, and V is the potential of the self-consistent field. The variables are separated in the Schrödinger equation based on the Hamiltonian in Eq. (F.2), and so the wave function for the system of electrons can be written as a product of single-electron wave functions. Symmetry properties must be imposed on this wave function. The state of a single electron is determined by the quantum numbers n, l, m, and σ, where n is the principal quantum number, l is the orbital angular momentum quantum number, m is the magnetic quantum number, and σ is the projection of the electron spin on a fixed direction. The quantities n, l, and m are integers satisfying the inequalities $n \geq 1$, $l \leq n - 1$, $-l \leq m \leq l$, and σ has the two possible values $\pm\frac{1}{2}$.

The energy of a single electron depends on the quantum numbers n and l only, which means that each electron state has a $2(2l + 1)$-fold degeneracy. The energy of a single electron increases with an increase in the n and l quantum numbers, so that states of the electron are enumerated beginning with the states of smallest n and l values. We must now take into account the Pauli principle, which requires that no more than one electron can be in a given state. This leads to a system of electron shells, each of which is characterized by fixed values of n and l, and each containing $2(2l + 1)$ substates.

This model of the atom is called the *shell model*. Internal shells are filled. Electrons of the outermost atomic shell have smaller binding energies than the inner electrons, and are called *valence electrons*.

The distribution of the atomic electrons in these shells is described by a standard notation that we summarize here. States with the values $l = 0, 1, 2, 3, 4, \ldots$ are labeled with the letters *s, p, d, f, g*, ... The principal quantum number is placed before the symbol that identifies the orbital angular momentum state, and the number of electrons that may occupy this shell is written as a superscript to the orbital angular momentum symbol. For example, the electron shell configuration for the ground state of the oxygen atom is written as

$$O\left(1s^2 2s^2 2p^4\right).$$

The notation signifies that two electrons are in the state with $n = 1,\ l = 0$; two more are in the state with $n = 2,\ l = 0$; and the state with $n = 2,\ l = 1$ is occupied by four electrons.

We shall now construct a wave function within the shell model: a wave function for the system of electrons formed from products of single-electron wave functions so composed as to obey the Pauli principle. This principle requires the wave function to change its sign upon the permutation of any two electrons. If all electron shells are filled, then only one combination of single-electron wave functions satisfies this condition. The wave function of an atom with k electrons is of the form

$$\begin{aligned}\Psi &= \widehat{P}\psi_{n_1 l_1 m_1}(\mathbf{r}_1)\,\eta_{\sigma_1}(1)\,\psi_{n_2 l_2 m_2}(\mathbf{r}_2)\,\eta_{\sigma_2}(2)\ldots \\ &\quad \times \ldots \psi_{n_k l_k m_k}(\mathbf{r}_k)\,\eta_{\sigma_k}(k).\end{aligned} \tag{F.3}$$

The operator $\widehat{P}$ includes all permutations of any two electrons, with each permutation changing the sign of the wave function. The total number of such permutations is

$$C_k^2 = k(k-1)/2,$$

so the wave function in Eq. (F.3) is the sum of C_k^2 products of k single-electron wave functions. In Eq. (F.3), $\psi_{n_i l_i m_i}(\mathbf{r}_i)\,\eta_{\sigma_i}(i)$ is the space part of the wave function for the ith electron with quantum numbers n_i, l_i, and m_i, and $\eta_{\sigma_i}(i)$ is the spin wave function for this electron with the spin projection σ_i on the fixed quantization axis. Equation (F.3) can be written in the form of a determinant for an atom with filled shells. For example, for the ground state of the helium atom we obtain

$$\Psi = \begin{vmatrix}\psi_{100}(\mathbf{r}_1)\,\eta_+(1) & \psi_{100}(\mathbf{r}_1)\,\eta_-(1) \\ \psi_{100}(\mathbf{r}_2)\,\eta_+(2) & \psi_{100}(\mathbf{r}_2)\,\eta_-(2)\end{vmatrix}, \tag{F.4}$$

where the notation coincides with that for Eq. (F.3), and labels a spin projection of $+\frac{1}{2}$ by $+$ and a spin projection of $-\frac{1}{2}$ by $-$. Expressions for atomic wave functions in the form of Eq. (F.4) are called *Slater determinants.*

APPENDIX G

VIBRATIONAL AND ROTATIONAL STATES OF A MOLECULE

The major difference between the masses of nuclei and the mass of the electron makes it possible to divide the problem of the description of molecular energy levels into two parts. We first determine electron energies at fixed positions of the nuclei, and then we consider the motion of the nuclei for a fixed state of the electrons. We denote by $U(r)$ the potential function for the interaction between nuclei. We restrict ourselves to the case of a diatomic molecule. The motion of the center of mass of the molecule is not of interest, and we confine our attention to the motion of the nuclei with respect to the center of mass.

Within the framework of the adiabatic approximation of quantum mechanics, the interaction potential $U(r)$ describes simultaneously the energy of the electron levels. The motion of the nuclei in this central potential reduces to a one-dimensional problem in terms of the effective potential

$$U(r) + \frac{\hbar^2}{2Mr^2}\widehat{\mathbf{K}}^2,$$

where $\widehat{\mathbf{K}} = \widehat{\mathbf{J}} - \widehat{\mathbf{L}}$ is the angular momentum of the nuclei, $\widehat{\mathbf{J}}$ is the total angular momentum of the molecule, and $\widehat{\mathbf{L}}$ is the total orbital angular momentum of the electrons. The second term in the expression for the effective potential is the centrifugal energy. The average over electron states of the centrifugal energy operator at a fixed distance r between the two nuclei is

$$\frac{\hbar^2}{2Mr^2}\langle\widehat{\mathbf{K}}^2\rangle = \frac{\hbar^2}{2Mr^2}\left[J(J+1) - 2\widehat{\mathbf{L}}\cdot\widehat{\mathbf{J}} + \widehat{\mathbf{L}}^2\right].$$

The last term in this expression depends only on the state of the electrons and can be included in the definition of $U(r)$. The same is also true for the second term, since the angular momentum $\widehat{\mathbf{K}}$ is perpendicular to the axis of the molecule, while the vector operator $\widehat{\mathbf{L}}$ is directed along this axis since it is the only physically explicit direction in the problem. Thus, the effective potential energy takes the form

$$U(r) + \frac{\hbar^2}{2Mr^2} J(J+1). \tag{G.1}$$

The motion of the nuclei in weak excited states can be considered as small oscillations with respect to the equilibrium position. A power series expansion about this equilibrium position gives the lowest order term

$$U_{\text{eff}} = U(r_c) + \frac{\hbar^2}{2I} J(J+1) + \frac{1}{2} U''(r_e)(r - r_e)^2, \tag{G.2}$$

where $I = Mr_e^2$ is the moment of inertia of the molecule, M is its mass, and r_e is the mean distance between the nuclei.

The third term in Eq. (G.2) is the potential of a one-dimensional harmonic oscillator. We can therefore give the energies of the vibrational-rotational levels as

$$E_{vJ} = \frac{\hbar^2}{2I} J(J+1) + \hbar\omega_e \left(v + \frac{1}{2}\right), \tag{G.3}$$

where v is the vibrational quantum number and J is the rotational quantum number. The quantity

$$\omega_e = \sqrt{U''(r_e)/M}$$

is the frequency of classical oscillations. The second term in Eq. (G.3) is the vibrational energy of the molecule, and the first term is the rotational energy. From the definition of ω_e, we see that it behaves as

$$\omega_e \sim 1/\sqrt{M}$$

in the atomic system of units. This quantity gives a measure of the intervals between vibrational levels. Since $M \gg 1$, this spacing is small as compared to the intervals between neighboring electron levels, which are of order unity in atomic units. Furthermore, since $I \sim M$, then the spacing of rotational levels can be estimated as

$$\frac{\hbar^2}{2I} \sim \frac{1}{M},$$

which is small compared to intervals between vibrational levels, which are themselves closely spaced by comparison with electron levels.

We conclude that vibrational motion splits an electron level into a manifold of levels, and each vibrational level is split into a manifold of rotational levels.

Now we consider the classification of molecular levels. If the spin S is zero, then the classification is the same as described above. However, if $S \neq 0$, then a relativistic interaction between spin and orbital angular momentum occurs. This interaction is small as compared to vibrational intervals, and so it is very small as compared to electron level intervals. However, the structure of rotational levels is usually so closely spaced that spin-orbit interactions can be more energetic than this rotational level splitting (excluding light molecules). We shall now consider the classification of such levels.

First, we neglect the rotation of the nuclei. In addition to the projection Λ of the electron orbital angular momentum onto the molecular axis and the quantum number v of the vibrational motion, we should also consider the projection of the total spin on the molecular axis. This quantity is usually given the designation Σ. Clearly, the value of Σ covers the range from $-S,\ -S+1,\ \ldots$ up to $\ldots,\ S-1,\ S$. The projection of the total angular momentum is $\Omega = \Lambda + \Sigma$. It can take the values

$$\Omega = \Lambda + S,\ \Lambda + S,\ \ldots\ \ldots,\ \Lambda - S.$$

An electron level with a projection of the orbital angular momentum Λ is split into $2S + 1$ sublevels, which differ in their Ω values. This splitting is called *fine structure*, in analogy to such splitting in atoms.

Now we consider the energies of the levels. Since the spin-orbit potential is proportional to $\widehat{\mathbf{L}} \cdot \widehat{\mathbf{S}}$, and the vector operator $\widehat{\mathbf{L}}$ is directed along the molecular axis, we shall write the spin-orbit interaction energy as $A(r)\Sigma$. For a fixed value of Λ, we obtain the electron energy as

$$U(r) + A(r)\Omega,$$

which means that the sublevels of the manifold are equidistant.

Now we describe the rotational levels. The centrifugal energy operator has the form

$$\frac{\hbar^2}{2Mr^2}\left(\widehat{\mathbf{J}} - \widehat{\mathbf{L}} - \widehat{\mathbf{S}}\right)^2.$$

When we average this quantity over electron states, we obtain the effective potential energy

$$U_{\rm eff} = U(r) + A(r)\Omega + \frac{\hbar^2}{2Mr^2}\left[\widehat{\mathbf{J}}^2 - 2\widehat{\mathbf{J}}\cdot\overline{\left(\widehat{\mathbf{L}} + \widehat{\mathbf{S}}\right)} + \overline{\widehat{\mathbf{L}}^2} + 2\overline{\widehat{\mathbf{L}}\cdot\widehat{\mathbf{S}}} + \overline{\widehat{\mathbf{S}}^2}\right]. \quad \text{(G.4)}$$

When we associate the operators with their quantum numbers, we can make the replacements

$$\widehat{\mathbf{J}}^2 = J(J+1), \qquad \overline{\widehat{\mathbf{L}}} = \Lambda\mathbf{n}, \qquad \overline{\widehat{\mathbf{S}}} = \Sigma\mathbf{n},$$

where $\mathbf{n}$ is the unit vector along the molecular axis. We also have the relationships

$$\overline{\left(\widehat{\mathbf{L}} + \widehat{\mathbf{S}}\right)}\cdot\mathbf{n} = \overline{\left(\widehat{\mathbf{J}} - \widehat{\mathbf{K}}\right)}\cdot\mathbf{n} = \overline{\widehat{\mathbf{J}}}\cdot\mathbf{n} = \Omega,$$

so Eq. (G.4) becomes

$$U_{\rm eff} = U(r) + A(r)\Omega + \frac{\hbar^2}{2Mr^2}\left[J(J+1) - 2\Omega^2 + \overline{\widehat{\mathbf{L}}^2} + 2\overline{\widehat{\mathbf{L}}\cdot\widehat{\mathbf{S}}} + \overline{\widehat{\mathbf{S}}^2}\right].$$

Averaging over electron states is done using wave functions that do not depend on spin, so we obtain

$$\overline{\widehat{\mathbf{S}}^2} = S(S+1).$$

Furthermore, the averaging of the angular momenta $\widehat{\mathbf{L}}$ and $\widehat{\mathbf{S}}$ can be done independently, so we find

$$\overline{\widehat{\mathbf{L}} \cdot \widehat{\mathbf{S}}} = \overline{\widehat{\mathbf{L}}} \cdot \overline{\widehat{\mathbf{S}}} = \Lambda\Sigma.$$

As above, the mean value $\overline{\widehat{\mathbf{L}}^2}$ can be incorporated into $U(r)$, since it does not depend on Σ or J. The terms proportional to Σ can be included in the term $A(r)\,\Omega$. The final expression we obtain for the effective potential is

$$U_{\text{eff}} = U(r) + A(r)\Omega + \frac{\hbar^2}{2Mr^2}\left[J(J+1) - 2\Omega^2\right].$$

The energies of the levels are obtained, as above, by means of an expansion over small oscillations of the nuclei to obtain

$$E = U_e + A_e\Omega + \hbar\omega_e\left(v + \frac{1}{2}\right) + \frac{\hbar^2}{2I}\left[J(J+1) - 2\Omega^2\right]. \qquad \text{(G.5)}$$

The constants in this expression are taken to be at equilibrium values of the internuclear distance.

APPENDIX H

EQUATION FOR THE DENSITY MATRIX

The density matrix is useful when a transition involves several states. The density matrix approach greatly simplifies the treatment of problems in which a relaxation process takes place among the states considered. In particular, the density matrix method is very useful for the description of relaxation due to collisions between particles.

The wave function of a system can be expressed as a superposition of stationary states by the expression

$$\Psi = \sum_{m} a_m \psi_m \exp\left(-i\frac{E_m t}{\hbar}\right), \tag{H.1}$$

where the ψ_m are eigenfunctions of the Hamiltonian H_0, and the E_m are its eigenvalues, that is,

$$H_0 \psi_m = E_m \psi_m.$$

The probability density is given by

$$\rho = \Psi^* \Psi = \sum_{m,n} \rho_{mn} \psi_m \psi_n^*, \tag{H.2}$$

where

$$\rho_{mn} = a_m a_n^* \exp(i\omega_{nm} t),$$

$$\omega_{nm} = (E_n - E_m)/\hbar.$$

The matrix ρ_{mn} is called the density matrix. According to its definition, the matrix elements satisfy the relation

$$\rho_{mn} = \rho_{nm}^{*}.$$

To derive the dynamical equation satisfied by the density matrix, we presume that transitions in a quantum system take place due to a perturbing potential V. The total Hamiltonian is then $\widehat{H} = \widehat{H_0} + V$. The Schrödinger equation for the wave function is

$$i\hbar \frac{\partial \Psi}{\partial t} = \left(\widehat{H_0} + V\right) \Psi. \tag{H.3}$$

When we substitute the expansion (H.1) into Eq. (H.3), multiply from the left by ψ_j^*, integrate over the variables that characterize the system, and use the orthogonality condition for the wave functions, we obtain the equation for the expansion coefficients,

$$i\dot{a}_j = \frac{i}{\hbar} \sum_m V_{jm} \exp\left(i\omega_{jm} t\right) a_m. \tag{H.4}$$

The equation for the density matrix elements follows from Eq. (H.4) as

$$\dot{\rho}_{jk} + i\omega_{jk}\rho_{jk} = \frac{i}{\hbar} \sum_n \left(V_{nk}\rho_{jn} - V_{jn}\rho_{nk}\right). \tag{H.5}$$

In particular, for the diagonal elements of the density matrix, we have

$$\dot{\rho}_{jj} = \frac{i}{\hbar} \sum_n \left(V_{nj}\rho_{jn} - V_{jn}\rho_{nj}\right). \tag{H.6}$$

Now we shall generalize Eq. (H.5) to apply to a density matrix appropriate when the states have a finite lifetime. Then the probability to find a system in the state m should be multiplied by the factor $\exp(-t/\tau_m)$, where τ_m is the lifetime of this state. This means that the probability amplitude a_m to find a system in the state m should be multiplied by $\exp\left(-t/2\tau_m\right)$. The density matrix ρ_{mn} acquires therefore the factor

$$\exp\left(-\frac{t}{2\tau_m} - \frac{t}{2\tau_n}\right),$$

and the quantity ω_{jk} in Eq. (H.5) should be replaced by

$$\omega_{jk} + \frac{i}{2\tau_m} + \frac{i}{2\tau_n}.$$

Equation (H.5) for the density matrix has now become

$$\dot{\rho}_{jk} + i\omega_{jk}\rho_{jk} + \left(\frac{1}{2\tau_j} + \frac{1}{2\tau_k}\right)\rho_{jk} = \frac{i}{\hbar} \sum_n \left(V_{nk}\rho_{jn} - V_{jn}\rho_{nk}\right), \tag{H.7}$$

and Eq. (H.6) for the diagonal elements of the density matrix is now

$$\dot{\rho}_{jj} + \frac{1}{\tau_j}\rho_{jj} = \frac{i}{\hbar} \sum_n \left(V_{nj}\rho_{jn} - V_{jn}\rho_{nj}\right). \tag{H.8}$$

The diagonal element of a density matrix is the probability to find a system in the state identified by the diagonal index. If transitions from other states to this state take place, then it should be accounted for in the equation for the density matrix. The quantity $1/\tau_{kj}$ is the transition rate from the state k to the state j. According to the definition of a state lifetime, we have

$$\frac{1}{\tau_k} = \sum_j \frac{1}{\tau_{kj}}.$$

Equation (H.8) for the diagonal density matrix elements thus takes the form

$$\dot{\rho}_{jj} + \frac{1}{\tau_j}\rho_{jj} = \sum_j \frac{1}{\tau_{kj}}\rho_{kk} + \frac{i}{\hbar}\sum_n \left(V_{nj}\rho_{jn} - V_{jn}\rho_{nj}\right). \tag{H.9}$$

Equation (H.8) for the off-diagonal matrix elements has not been changed.

We continue to generalize by specifying a part of the density matrix determined by collisions between particles. Each collision constitutes a strong interaction between colliding particles and makes a fundamental change in the state of the molecule. These collisions are random, however, and collision times are small compared to times for radiative transitions. We may therefore disregard the time duration of a collision. We now convert these qualitative considerations into terms in the density matrix equation arising from collisions.

We take τ_q to be the time at which some particular collision occurs between the atom being examined and the surrounding atoms. The parameters defining the collision are the impact parameter, the relative velocity of the collision, the angle between the relative velocity vector and a fixed direction in space that may be a direction of an external field or the direction of observation. We characterize the transitions resulting from the collision by a transition S matrix, so that

$$a_k\left(\tau_q + \Delta t\right) = \sum_n (Snk)\, a_n\left(\tau_q - \Delta t\right),$$

where Δt is a time much larger than a collision time for the atoms but much less than a typical time for an alteration of the $a_k(t)$ amplitudes under the action of a radiation field or other external field. The change in the amplitude during a time t as a result of collisions is

$$a_k(t) - a_k(0) = \sum_n (S_{nk} - \delta_{nk})\, a_n(0).$$

Hence the change in the density matrix due to the collision is

$$\begin{aligned}\rho_{jk}(t) - \rho_{jk}(0) &= a_j(t)\, a_k^*(t) - a_j(0)\, a_k^*(0)\\ &= \sum_{m,n}\left(S_{mj}S_{nk}^* - \delta_{mj}\delta_{nk}\right)\rho_{mn}.\end{aligned}$$

This change takes place in a time much less than typical times for radiation processes. When we take the limit in which the collision time goes to zero, we obtain the time derivative of the density matrix due to collisions,

$$\left(\frac{d\rho_{jk}}{dt}\right)_{\text{coll}} = \sum_{m,n} \left(S_{mj}S_{nk}^{*} - \delta_{mj}\delta_{nk}\right)\rho_{mn}\delta\left(t - \tau_q\right), \tag{H.10}$$

where τ_q is the time at which a collision occurs.

Since collisions take place chaotically and abruptly compared to radiative processes, we can average Eq. (H.10) over collision parameters. In particular, if the collision between the radiating atom and the perturbing atom occurs according to the laws of classical mechanics, then the collision probability during the time dt with the impact parameter in the range $[\rho, \rho + d\rho]$ is given by

$$Nv 2\pi\rho\, d\rho\, dt,$$

where N is the density of perturbing atoms, and v is the relative velocity for collisions between the radiating and perturbing atoms. When we substitute this relation into Eq. (H.10) and integrate over collision time, we obtain

$$\left(\frac{d\rho_{jk}}{dt}\right)_{\text{coll}} = \sum_{m,n}\left\langle Nv\int_0^{\infty}\left(S_{mj}S_{nk}^{*} - \delta_{mj}\delta_{nk}\right)2\pi\rho\, d\rho\, \rho_{mn}\right\rangle,$$

where the angle bracket symbolizes averaging over relative velocities of the atoms. When the collision contribution is inserted into Eq. (H.7) for the density matrix, we obtain

$$\frac{d\rho_{jk}}{dt} + \left(i\omega_{jk} + \frac{1}{2\tau_j} + \frac{1}{2\tau_k}\right)\rho_{jk} = \frac{i}{\hbar}\sum_n\left(V_{nk}\rho_{jn} - V_{jn}\rho_{nk}\right) + \left\langle Nv\int_0^{\infty}2\pi\rho\, d\rho\left(S_{mj}S_{nk}^{*} - \delta_{mj}\delta_{nk}\right)\rho_{mn}\right\rangle. \tag{H.11}$$

We now examine the special case where there is no inelastic scattering. All possible channels are presumed to have elastic scattering only. The transition S matrix is then of the simple form

$$S_{mj}(\rho) = \delta_{mj}\exp\left(2i\delta_j\right),$$

where $\delta_j(\rho)$ is the scattering phase for the channel j. Equation (H.11) simplifies to

$$\frac{d\rho_{jk}}{dt} + \left(i\omega_{jk} + \frac{1}{2\tau_j} + \frac{1}{2\tau_k} + \operatorname{Re}\nu_{jk} + i\operatorname{Im}\nu_{jk}\right)\rho_{jk} = \frac{i}{\hbar}\sum_n\left(V_{nk}\rho_{jn} - V_{jn}\rho_{nk}\right), \tag{H.12}$$

where the notation has been introduced that

$$\nu_{jk} = \left\langle Nv\int_0^{\infty}2\pi\rho\, d\rho\left\{1 - \exp\left[2i\left(\delta_j - \delta_k\right)\right]\right\}\right\rangle.$$

The equation for the diagonal elements of the density matrix is not affected by collisions in this case and continues to be expressed by Eq. (H.9).

APPENDIX I

ATOMIC UNITS AND MEASURES OF INTENSITY

Atomic units greatly simplify the appearance of expressions in atomic physics, and these units are thus widely used. They are due to Hartree. The basic atomic units are summarized, and then various measures of electromagnetic field quantities are introduced in terms of atomic units, and relations among these field quantities are summarized.

I.1 ATOMIC UNITS

Atomic units are specified by the requirement that

$$\hbar = 1, \qquad m = 1, \qquad e^2 = 1, \tag{I.1}$$

where m and e^2 refer, respectively, to the mass of the electron and the squared charge of the electron. These units then are applied in terms of the properties of the simple hydrogen atom based on infinite mass of the atomic nucleus. A single atomic unit of the most important physical quantities is then:

Unit of length: Bohr radius $= a_0 = \hbar^2/\left(me^2\right) \equiv 1$ a.u. $= 5.29177249 \times 10^{-11}$ m.

Unit of velocity: Velocity of an electron in the first Bohr orbit $= v_0 = e^2/\hbar \equiv$ 1 a.u. $= 2.18769142 \times 10^6$ m/s.

Unit of time: Time for an electron to travel a distance a_0 at the velocity $v_0 = \tau_0 = a_0/v_0 = \hbar^3/me^4 \equiv 1$ a.u. $= 2.41888433 \times 10^{-17}$ s.

Unit of frequency: Inverse of the unit of time $= \nu_0 = 1/\tau_0 = me^4/\hbar^3 \equiv 1$ a.u. $=$ $4.13413732 \times 10^{16}$ Hz.

Unit of energy: Twice the binding energy of hydrogen $= E_0 = 2R_\infty = me^4/\hbar^2 \equiv$ 1 a.u. $= 27.2113962$ eV.

I.2 ELECTROMAGNETIC FIELD QUANTITIES

I.2.1 Basic Field Quantities

Unit of electric field strength: $\mathcal{E}_0 = E_0/a_0 = m^2e^5/\hbar^4 \equiv 1$ a.u. $= 5.14220826 \times 10^{11}$ V/m.

Unit of energy flux (intensity): $I_0 = cE_0^2/(8\pi) = 3.50944758 \times 10^{16}$ W/cm^2.

I.2.2 Convenient Relations

Angular frequency in a.u. and wavelength in nm: $\omega(\text{a.u.}) = 45.5633526/\lambda(\text{nm})$.

Velocity of light and the fine structure constant: $c = 1/\alpha$.

I.2.3 Field Measures Useful in Strong-Field Atomic Physics

To set the convention used for the amplitudes, we shall regard a monochromatic linearly polarized field as being of the form

$$\mathbf{E} = E_0 \boldsymbol{\epsilon} \sin(\omega t - \mathbf{k} \cdot \mathbf{r}), \qquad \boldsymbol{\epsilon} \cdot \boldsymbol{\epsilon} = 1; \tag{I.2}$$

and a circularly polarized monochromatic field as being of the form

$$\mathbf{E} = \frac{E_0}{2^{1/2}} \left[\mathbf{e}_x \sin(\omega t - \mathbf{k} \cdot \mathbf{r}) \pm \mathbf{e}_y \cos(\omega t - \mathbf{k} \cdot \mathbf{r}) \right], \tag{I.3}$$

$$\mathbf{e}_x \cdot \mathbf{e}_x = \mathbf{e}_y \cdot \mathbf{e}_y = 1, \qquad \mathbf{e}_x \cdot \mathbf{e}_y = 0.$$

In Eq. (I.3), the presumption is that wave propagation is along the z axis, and $\mathbf{e}_x$, $\mathbf{e}_y$ are unit vectors along the x and y axes, respectively. Whereas the use of the amplitude E_0 in Eq. (I.2) is universal, the definition employed in Eq. (I.3) is not. Some authors omit the factor $2^{-1/2}$. Equations (I.2) and (I.3) have the amplitude relations

$$|\mathbf{E}| = \begin{cases} E_0, & \text{linear} \\ E_0/2^{1/2}, & \text{circular} \end{cases},$$

$$\langle \mathbf{E}^2 \rangle = \begin{cases} \frac{1}{2}|\mathbf{E}|^2 = \frac{1}{2}E_0^2, & \text{linear} \\ |\mathbf{E}|^2 = \frac{1}{2}E_0^2, & \text{circular} \end{cases}, \tag{I.4}$$

where the angle bracket indicates a time average over a period.

Irradiance, or Energy Flux

A physical quantity frequently employed to measure the "strength" of the field is often expressed in units of Watts/cm^2, or Joules/(s · cm^2). This is a flux of energy, also known as irradiance. It has, however, now become quite common to refer to this quantity simply as the intensity. We shall here use the designation I for the irradiance, given by the amplitude of the Poynting vector, or

$$I \text{ (Gaussian units)} = uc = \frac{c}{4\pi}\left\langle \mathbf{E}^2 \right\rangle = \frac{c}{8\pi} E_0^2.$$

The atomic unit of irradiance I_0, as indicated above, is that energy flux that corresponds to $E_0 = 1$ a.u., so that $I_0 = c/(8\pi) = 3.50944758 \times 10^{16}\,\mathrm{W/cm^2}$.

When all quantities are in atomic units, then

$$I = 2\left\langle \mathbf{E}^2 \right\rangle = E_0^2.$$

Ponderomotive Energy

The ponderomotive energy is that energy that a free charged particle possesses because of its oscillation, or "quiver," motion when in an electromagnetic field. It is exactly the kinetic energy of such oscillatory motion as expressed in the frame of reference in which the center of mass (c.m.) of the charged particle is, on a time average, at rest. If a charged particle should emerge adiabatically from a region in which there is an electromagnetic field into a region free of the field, this ponderomotive energy is converted into a directed kinetic energy. The ponderomotive energy U_p is thus sometimes called the ponderomotive potential. In atomic units,

$$U_p = \frac{I}{4\omega^2} = \frac{\left\langle \mathbf{E}^2 \right\rangle}{2\omega^2} = \left(\frac{E_0}{2\omega}\right)^2.$$

Free-Electron Amplitude of Motion

The amplitude of motion of an electron in a plane wave field is often designated α_0. As expressed in the frame of reference in which the c.m. of the electron is at rest on the average, this motion is a rectilinear oscillation along the polarization direction for a linearly polarized plane wave of moderate intensity, with

$$\alpha_0 = \frac{I^{1/2}}{\omega^2} = \frac{2}{\omega} U_p^{1/2} = \frac{E_0}{\omega^2}.$$

This result is actually the limit of the true relativistic motion of the electron, where the relativistic nature of the motion will arise as a result of the intensity of the field. The relativistic motion is in the form of a figure 8, with the axis of the lobes along the polarization direction, and the plane of the figure 8 determined by the polarization direction and the direction of propagation.

For a circularly polarized electromagnetic field, the electron motion in the simplest frame of reference is a circle executed in the plane perpendicular to the direction of propagation. This circle has the radius

$$\alpha_0 = \frac{1}{\omega^2}\left(\frac{I}{2}\right)^{1/2} = \frac{(2U_p)^{1/2}}{\omega} = \frac{E_0}{2^{1/2}\omega^2}.$$

Dimensionless Intensity Parameters

The Keldysh parameter, or adiabaticity parameter, is actually an inverse intensity parameter. It is given as

$$\gamma = \frac{\omega (2E_i)^{1/2}}{E_0} = \frac{\omega (2E_i)^{1/2}}{I^{1/2}} = \frac{E_i^{1/2}}{(2U_p)^{1/2}},$$

where E_i is the ionization potential of the atom.

The inverse square of the adiabaticity parameter is directly an intensity parameter, which can be defined *ab initio* as twice the ratio of the ponderomotive energy to the ionization potential of the atom. When used in this form, it is generally designated z_1, where

$$z_1 \equiv \frac{2U_p}{E_i} = \frac{I}{2\omega^2 E_i} = \frac{E_0^2}{2\omega^2 E_i} = \begin{cases} (\omega\alpha_0)^2 / (2E_i), & \text{linear} \\ (\omega\alpha_0)^2 / E_i, & \text{circular} \end{cases}.$$

This quantity has been called the *bound-state intensity parameter.*

Another intensity parameter is the ratio of the ponderomotive energy of the electron to the energy of a single photon of the field. That is, it is given by

$$z \equiv \frac{U_p}{\omega} = \frac{I}{4\omega^3} = \frac{E_0^2}{4\omega^3} = \begin{cases} \omega\alpha_0^2/4, & \text{linear} \\ \omega\alpha_0^2/2, & \text{circular} \end{cases}.$$

This quantity has been called the *intensity parameter for the final continuum state*, and also the *nonperturbative intensity parameter*, since it is the quantity that is the primary indicator of when perturbation theory will fail.

There is a free-electron parameter that always arises in the description of the interaction of a free electron with an electromagnetic field. There is no universally accepted symbol for this parameter, although many authors have defined the very same quantity. We here designate it z_f. It is most readily defined as twice the ratio of the ponderomotive energy of the electron in the field to the rest energy of the electron, or

$$z_f \equiv \frac{2U_p}{c^2} = \frac{I}{2(\omega c)^2} = \frac{1}{2}\left(\frac{E_0}{\omega c}\right)^2 = \begin{cases} \frac{1}{2}(\omega\alpha_0/c)^2, & \text{linear} \\ (\omega\alpha_0/c)^2, & \text{circular} \end{cases}.$$

By its definition, this parameter plainly is a measure of the appearance of relativistic effects in the field-induced motion of an electron. It is the *free-electron intensity parameter.*

APPENDIX J

PROPERTIES OF THE GENERALIZED BESSEL FUNCTION

The generalized Bessel function $J_n(u, v)$ arises in connection with Volkov solutions for an electron in a linearly polarized electromagnetic field. Because the Volkov wave function is an exact solution of the Schrödinger equation for a free electron in a superposition of unidirectional plane wave fields, it has been widely employed in the treatment of strong-field problems whenever at least one of the states in the problem describes an unbound electron. The function $J_n(u, v)$ is thus frequently encountered in the strong-field literature, but its properties are not to be found in any of the standard reference works on transcendental functions. The principal properties of $J_n(u, v)$ are summarized below.

It will be assumed that n is an integer, and u and v are real.

Integral representation:

$$J_n(u, v) = \frac{1}{2\pi} \int_{-\pi}^{\pi} d\theta \exp\left[i(u \sin\theta + v \sin 2\theta - n\theta)\right].$$

Series representation:

$$J_n(u, v) = \sum_{k=-\infty}^{\infty} J_{n-2k}(u) J_k(v).$$

Generating function:

$$\exp\left[i(u \sin\theta + v \sin 2\theta)\right] = \sum_{n=-\infty}^{\infty} e^{in\theta} J_n(u, v).$$

Sums of generalized Bessel functions:

$$\sum_{n=-\infty}^{\infty} J_n(u, v) = 1,$$

$$\sum_{n=-\infty}^{\infty} (-1)^n J_n(u, v) = 1,$$

$$\sum_{n=-\infty}^{\infty} i^n J_n(u, v) = e^{iu}.$$

Recursion relations:

$$J_{n-1}(u, v) - J_{n+1}(u, v) = 2\frac{\partial}{\partial u} J_n(u, v),$$

$$J_{n-2}(u, v) - J_{n+2}(u, v) = 2\frac{\partial}{\partial v} J_n(u, v),$$

$$2nJ_n(u, v) = u\left[J_{n-1}(u, v) + J_{n+1}(u, v)\right] + 2v\left[J_{n-2}(u, v) + J_{n+2}(u, v)\right],$$

$$\frac{\partial}{\partial v} J_n(u, v) = 2\frac{\partial}{\partial u} J_{n-1}(u, v) + 2\frac{\partial}{\partial u} J_{n+1}(u, v).$$

Symmetry relations:

$$J_n(-u, v) = (-1)^n J_n(u, v),$$

$$J_n(u, -v) = (-1)^n J_{-n}(u, v).$$

Sums of binary products:

$$\sum_{k=-\infty}^{\infty} J_{n\mp k}(u, v) J_k\left(u', v'\right) = J_n\left(u \pm u', v \pm v'\right),$$

$$\sum_{k=-\infty}^{\infty} \left[J_k(u, v)\right]^2 = 1,$$

$$\sum_{k=-\infty}^{\infty} (-1)^k \left[J_{-k}(u, v)\right]^2 = J_0(2u),$$

$$\sum_{k=-\infty}^{\infty} J_k(u, v) J_{-k}(u, v) = J_0(2u, 2v),$$

$$\sum_{k=-\infty}^{\infty} (-1)^k J_k(u, v) J_{-k}(u, v) = 0.$$

Limiting cases:

$$J_n(u, 0) = J_n(u),$$

$$J_n(0, v) = \begin{cases} J_{n/2}(v), & n \text{ even} \\ 0, & n \text{ odd} \end{cases},$$

$$J_n(u, v) \approx J_n(u) + \tfrac{1}{2} v \left[J_{n-2}(u) - J_{n+2}(u)\right], \qquad |v| \ll 1,$$

$$J_n(u, v) \approx \left[1 - \frac{u^2}{4}\left(1 - \frac{n}{2v}\right) J_{n/2}(v)\right], \qquad |u| \ll 1, \quad n \text{ even},$$

$$J_n(u, v) \approx \tfrac{1}{2} u \left[J_{(n-1)/2}(v) - J_{(n+1)/2}(v)\right], \qquad |u| \ll 1, \quad n \text{ odd},$$

$$\lim_{\epsilon \to 0} J_n\left(\epsilon^{1/2}\mu, \epsilon\nu\right)$$

$$= \begin{cases} \lim\limits_{\epsilon \to 0} \left(\dfrac{\epsilon\nu}{2}\right)^{n/2} \displaystyle\sum_{k=0}^{n/2} \dfrac{\left(\mu^2/2\nu\right)^k}{(2k)!\,(n/2 - k)!}, & n \text{ even} \\ \lim\limits_{\epsilon \to 0} \left(\dfrac{\epsilon\nu}{2}\right)^{n/2} \displaystyle\sum_{k=0}^{(n-1)/2} \dfrac{\left(\mu^2/2\nu\right)^{k+1/2}}{(2k+1)!\,\left[(n-1)/2 - k\right]!}, & n \text{ odd} \end{cases}$$

APPENDIX K

IONIZATION POTENTIALS

Ionization potentials of atoms and positive ions are given in Table K.1. The values I_0, I_1, I_2, and I_3 are, respectively, the ionization potentials of the neutral atom and the singly, doubly, and triply ionized atoms. Potentials are given in eV. The values are those listed in the *Handbook of Chemistry and Physics* (Ref.11).

TABLE K.1. Ionization Potentials of Atoms and Positive Ions

Z	Atom	I_0, eV	I_1, eV	I_2, eV	I_3, eV
1	H	13.598			
2	He	24.587	54.418		
3	Li	5.392	75.640	122.45	
4	Be	9.323	18.211	153.90	217.72
5	B	8.298	25.155	37.931	259.38
6	C	11.260	24.383	47.89	64.49
7	N	14.534	29.601	47.45	77.47
8	O	13.618	35.117	54.936	77.414
9	F	17.423	34.971	62.71	87.14
10	Ne	21.565	40.963	63.45	97.12
11	Na	5.139	47.286	71.620	98.91
12	Mg	7.646	15.035	80.144	109.27
13	Al	5.986	18.829	28.448	119.99
14	Si	8.152	16.346	33.493	5.142
15	P	10.487	19.770	30.203	51.444
16	S	10.360	23.338	34.79	47.222

(continued)

TABLE K.1. Ionization Potentials of Atoms and Positive Ions

Z	Atom	I_0, eV	I_1, eV	I_2, eV	I_3, eV
17	Cl	12.968	23.814	39.61	53.47
18	Ar	15.760	27.630	40.74	59.81
19	K	4.341	31.63	45.81	60.91
20	Ca	6.113	11.872	50.913	67.3
21	Sc	6.561	12.800	24.757	73.49
22	Ti	6.828	13.58	27.49	43.27
23	V	6.746	14.66	29.31	46.71
24	Cr	6.767	16.49	31.0	49.2
25	Mn	7.434	15.640	33.67	51.2
26	Fe	7.902	16.188	30.65	54.8
27	Co	7.88	17.083	33.5	51.3
28	Ni	7.640	18.169	35.19	54.9
29	Cu	7.726	20.292	36.84	57.4
30	Zn	9.394	17.964	39.72	59.4
31	Ga	5.999	20.51	30.7	64.
32	Ge	7.900	15.935	34.2	45.7
33	As	9.815	18.63	28.4	50.1
34	Se	9.752	21.19	30.82	42.95
35	Br	11.814	21.81	35.90	47.3
36	Kr	14.000	24.360	36.95	52.5
37	Rb	4.177	27.285	39.2	52.6
38	Sr	5.695	11.030	42.88	56.28
39	Y	6.217	12.24	20.525	60.60
40	Zr	6.634	13.13	23.0	34.34
41	Nb	6.759	14.32	25.0	38.3
42	Mo	7.092	16.16	27.2	46.4
43	Tc	7.28	15.26	29.5	
44	Ru	7.361	16.76	28.5	
45	Rh	7.46	18.08	31.1	
46	Pd	8.337	19.43	32.9	
47	Ag	7.576	21.49	34.8	
48	Cd	8.994	16.908	37.48	
49	In	5.786	18.87	28.0	54
50	Sn	7.344	14.632	30.50	40.74
51	Sb	8.64	16.53	25.32	44.16
52	Te	9.010	18.6	27.96	37.41
53	I	10.451	19.131	33.0	
54	Xe	12.130	21.21	32.1	45
55	Cs	3.894	23.15	33.4	46
56	Ba	5.212	10.004	35.8	47
57	La	5.577	11.1	19.18	49.9
58	Ce	5.539	10.8	20.20	39.76
59	Pr	5.464	10.6	21.62	38.98

(*continued*)

TABLE K.1. Ionization Potentials of Atoms and Positive Ions

Z	Atom	I_0, eV	I_1, eV	I_2, eV	I_3, eV
60	Nd	5.525	10.7	22.1	40.4
61	Pm	5.55	10.9	22.3	41.1
62	Sm	5.644	11.1	23.4	41.4
63	Eu	5.670	11.24	24.9	42.7
64	Gd	6.150	12.1	20.6	44.0
65	Tb	5.864	11.5	21.9	39.8
66	Dy	5.939	11.7	22.8	41.4
67	Ho	6.022	11.8	22.8	42.5
68	Er	6.108	11.9	22.7	42.7
69	Tm	6.184	12.1	23.7	42.7
70	Yb	6.254	12.18	25.05	43.6
71	Lu	5.426	13.9	20.96	45.25
72	Hf	6.825	14.9	23.3	33.3
73	Ta	7.89			
74	W	7.98			
75	Re	7.88			
76	Os	8.73			
77	Ir	9.05			
78	Pt	8.96	18.56		
79	Au	9.226	20.5	34	43
80	Hg	10.438	18.76	34.2	46
81	Tl	6.108	20.43	29.83	
82	Pb	7.417	15.032	31.94	42.32
83	Bi	7.289	16.7	25.56	45.3
84	Po	8.417			
85	At	9.0			
86	Rn	10.75			
87	Fr	4.0			
88	Ra	5.278	10.15		
89	Ac	5.2	11.75	20	
90	Th	6.1	11.5	20	28.7
91	Pa	5.89			
92	U	6.194	11.9	20	37
93	Np	6.266			
94	Pu	6.06			
95	Am	6.0			
96	Cu	6.02			
97	Bk	6.23			
98	Cf	6.30			
99	Es	6.42			
100	Fm	6.50			
101	Md	6.58			
102	No	6.65			

Table K.2 gives electron binding energies (electron affinities) in negative ions, labeled as EA in the table. The accuracy to which these affinities is known is indicated in the table by the code: A–the accuracy is better than 1%, B–the accuracy is between 1 and 3%, C–the accuracy is between 3 and 10%, D–the accuracy is lower than 10%. The entry "no" means that the stable negative ion of this element does not exist.

TABLE K.2. Electron Binding Energies in Negative Ions

Z	Ion, term	Shell	EA, (eV)
1	$H^-(^1S)$	$1s^2$	0.75416 (A)
2	He^-	$2s$	No
3	$Li^-(^1S)$	$2s^2$	0.618 (A)
4	Be^-	$2p$	No
5	$B^-(^3P)$	$2p^2$	0.28 (C)
6	$C^-(^4S)$	$2p^3$	1.2629 (A)
6	$C^-(^2D)$	$2p^3$	0.035 (C)
7	N^-	$2p^4$	No
8	$O^-(^2P)$	$2p^5$	1.4611 (A)
9	$F^-(^1S)$	$2p^6$	3.4012 (A)
10	Ne^-	$3s$	No
11	$Na^-(^1S)$	$3s^2$	0.5479 (A)
12	Mg^-	$3p$	No
13	$Al^-(^3P)$	$3p^2$	0.441 (A)
13	$Al^-(^1D)$	$3p^2$	0.332 (B)
14	$Si^-(^4S)$	$3p^3$	1.394 (A)
14	$Si^-(^2D)$	$3p^3$	0.526 (B)
14	$Si^-(^2P)$	$3p^3$	0.034 (C)
15	$P^-(^3P)$	$3p^4$	0.7465 (A)
16	$S^-(^2P)$	$3p^5$	2.0771 (A)
17	$Cl^-(^1S)$	$3p^6$	3.6127 (A)
18	Ar^-	$4s$	No
19	$K^-(^1S)$	$4s^2$	0.5015 (A)
20	$Ca^-(^2P)$	$4s^24p$	0.018 (D)
21	$Sc^-(^1D)$	$3d4p$	0.19 (C)
21	$Sc^-(^3D)$	$3d4p$	0.04 (D)
22	$Ti^-(^4F)$	$3d^3$	0.08 (D)
23	$V^-(^5D)$	$3d^4$	0.53 (C)
24	$Cr^-(^6S)$	$3d^5$	0.67 (B)
25	Mn^-	$3d^6$	No
26	$Fe^-(^4F)$	$3d^7$	0.151 (B)
27	$Co^-(^3F)$	$3d^8$	0.662 (A)
28	$Ni^-(^2D)$	$3d^9$	1.15 (A)
29	$Cu^-(^1S)$	$3d^{10}$	1.23 (A)
30	Zn^-	$4p$	No
31	$Ga^-(^2P)$	$4p^2$	0.5 (D)

(continued)

TABLE K.2. Electron Binding Energies in Negative Ions

Z	Ion, term	Shell	EA, (eV)
32	$Ge^-(^4S)$	$4p^3$	1.233 (A)
33	$As^-(^3P)$	$4p^4$	0.80 (C)
34	$Se^-(^2P)$	$4p^5$	2.0207 (A)
35	$Br^-(^1S)$	$4p^6$	3.3636 (A)
36	Kr^-	$5s$	No
37	$Rb^-(^1S)$	$5s^2$	0.4859 (A)
38	$Sr^-(^2P)$	$5s^25p$	0.026 (D)
39	$Y^-(^1D)$	$4d5p$	0.31 (C)
39	$Y^-(^1D)$	$4d5p$	0.16 (D)
40	$Zr^-(^4F)$	$4d^3$	0.43 (B)
41	$Nb^-(^5D)$	$4d^4$	0.89 (B)
42	$Mo^-(^6S)$	$4d^5$	0.75 (B)
43	$Tc^-(^5D)$	$4d^6$	1.0 (D)
44	$Ru^-(^4F)$	$4d^7$	1.5 (D)
45	$Rh^-(^3F)$	$4d^8$	1.14 (A)
46	$Pd^-(^2D)$	$4d^9$	0.56 (B)
47	$Ag^-(^1S)$	$4d^{10}$	1.30 (A)
48	Cd^-	$5p$	No
49	$In^-(^3P)$	$5p^2$	0.3 (D)
50	$Sn^-(^4S)$	$5p^3$	1.112 (A)
51	$Sb^-(^3P)$	$5p^4$	1.05 (B)
52	$Te^-(^2P)$	$5p^5$	1.9708 (A)
53	$I^-(^1S)$	$5p^6$	3.0590 (A)
54	Xe^-	$6s$	No
55	$Cs^-(^1S)$	$6s^2$	0.4716 (A)
56	Ba^-	$5d$	No
57	$La^-(^3F)$	$5d^2$	0.5 (D)
73	$Ta^-(^5D)$	$5d^4$	0.32 (C)
74	$W^-(^6S)$	$5d^5$	0.816 (A)
75	$Re^-(^5D)$	$5d^6$	0.15 (D)
76	$Os^-(^4F)$	$5d^7$	1.4 (D)
77	$Ir^-(^3F)$	$5d^8$	1.57 (A)
78	$Pt^-(^2D)$	$5d^9$	2.128 (A)
79	$Au^-(^1S)$	$5d^{10}$	2.3086 (A)
80	Hg^-	$6p$	No
81	$Tl^-(^3P)$	$6p^2$	0.5 (D)
82	$Pb^-(^4S)$	$6p^3$	0.364 (A)
83	$Bi^-(^3P)$	$6p^4$	0.95 (B)
84	$Po^-(^2P)$	$6p^5$	1.6 (D)

APPENDIX L

PARAMETERS OF THE HYDROGEN ATOM

L.1 ANGULAR WAVE FUNCTIONS OF THE HYDROGEN ATOM

l	m	$\mathrm{Y}_{lm}(\theta, \varphi)$
0	0	$\frac{1}{\sqrt{4\pi}}$
1	0	$\sqrt{\frac{3}{4\pi}} \cos\theta$
1	± 1	$\mp\sqrt{\frac{3}{8\pi}} \sin\theta \exp(\pm i\varphi)$
2	0	$\sqrt{\frac{5}{16\pi}} \left(3\cos^2\theta - 1\right)$
2	± 1	$\mp\sqrt{\frac{15}{8\pi}} \sin\theta\cos\theta \exp(\pm i\varphi)$
2	± 2	$\frac{1}{2}\sqrt{\frac{15}{8\pi}} \sin^2\theta \exp(\pm 2i\varphi)$
3	0	$\frac{1}{4}\sqrt{\frac{7}{\pi}} \left(5\cos^3\theta - 3\cos\theta\right)$
3	± 1	$\mp\frac{1}{8}\sqrt{\frac{21}{\pi}} \sin\theta \left(5\cos^2\theta - 1\right) \exp(\pm i\varphi)$
3	± 2	$\frac{1}{4}\sqrt{\frac{105}{2\pi}} \sin^2\theta\cos\theta \exp(\pm 2i\varphi)$
3	± 3	$\mp\frac{1}{8}\sqrt{\frac{35}{\pi}} \sin^3\theta \exp(\pm 3i\varphi)$
4	0	$\sqrt{\frac{9}{4\pi}} \left(\frac{35}{8}\cos^4\theta - \frac{15}{4}\cos^2\theta + \frac{3}{8}\right)$
4	± 1	$\mp\frac{3}{4}\sqrt{\frac{5}{4\pi}} \sin\theta\cos\theta \left(7\cos^2\theta - 3\right) \exp(\pm i\varphi)$
4	± 2	$\frac{3}{8}\sqrt{\frac{5}{2\pi}} \sin^2\theta \left(7\cos^2\theta - 10\right) \exp(\pm 2i\varphi)$
4	± 3	$\mp\frac{3}{4}\sqrt{\frac{35}{4\pi}} \sin^3\theta\cos\theta \exp(\pm 3i\varphi)$
4	± 4	$\frac{3}{8}\sqrt{\frac{35}{8\pi}} \sin^4\theta \exp(\pm 4i\varphi)$

L.2 RADIAL WAVE FUNCTIONS OF THE HYDROGEN ATOM

State	R_{nl}
$1s$	$2\exp(-r)$
$2s$	$\frac{1}{\sqrt{2}}\left(1-\frac{r}{2}\right)\exp\left(-\frac{r}{2}\right)$
$2p$	$\frac{1}{\sqrt{24}}r\exp\left(-\frac{r}{2}\right)$
$3s$	$\frac{2}{3\sqrt{3}}\left(1-\frac{2r}{3}+\frac{2r^2}{27}\right)\exp\left(-\frac{r}{3}\right)$
$3p$	$\frac{2}{27}\sqrt{\frac{2}{3}}r\left(1-\frac{r}{6}\right)\exp\left(-\frac{r}{3}\right)$
$3d$	$\frac{4}{81\sqrt{30}}r^2\exp\left(-\frac{r}{3}\right)$
$4s$	$\frac{1}{4}\left(1-\frac{3r}{4}+\frac{r^2}{8}-\frac{r^3}{192}\right)\exp\left(-\frac{r}{4}\right)$
$4p$	$\frac{1}{16}\sqrt{\frac{5}{3}}r\left(1-\frac{r}{4}+\frac{r^2}{80}\right)\exp\left(-\frac{r}{4}\right)$
$4d$	$\frac{1}{64\sqrt{5}}r^2\left(1-\frac{r}{12}\right)\exp\left(-\frac{r}{4}\right)$
$4f$	$\frac{1}{768\sqrt{35}}r^3\exp\left(-\frac{r}{4}\right)$

L.3 ALGEBRAIC EXPRESSIONS FOR EXPECTATION VALUES OF POWERS OF THE RADIAL COORDINATE IN THE HYDROGEN ATOM: $\langle r^N \rangle = \int_0^\infty r_{nl}^2(r)\, r^{n+2}\, dr$

Parameter	Expression
$\langle r \rangle$	$\frac{1}{2}\left[3n^2 - l(l+1)\right]$
$\langle r^2 \rangle$	$\frac{n^2}{2}\left[5n^2 + 1 - 3l(l+1)\right]$
$\langle r^3 \rangle$	$\frac{n^2}{8}\left[35n^2\left(n^2 - 1\right) - 30n^2(l+2)(l-1) + 3(l+2)(l+1)l(l-1)\right]$
$\langle r^4 \rangle$	$\frac{n^4}{8}\left[63n^4 - 35n^2\left(2l^2 + 2l - 3\right) + 5l(l+1)\left(3l^2 + 3l - 10\right) + 12\right]$
$\langle r^{-1} \rangle$	$\frac{1}{n^2}$
$\langle r^{-2} \rangle$	$\frac{1}{n^3\left(l + \frac{1}{2}\right)}$
$\langle r^{-3} \rangle$	$\frac{1}{n^3(l+1)\left(l + \frac{1}{2}\right)l}$
$\langle r^{-4} \rangle$	$\frac{3n^2 - l(l+1)}{2n^5\left(l + \frac{3}{2}\right)(l+1)\left(l + \frac{1}{2}\right)l\left(l - \frac{1}{2}\right)}$

L.4 EXPECTATION VALUES OF POWERS OF THE RADIAL COORDINATE IN THE HYDROGEN ATOM (IN ATOMIC UNITS)

State	$\langle r \rangle$	$\langle r^2 \rangle$	$\langle r^3 \rangle$	$\langle r^4 \rangle$	$\langle r^{-1} \rangle$	$\langle r^{-2} \rangle$	$\langle r^{-3} \rangle$	$\langle r^{-4} \rangle$
$1s$	1.5	3	7.5	22.5	1	2	—	—
$2s$	6	42	330	2880	0.25	0.25	—	—
$2p$	5	30	210	1680	0.25	0.0833	0.0417	0.0417
$3s$	13.5	207	3442	6.136×10^4	0.111	0.0741	—	—
$3p$	12.5	180	2835	4.420×10^4	0.111	0.0247	0.0123	0.0137
$3d$	10.5	126	1701	2.552×10^4	0.111	0.0148	0.0247	5.49×10^{-3}
$4s$	24	648	18720	5.702×10^5	0.0625	0.0312	—	—
$4p$	23	600	16800	4.973×10^5	0.0625	0.0104	5.21×10^{-3}	5.49×10^{-4}
$4d$	21	504	13100	3.629×10^5	0.0625	0.00625	1.04×10^{-3}	2.60×10^{-4}
$4f$	18	360	7920	1.901×10^5	0.0625	0.00446	3.72×10^{-4}	3.7×10^{-5}

L.5 OSCILLATOR STRENGTHS AND LIFETIMES FOR RADIATIVE TRANSITIONS IN THE HYDROGEN ATOM

Transition	f_{if}	τ_{fi}, ns
$1s - 2p$	0.4162	1.6
$1s - 3p$	0.0791	5.4
$1s - 4p$	0.0290	12.4
$1s - 5p$	0.0139	24
$2s - 3p$	0.4349	5.4
$2s - 4p$	0.1028	12.4
$2s - 5p$	0.0419	24
$2p - 3s$	0.014	160
$2p - 3d$	0.696	15.6
$2p - 4s$	0.0031	230
$2p - 4d$	0.122	26.5
$2p - 5s$	0.0012	360
$2p - 5d$	0.044	70
$3s - 3p$	0.484	12.4
$3s - 5p$	0.121	24
$3p - 4s$	0.032	230
$3p - 4d$	0.619	36.5
$3p - 5s$	0.007	360
$3p - 5d$	0.139	70
$3d - 4p$	0.011	12.4
$3d - 4f$	1.016	73
$3d - 5p$	0.0022	24
$3d - 5f$	0.156	140
$4s - 5p$	0.545	24
$4p - 5s$	0.053	360
$4p - 5d$	0.610	70
$4d - 5p$	0.028	24
$4d - 5f$	0.890	140
$4f - 5d$	0.009	70
$4f - 5g$	1.345	240

REFERENCES

[1] W. Heitler, *Quantum Theory of Radiation*, Oxford University Press, Oxford, 1954.

[2] R. Loudon, *Quantum Theory of Light*, Oxford University Press, Oxford, 1983.

[3] H. A. Bethe and E. E. Salpeter, *Quantum Mechanics of One- and Two-Electron Atoms*, Plenum, New York, 1977.

[4] A. B. Migdal and V. P. Krainov, *Approximation Methods in Quantum Mechanics*, Benjamin, New York, 1969.

[5] V. B. Berestetskii, E. M. Lifshitz, and L. P. Pitaevskii, *Relativistic Quantum Theory*, Pergamon, Oxford, 1971.

[6] J. J. Sakurai, *Advanced Quantum Mechanics*, Addison-Wesley, Reading, MA, 1967.

[7] J. D. Jackson, *Classical Electrodynamics*, Wiley, New York, 1975.

[8] L. D. Landau and E. M. Lifshitz, *Quantum Mechanics: Nonrelativistic Theory*, Pergamon, Oxford, 1977.

[9] A. R. Edmonds, *Angular Momentum in Quantum Mechanics*, Princeton University Press, Princeton, 1960.

[10] M. E. Rose, *Elementary Theory of Angular Momentum*, Wiley, New York, 1957.

[11] *Handbook of Chemistry and Physics*, 76th ed., CRC Press, Cleveland, 1995/96.

BIBLIOGRAPHY

A. I. Akhiezer and V. B. Berestetskii, *Quantum Electrodynamics*, Interscience, New York, 1965.

L. Allen and J. H. Eberly, *Optical Resonance and Two-Level Atoms*, Wiley, New York, 1975.

S. Bashkin and J. O. Stoner, *Atomic Energy Levels and Grotrian Diagrams, vol. I: Hydrogen I–Phosphorus XV*, North-Holland, Amsterdam, 1975.

S. Bashkin and J. O. Stoner, *Atomic Energy Levels and Grotrian Diagrams, vol. II: Sulphur I–Titanium XXII*, North-Holland, Amsterdam, 1978.

S. Bashkin and J. O. Stoner, *Atomic Energy Levels and Grotrian Diagrams, vol. III: Vanadium I–Chromium XV*, North-Holland, Amsterdam, 1981.

S. Bashkin and J. O. Stoner, *Atomic Energy Levels and Grotrian Diagrams, vol. IV: Manganese I–XXV*, North-Holland, Amsterdam, 1982.

K. Blum, *Density Matrix Theory and Applications*, Plenum, New York, 1981.

M. Born and E. Wolf, *Principles of Optics*, Pergamon, Oxford, 1970.

B. H. Bransden, *Atomic Collision Theory*, Benjamin Cummings, Reading, MA, 1983.

B. H. Bransden and C. J. Joachain, *Physics of Atoms and Molecules*, Longman, Harlow, UK, 1983.

P. G. Burke, *Potential Scattering in Atomic Physics*, Plenum, New York, 1977.

C. Cohen-Tannoudji, *Atom-Photon Interactions*, Wiley, New York, 1992.

E. U. Condon and H. Odabasi, *Atomic Structure*, Cambridge University Press, Cambridge, 1980.

E. U. Condon and G. H. Shortley, *Theory of Atomic Spectra,* Cambridge University Press, Cambridge, 1967.

N. B. Delone, *Interaction of Laser Radiation with Matter*, Editions Frontières, Gif-sur-Yvette, 1993.

N. B. Delone and V. P. Krainov, *Atoms in Strong Light Fields*, Springer-Verlag, Berlin, 1985.

N. B. Delone and V. P. Krainov, *Fundamentals of Nonlinear Optics of Atomic Gases*, Wiley, New York, 1988.

N. B. Delone and V. P. Krainov, *Multiphoton Processes in Atoms*, Springer-Verlag, Berlin, 1994.

J. H. Eberly, J. Javanainen, and K. Rzążewski, Above-Threshold Ionization, *Physics Reports* **204**, 332–383 (1991).

F. H. M. Faisal, *Theory of Multiphoton Processes*, Plenum, New York, 1986.

U. Fano and L. Fano, *Physics of Atoms and Molecules*, University of Chicago Press, Chicago, 1972.

U. Fano and A. R. P. Rau, *Atomic Collisions and Spectra*, Academic, Orlando, 1986.

H. Friedrich, *Theoretical Atomic Physics*, Springer-Verlag, Berlin, 1991.

S. Gasiorowicz, *Quantum Physics*, Wiley, New York, 1974.

M. Gavrila, *Atoms in Intense Laser Fields*, Academic, New York, 1992.

G. Herzberg, *Atomic Spectra and Atomic Structure*, Dover, New York, 1944.

L. D. Landau and E. M. Lifshitz, *Classical Theory of Fields*, Pergamon, Oxford, 1979.

I. Lindgren and J. Morrison, *Atomic Many-Body Theory*, Springer-Verlag, Berlin, 1985.

W. H. Louisell, *Quantum Statistical Properties of Radiation*, Wiley, New York, 1973.

A. Messiah, *Quantum Mechanics, Vol. I*, Wiley, New York, 1965.

A. Messiah, *Quantum Mechanics, Vol. II*, Wiley, New York, 1966.

P. W. Milonni and J. H. Eberly, *Lasers*, Wiley, New York, 1988.

M. H. Mittleman, *Theory of Laser-Atom Interaction*, Plenum, New York, 1993.

R. G. Newton, *Scattering Theory of Waves and Particles*, Springer-Verlag, Berlin, 1982.

A. A. Radzig and B. M. Smirnov, *Reference Data on Atoms, Molecules, and Ions*, Springer-Verlag, Berlin, 1985.

H. R. Reiss, S-Matrix and Keldysh Techniques for Strong-Field Processes, *Progress in Quantum Electronics*, **16**, 1–71 (1992).

M. E. Rose, *Multipole Fields*, Wiley, New York, 1955.

Y. R. Shen, *Principles of Nonlinear Optics*, Wiley, New York, 1984.

I. I. Sobelman, *Atomic Spectra and Radiative Transitions*, Springer-Verlag, Berlin, 1979.

A. F. Starace, *Theory of Atomic Photoionization* in *Encyclopedia of Physics, Vol. 31*, Springer-Verlag, Berlin, 1982, pp. 1–121.

C. H. Townes and A. L. Schawlow, *Microwave Spectroscopy*, McGraw-Hill, New York, 1955.

INDEX